GIS in Land and Prop
Management

CW01572547

GIS in Land and Property Management

Peter Wyatt and Martin Ralphs

Routledge
Taylor & Francis Group

LONDON AND NEW YORK

First published 2003 by Spon Press

This edition published by Routledge
2 Park Square, Milton Park, Abingdon, Oxon OX14 4RN

Simultaneously published in the USA and Canada
by Routledge
711 Third Avenue, New York NY 10017

Routledge is an imprint of the Taylor & Francis Group, an informa business

© 2003 Peter Wyatt and Martin Ralphs

Typeset in Sabon by
Newgen Imaging Systems (P) Ltd, Chennai, India

All rights reserved. No part of this book may be reprinted or
reproduced or utilised in any form or by any electronic,
mechanical, or other means, now known or hereafter
invented, including photocopying and recording, or in any
information storage or retrieval system, without permission in
writing from the publishers.

British Library Cataloguing in Publication Data
A catalogue record for this book is available
from the British Library

Library of Congress Cataloging in Publication Data
Wyatt, Peter, 1968–
    GIS in land and property management / Peter Wyatt and
Martin Ralphs.
        p. cm.
    Includes bibliographical references and index.
    1. Real estate business – Data processing. 2. Land use – Data
processing. 3. Geographic information systems.
I. Ralphs, Martin. II. Title
    HD1380 .W93 2002
    333.3'0285–dc21   2002153755

ISBN 978-0-415-24065-9 (pbk)
ISBN 978-0-415-24064-2 (hbk)

This book is dedicated to Jemma, Sam and Tom (PW).

To my wife Sara, with heartfelt thanks for her patience and support throughout this project (MR).

Contents

List of figures		x
List of tables		xv
Foreword		xvi
Preface		xviii
Acknowledgements		xxiii
List of abbreviations		xxv

PART I
Geographical Information Systems — 1

1 Setting the scene — 3

Definitions and key terminology 4
The nuts and bolts of GIS technology 11

2 Maps and mapping — 17

An introduction to spatial referencing 17
The map: a window on the world 27
Bringing geographical information into the computer 35

3 Mapping and analysis using GIS — 54

Mapping and presenting information with GIS 54
Using geographical relationships in spatial analysis 64
Visualisation techniques 80

PART II
GIS applications in land and property management — 85

4 Mapping, land information systems and conveyancing — 87

Mapping property information 88
Land registration and land information systems 96
Property marketing and conveyancing 111

5 Property management   121

   Local authority property management 121
   Large landowners 134
   Facilities management 140
   Rural land management 148

6 Planning and development   159

   Planning 160
   Property development 183
   Urban design, 3D modelling and interaction 190

7 Retail and financial market research   200

   Geo-demographic analysis 201
   Retail location planning 202
   Office location planning 217
   Insurance and finance 231

8 Property market analysis   239

   Visualisation of property data 239
   Thematic mapping of property data 246
   Property market analysis 258
   Property research 267

PART III
GIS issues in land and property management   **281**

9 Information management   283

   Data quality and liability 285
   Data access 298
   Data standards 308

10 Implementing and managing GIS   331

   Project-led GIS implementation 331
   Corporate GIS implementation 335
   *Implementation issues for national land and property
     management initiatives 358*
   *Organisation and administration issues for
     GIS implementation 360*

11 Future prospects 367

The use of GIS in land and property management 367
New 'information' markets in land and
  property management 371
National initiatives for land and property information 374
New methods of data visualisation and exploration 376
GIS and the Internet 377
A mobile future 378

Glossary 382
Index 385

Figures

1.1	An indicative GIS hardware configuration	12
1.2	Digital aerial photograph of Oxford Circus and its environs	14
2.1	Latitude, longitude and height measurements on the sphere	18
2.2	Cross section of the Earth as a spheroid	19
2.3	Latitude and longitude measured on the spheroid	20
2.4	Parallels and meridians on the sphere	21
2.5	Three families of map projections: planar, cylindrical and conic	22
2.6	The grid co-ordinate system	23
2.7	Orientation of the Transverse Mercator projection for the British National Grid	25
2.8	The relationship between true north and grid north	25
2.9	The National Grid with the true and false origins	26
2.10	Example of a scale bar	28
2.11	Point, line and area representations	28
2.12	An example of surface representation	30
2.13	Topographic mapping from the OS 1:50,000 Landranger series of maps of Great Britain	33
2.14	Choropleth map depicting relative scores on the Townsend Index of Deprivation	34
2.15	An example of an area class map showing soil types	35
2.16	Contour mapping of height on Mt Saint Helens	36
2.17	Representation of point features	40
2.18	Representation of lines in a GIS	41
2.19	Example of a link and node vector data structure based on Figure 2.18	42
2.20	Storing polygons using the link and node structure	44
2.21	The raster data structure	46
2.22	An example of high resolution and low resolution raster data	47
2.23	Raster orientation in relation to true north	48
2.24	Positioning on a raster grid	48
2.25	Spatial referencing on a raster grid	49

2.26	Comparison of vector and raster representations	50
3.1	An example of single feature mapping	56
3.2	Shading of a single feature type according to category	57
3.3	Proportional symbolisation of power generation facilities	58
3.4	Choropleth mapping of student GCSE performance	59
3.5	Classification of deprivation using the natural breaks method	62
3.6	Classification of deprivation using the equal intervals method	62
3.7	Classification of deprivation using the standard deviation method	63
3.8	Classification of deprivation using the quantiles method	63
3.9	Measurement of straight line distance	65
3.10	A simple geographical query	66
3.11	Visualising the results of a database query on the map	66
3.12	Flood risk in a river valley	67
3.13	Circular, rectangular and polygonal buffer zones around point, line and area features	68
3.14	An example to illustrate some simple set concepts	70
3.15	Siting a new newsagent in a suburban centre using buffer zones and overlay techniques (1 of 4)	72
3.16	Overlay and buffer generation (2 of 4)	72
3.17	Overlay and buffer generation (3 of 4)	73
3.18	Overlay and buffer generation (4 of 4)	74
3.19	An example of combinatory geographical overlay	74
3.20	Combinatory overlay requiring attribute integration	75
3.21	Mapping business densities in East London	81
3.22	The same business density data set viewed in three dimensions	81
3.23	3D visualisation of the built environment	82
4.1	John Snow's map of the Broad Street pump cholera outbreak	88
4.2	OS Landline map data	89
4.3	OS Mastermap	90
4.4	Seed points for property interests on each floor of a building	91
4.5	The relationship between OS map data and the NLPG	93
4.6	BLPU and UPRN	93
4.7	BS7666 gazetteer application	94
4.8	Register of Title	97
4.9	Title Plan	99
4.10	Multi-purpose cadastre	103
4.11	NLIS hub	108
4.12	Selecting a state	114
4.13	Selecting an area within a state	114

4.14	Choosing a county	115
4.15	Selecting an area within a county	115
4.16	Choosing a specific neighbourhood	116
4.17	Viewing a picture of a selected property	116
4.18	NLIS conveyancing pilot application	118
5.1	Property information required for local authority property management	125
5.2	Birmingham City Council property map	132
5.3	Floor 3 rooms classified by departmental responsibility	141
5.4	Rooms on floor 2 classified by use	142
5.5	Rooms on floor 1 classified by occupancy level	142
5.6	Rooms on floor 2 used by the Pathology Department and have a digital door-lock	143
5.7	Rooms on floor 3 used by the Ante-Natal Department and Ante-Natal Clinic	144
5.8	Areas of Ancient Woodland	151
5.9	Character Areas	152
5.10	Natural Areas	153
5.11	Extract from the Land Cover of Great Britain	156
6.1	The construction of an index of town centre activity	162
6.2	Contours representing levels of town centre activity	163
6.3	3D visualisation of the index of town centre activity	163
6.4	Area of town centre activity	164
6.5	NLUD Phase 1 sites in the Cotswolds	168
6.6	Spittelmarkt, Berlin	190
6.7	Bristol Harbourside, modelled in MapInfo GIS	191
6.8	Bristol Harbourside, plan view of 3D model	192
6.9	Bristol Harbourside, virtual reality image	192
6.10	Building outlines and ring-road for Wolverhampton	194
6.11	3D visualisation of Figure 6.10	194
6.12	GIS-based photographic image of Oxford Street in London	195
6.13	Ringed building from Figure 6.12 has been repositioned	196
7.1	A 500-metre drive-time defined by Euclidean distance	203
7.2	A 500-metre drive-time defined by road network distance	203
7.3	How census tract level demographic data are apportioned using area ratios	210
7.4	Drive-times based on postcode sectors	212
7.5	Retail catchment areas around Neston (1)	213
7.6	Retail catchment areas around Neston (2)	213
7.7	Rental growth by local authority area	215
7.8	Normalisation of rental growth figures	215
7.9	Composite map showing a ranking of suitable locations for a new shopping centre	216
7.10	Mapping the competition, catchment and key client targets	220
7.11	Office location and staff residences	222

7.12 Office location and staff residences (excluding employees
      not commuting from within the local area)   223
7.13 The 20-, 40- and 60-minute peak drive-times to the
      Cherwick office   223
7.14 The 20-, 40- and 60-minute peak drive-times to the
      Breconsfield office   224
7.15 The 20-, 40- and 60-minute peak drive-times to the
      Grayford office   224
7.16 Existing morning peak travel-to-work times   225
7.17 Overlap between the three main offices of 1-hour peak
      drive-times   226
7.18 The 20-, 40- and 60-minute peak drive-times to the
      proposed Sloughton site   227
7.19 The 20-, 40- and 60-minute peak drive-times to the
      proposed Blockworth site   227
7.20 Morning peak travel-to-work times for proposed offices   228
7.21 Average travel-to-work time   228
8.1   Luton drive-time map   241
8.2   Luton key industrial locations   241
8.3   Primary catchment area around Neston   242
8.4   Ipswich town centre – retail provision   243
8.5   Ipswich and competing centres   243
8.6   Street-based property map for Lille, France   244
8.7   Distribution of deals by rental levels   245
8.8   Average house prices in the third quarter of 1996   247
8.9   Average house prices in the third quarter of 1997   247
8.10 Average house prices in the third quarter of 1998   248
8.11 Average house prices in the third quarter of 1999   248
8.12 Average house prices in the third quarter of 2000   249
8.13 Average house prices in London in the third quarter
      of 2000   250
8.14 Property investments in Central London classified by
      capital value   251
8.15 Property investments in Central London classified by
      total return   252
8.16 Density of retail property investments in the IPD databank   252
8.17 Total return on property investments in south-east England
      in 1996   253
8.18 Total return on property investments in south-east England
      in 1997   254
8.19 Total return on property investments in south-east England
      in 1998   254
8.20 Core areas in Leeds city centre   255
8.21 Motorway buffers around London   256
8.22 Average movement on each pavement   259

8.23   Pedestrian and bicycle observation points   261
8.24   Pedestrian routes through the Sidgwick Campus   261
8.25   Cycle routes through the Sidgwick site   262
8.26   An example of observation points for VGA   263
8.27   An example of the visibility measure inside a building   264
8.28   An example of accessibility analysis inside a building   265
8.29   Surveillance from building entrances   265
8.30   Barbican, overlaid by isovals of 1939 land values   268
8.31   Barbican, overlaid by isovals of 1969 land values   268
8.32   Relative shop values in Lewes, east Sussex   269
8.33   Rateable values per street block   270
8.34   A taxonomy of factors that influence property value   271
8.35   GIS-based value map of the central retail area of
       Horsham in west Sussex   276
9.1   2001 general election results by parliamentary
       constituency   296
9.2   Nodal structure for a national LIS/GIS   312
9.3   Logical data model for an LPG   318
9.4   Logical data model for the Bristol LPG   321
9.5   Seed points for property interests on each floor of a
       building   324
9.6   Core spatial units in the UKSGB   325

Tables

3.1 Upper boundaries of classes for different classification
    procedures   64
4.1 Potential applications of NLIS   107
5.1 GIS procurement in local government in 2000   122
5.2 Local authority information requirements   133
5.3 Building facilities databases   144
6.1 NLUD sites within the administrative
    boundary of Cotswold District Council   166
6.2 Local authority planning responsibilities   170
6.3 Local government applications of GIS   171
6.4 Volume of data held and used by Birmingham City Council   174
6.5 Development sieve analysis   186
7.1 Drive-time matrix of all staff to each site option within
    drive-time bands – numbers   229
7.2 Drive-time distribution of all staff to each site option within
    drive-time bands – percentages   229
7.3 Disqualifying criteria   232
9.1 Aggregated and disaggregated property data   290
9.2 Summary of process for assessing quality of geographical
    data   297
9.3 Data sets that need not be registered or are exempt from
    the Data Protection Act   302
9.4 Important government departments as far as land and
    property data are concerned   305
10.1 GIS implementation issues   345
10.2 GIS course content   363
10.3 Skills matrix   364
11.1 How geo-referenced property information is used in the
    property cycle   369

Foreword

Land has many dimensions – social, cultural, functional and economic. It is the space in which we live, underpinning all human activity – even a journey through virtual reality requires some physical space for the computer that generates it and the traveller who experiences it. It is a basis for production and a commodity that can be used to generate wealth through investment. It is the most valuable item that the majority of people are ever likely to possess.

Yet, our whole approach to land has been fragmented between different professions and different perceptions. A land surveyor sees it as something to measure while an anthropologist sees it as a factor in human evolution. A lawyer sees it as a set of abstract property rights while an accountant sees it as a financial element on a company's balance sheet. Those who estimate its market value claim that the three most important factors in determining its worth are 'location, location and location' and yet until recently there has been little scientific analysis and evaluation of this claim. Why is it that we understand so little about the multitude of factors that influence one of our most prized possessions? More importantly, how can we improve our knowledge about land and the workings of the land market?

Many of the answers lie in this book and in Geographical Information Systems (GIS) technology that helps us to analyse land and its many interrelated components. The marriage between GIS and real property management has been long overdue. Land and property data have been like a sleeping giant waiting to be aroused but a new day is now dawning in which the open flow of land-related data is leading to a new and more precise understanding of the world about us. In focusing on the application of GIS to land and property management this book seeks to demystify the technology and review its actual and potential applications. As such it should have a much wider appeal than just to the specialist. Although it focuses much upon developments in the UK the underlying message is global for there are many countries, especially those in economic transition to a market-based economy, that are struggling with the problems of developing a land market.

Both authors have considerable research and practical experience. They have brought together the strands of an unfolding story in which one can at present catch merely a glimpse of what is to come. Their contribution is timely and significant and their efforts are to be commended.

Peter Dale, OBE
Honorary President, International Federation of Surveyors
2002

Preface

Land is a fundamental resource. The proper functioning of society and the economy depends upon its efficient allocation and optimum use. Economic output depends on the proper use of land resources. The effective use of land is in turn dependent upon the proper functioning of the land market and the efficient dissemination of and access to information about it. The representative bodies of key players in the land and property market agree that in the UK the proper management of land resources and the efficient working of the property market are hindered by a lack of good quality information. Better data, made widely available, will result in better quality decision-making about land and property.

Land is a resource fixed in locational terms. Unlike labour and capital, one unit of land is not directly substitutable for another because each unit is unique at least in terms of its geographical location. Consequently the locational aspect of land and property information is of vital importance. Herein lies the significance of information systems that are able to handle the locational attributes of property information and that can relate that information to standard property management data sets.

Land and property advisors are beginning to recognise the opportunities that new technologies such as Geographical Information Systems (GIS) can offer. Often, when considering the impact of new technology, advisors have tended to think of the application of particular types of software to discrete tasks. Increasingly though, the potential of the digital property information system as a valuable decision support tool is being recognised. Such systems allow for rapid access to data about land and property whenever and wherever they are required and the formulation of complex resource management questions. As such, they present the property information manager with a decision support tool that can be applied to a wide range of diverse applications.

The aims of this book are to introduce the use of GIS as a tool for land and property management by providing an insight into ways in which GIS is being used by property people and examining the issues involved when using GIS as a decision aid for land and property management. The book introduces GIS as a technology, describes current practical developments and examines the key issues in the application of GIS to land and property management. The book is divided into three parts which have the following objectives:

1   To introduce digital mapping and GIS, together with a brief history of the development of GIS and LIS, all with an emphasis on property.
2   To describe the spectrum of GIS applications in land and property management with selected case studies to provide detail for key application areas.
3   To examine the issues drawn out from the above and provide guidance and recommendations for future implementation and use of GIS in land and property management.

Part I

Geographical Information Systems are a relatively new technology in computing terms. Its origins lie in the development of computer graphics, computer-assisted cartography and automated mapping techniques during the 1960s and 1970s and the parallel development of computerised information management applications using database approaches. In the 1980s, the development of the IBM Personal Computer and the Apple Macintosh began a revolution in user-friendly and affordable computing and information storage technology. Since that time we have seen rapid acceleration in the speed and power of computer processing and graphics techniques, huge expansion of data storage capacity and an increasing sophistication in the user interfaces for programs and operating systems. In 1990, a software package called AutoRoute, which uses GIS techniques to identify and measure road trips in terms of distance, time and cost, was the second biggest selling computer software package in that year. Since the early 1990s, the GIS sector has experienced very high rates of growth as the technology has moved out of the specialist computer laboratory and onto the desktop. GIS has been adopted across a range of application areas from infrastructure management through environmental resource monitoring to social policy formulation as users have become aware of its potential for managing and bringing together their information holdings, many of which include a geographical component.

This part of the book introduces the concepts and key terminology of GIS and considers how the technology can be used to represent and interrogate geographical information. It addresses common GIS functionality and explores the use of the technology for the query and inventory of resource information before progressing to issues of data modelling and visualisation. It discusses GIS and Land Information Systems and the differences between them and, as part of its review of GIS technology, introduces the Internet and new communications technologies as new resources for handling and delivering geographical information. All of the elements covered are framed within the context of land and property management and the specific characteristics of this application area.

Part II

Land and property is a diverse resource ranging from rural land and agricultural property through to urban and residential land uses and infrastructure for transport and service delivery. It is unsurprising then that the management of land and property involves a similarly diverse range of activities. Unfettered access to accurate and comprehensive information is vital to good property management decision-making. Part II illustrates how GIS is being used to help collect, store, manipulate and present land and property information as an aid to management decisions. The chapters focus on those users who have invested in GIS to help manage and analyse property information to help inform property decision-making at the operational and strategic levels. They consider, therefore, property market agents, landowners, local authorities, lenders and insurers rather than environmental agencies and market research companies whose interest in property is often secondary to business or customer-focused decision-making. GIS applications are considered from a business perspective in Birkin *et al.* (1996) and Grimshaw (1995).

Markets exist to transfer ownership of interests in land and property. Some markets operate more efficiently and openly than others. The property market has no centralised trading place and involves the conveyance of unique 'products' that differ legally, physically and always geographically. Expert advice is therefore often required during the transaction process. The proper functioning of any market depends on ready access to market information and the property industry has been criticised for its opaqueness. The government is keen to improve the speed at which property transactions take place and are therefore currently reviewing the process in England and Wales. The private sector also perceives a need to introduce initiatives that improve the property transaction process: GIS is being used as a tool at the marketing and conveyancing stages of a property transaction.

Many owners and occupiers of and investors in property have portfolios of interests and seek to manage them collectively. Where these interests are located in close proximity GIS is often seen as a useful management tool. Consequently local authorities pioneered the application of GIS in land and property management. This is because many local government services require the collection and management of property data within discrete geographical areas. GIS is a tool well suited to the management of such data. Similarly, utilities use GIS to plot cable and pipe networks, plan excavations and assist in fault-finding.

Local authorities are also using GIS to assist their statutory planning functions. Most development decisions require planning permission and local authorities are required to produce development plans. The use of GIS in planning departments is now widespread in the UK and there are good examples of GIS being used for, *inter alia*, plan-making, land charge searches, development control and planning constraint notification. Developers and, in particular, location-dependent businesses such as retailers, are also using GIS to assist in site selection, store location, branch network distribution and performance measurement. Here, land and property data are combined with population demographics and business trading data within a GIS to create a powerful decision aid. Some of the more innovative and sophisticated applications of GIS can be found in this field.

Increasingly, lenders and insurers are looking to use GIS to help calculate their risk exposure in lending decisions and premium calculations, respectively. In terms of secured lending, property is probably the most prevalent form of collateral and insurers have become much more aware of the localised risk factors that can affect building insurance claims – recent flooding has heightened this awareness. The use of GIS to analyse environmental data such as geology, historic uses of previously developed land and flood risk is of particular importance to lenders and insurers.

Finally, with regard to property market analysis, location has long been recognised as the primary driver of property decisions but has been neglected in terms of sophisticated geographical analysis. This has been blamed on a lack of appropriate and easily understood analysis techniques and a paucity of data on which to perform such analysis. The influence of location on property value, development viability and business success has been left to the practical experience and local market knowledge of the advisor. It is not unusual to see paper maps on display in the offices of property advisors, often colour-coded, showing the location of important places and circumscribed by radii representing drive-times. These are paper-based GIS but computerised GIS are starting to be used for market analysis rather than simple data storage and geographical display. The analytical power of GIS is beginning to be utilised in land and property management.

Part III

There are important information management issues that need to be addressed if GIS is to become a vital tool for land and property management. The final part of the book addresses these issues.

Chapter 9 focuses on issues of data and information that affect the use of GIS. This concentrates on two broad areas. First, legal and administrative issues of information management including copyright and data protection legislation, problems of access to data, restrictions on use and licensing agreements. Second, technical issues of information management are considered, such as data quality and reliability, the use and adoption of standards and the management of error in data. Chapter 10 explores factors affecting GIS implementation and different strategies for GIS development, drawing on examples to illustrate the different approaches that have been taken. The main function of this chapter is to draw attention to the issues involved in implementation and to successful strategies for taking it forward.

In the information age data can easily be assembled, shared and traded irrespective of place. The National Land Information Service (NLIS) initiative has lobbied for public access to information on the ownership, use and price of land and property in the UK. Chapter 11 looks at the future prospects for the use of GIS in land and property management. It considers the embryonic NLIS and looks at some of the cutting edge developments in GIS technology that might impact upon the GIS property professional.

References

Birkin, M., Clarke, G., Clarke, M. and Wilson, A. (1996) *Intelligent GIS: Location Decisions and Strategic Planning*, GeoInformation International, Cambridge, UK.

Grimshaw, D. (2000) *Bringing Geographical Information Systems into Business* (2nd edition), John Wiley & Sons, New York.

Acknowledgements

The material for this book has come from a wide range of sources, the supply of which depends upon the generosity of a great many GIS and property people. We are very grateful to those who helped make this book more colourful through the use of examples and case studies, particularly

- Andy Coote, Nick Chappallaz and Peter Beaumont at ESRI (UK)
- Dave Roberts at Innogistic GIS
- Nick Land at the Ordnance Survey
- Her Majesty's Land Registry
- Jake Desyllas at The Intelligent Space Partnership
- Professor Peter Dale and Mark Thurstain-Goodwin at University College London
- Andrew Larner at the Improvement and Development Agency
- John Allinson and John Counsell at the University of the West of England
- Roger Monk at Birmingham City Council
- Tony Key and Dylan McBurney at IPD
- Matthew Bush and Tom Whittington at FPDSavills
- Marion Murphy at Jones Lang Lasalle
- Mark Teale at CB Hillier Parker
- Phil Hammond at Property Market Analysis
- Tony Black and Nick Griffiths at Intelligent Addressing
- Michael Nicholson at Property Intelligence
- Jennifer MacLellan at the Barking and Dagenham London Borough Council
- Bruce Yeoman, Independent Consultant
- George Griffith at the University of Bristol Healthcare Trust
- Andrew Smith at Henderson Global Investors

Figures 2.13, 3.5, 3.6, 3.7, 3.8, 3.9, 3.10, 3.11, 3.15, 3.16, 3.17, 3.18, 3.21, 3.22, 4.2, 4.3, 4.7, 4.8, 4.9, 4.18, 5.2, 5.8, 5.9, 5.10, 6.2, 6.3, 6.4, 6.5, 6.7, 6.8, 6.9, 6.10, 6.11, 7.1, 7.2, 7.7, 7.9, 8.4, 8.5, 8.9, 8.10, 8.11,

8.12, 8.13, 8.14, 8.15, 8.16, 8.17, 8.19, 8.20, 8.22, 8.36 reproduced from Ordinance Survey material on behalf of the Controller of Her Majesty's Stationery Office © Crown Copyright MC100038806.

Figures 7.10, 7.13, 7.14, 7.15, 7.17, 7.18, 7.19 Drive-time boundaries were produced using MapInfo Drivetime® software.

Abbreviations

AGI	Association for Geographical Information
BCC	Bristol City Council
BGS	British Geological Survey
BLPU	Basic Land and Property Unit
BS7666	British Standard 7666
CAD	Computer Aided Design
CAFM	Computer Aided Facilities Management
CASA	Centre for Advanced Spatial Analysis
CBD	Central Business District
DEM	Digital Elevation Model
DETR	Department of the Environment, Transport and the Regions
DNF	Digital National Framework
DoE	Department of Environment
DSS	Decision Support System
DTLR	Department for Transport, Local Government and Regions
ESMR	Essex sites and Monuments Records
GIS	Geographical Information Systems
GPS	Global Positioning System
GSDI	Global Spatial Data Infrastructure
HMLR	Her Majesty's Land Registry
IACS	Integrated Administration and Control System
ICT	Information and Communications Technology
IDeA	Improvement and Development Agency
IGGI	Interdepartmental Group on Geographical Information
ISO	International Standards Organisation
ITT	Invitation To Tender
JLL	Jones Lang Lasalle
LGIH	Local Government Information House
LIS	Land Information System
LLPG	Local Land and Property Gazetteer
LPG	Land and Property Gazetteer
LPI	Land and Property Identifier
LVRS	Location Value Response Surface

MAFF	Ministry of Agriculture Fisheries and Food (now part of DEFRA – Department for the Environment, Farming and Rural Affairs)
NGDF	National Geospatial Data Framework
NLIS	National Land Information Service
NLPG	National Land and Property Gazetteer
NLUD	National Land Use Database
NNR	National Nature Reserve
NPV	Net Present Value
NSDI	National Spatial Data Infrastructure
ODPM	Office of the Deputy Prime Minister
OPD	Occupier Property Databank
OS	Ordnance Survey
PACE	Property Advisors to the Civil Estate
PAF	Postcode Address File
PAON	Primary Addressable Object Name
PFI	Private Finance Initiative
PROMIS	Property Market Information Service
RFP	Request for Proposal
RICS	Royal Institution of Chartered Surveyors
RTPI	Royal Town Planning Institute
SAC	Special Area of Character
SAON	Secondary Addressable Object Name
SDTS	Spatial Data Transfer Standard
SGB	Standard Geographic Base
SPA	Special Protection Area
SSSI	Site of Special Scientific Interest
UBHT	University of Bristol Healthcare Trust
UPRN	Unique Property Reference Number
URBED	Urban and Economic Development Group
VENUE	Virtual Environments for Urban Environments
VO	Valuation Office
VRML	Virtual Reality Modelling Language
2D	Two-Dimensional
3D	Three-Dimensional

Part I
Geographical Information Systems

1 Setting the scene

Introduction

A primary goal of the study of geography is to try and increase our under-standing of the world. One of the most important facilities in the scientific toolkit of the geographer is the ability to collect, synthesise, analyse and visualise data. Through the exploration of data we can test our theoretical understanding or apply existing knowledge in specific cases to try and make well-informed decisions about the management and use of limited natural and socio-economic resources.

This book is about geography. More specifically, it is about how modern techniques based largely around the use of computer systems for informa-tion processing can be brought to bear on one particular aspect of geography: the management of land and property resources. This area of activity has been central to human society since the development of permanent settle-ments. Records of land and property information can be found on the clay tablet maps of the early civilisations of the Indus Valley, Sumeria and Mesopotamia. The Egyptians used large-scale mapping for resource man-agement on the Nile flood plain. Later, the Romans created large-scale maps of land ownership based on the measurement of distances and angles (Dorling and Fairbairn, 1997) while the masters of feudal Europe created complex inventories of rights over land resources, the most famous being the Domesday Book of William the Conqueror.

In the twenty-first century we have access to more information about land and property than ever before. Space-borne satellite sensor platforms collect enormous volumes of information on a daily basis at ever increasing levels of detail describing the land cover of our planet. National governments hold computerised databases of land inventory information and commercial data suppliers provide wide-ranging information resources about land value and land use. Employing this information effectively and maximising its benefits are vital to the successful future development of our society.

Computerised approaches to geographical exploration, sometimes referred to as 'automated geography' (DeMers, 2000) can be brought to bear to assist those who need to maximise the benefits of modern property information resources. Part I of this book is an introduction to computerised approaches to the processing of information about land and property, with particular focus on the core technology of Geographical Information Systems (GIS) and the capabilities of these systems in handling, analysing and visualising land and property information.

It is timely for property professionals in the UK to increase their awareness of the opportunities that GIS can offer. The property sector now regards information as a critical corporate resource and managing and analysing this resource has become a high priority. The application of GIS technology to the management and analysis of property information brings forward numerous opportunities. Databases of property information can be created to which access is sold; new analytical skills can be learnt to meet the changing demands of clients and new business opportunities – made possible by technology and information access – can be sought.

There are already numerous introductory texts of good quality available about the technologies and capabilities of GIS and it is not our purpose in this book to reinvent the wheel by presenting broadly similar material. Instead, we have sought to summarise the basics of the technology and its capabilities in this first part and in the later parts to present more detail about the specific relevance of GIS methods to the application area of land and property management.

Definitions and key terminology

We begin our exploration by introducing and discussing some of the fundamental terms that we will be using throughout this book and providing some background on these concepts.

Land and property

'Land' refers to the physical resource represented by the surface of the earth. For human beings, land is a fundamental resource that is essential for all of the activities that we perform. A useful definition is provided by Dale and McLaughlin (1988) who consider that land 'encompasses all those things directly associated with the surface of the earth, including those areas covered by water. It includes a myriad of physical and abstract attributes from the right to light or build upon the land to ground water and minerals and the rights to use and exploit them'.

In economic terms, land is a factor of production. Land has utility, which might be the growing of crops, the grazing of livestock or the extraction of minerals. Such utility may be enhanced by constructing buildings that allow the environment to be adapted for other uses such as living accommodation, the manufacture and sale of goods or the provision of services. The term 'property' or 'real estate' refers to the combination of land and buildings. Hence 'property management' can be defined as the set of activities whereby land resources are put to good effect and includes the following processes:

- conveyancing, that is, the transfer of an interest in property from one owner to another;
- valuation, that is an estimate of the financial worth of an interest in property to a particular person for a specific purpose;
- development or redevelopment of property;
- management and maintenance of property;
- formation and implementation of land use planning policies and the monitoring of land use;
- environmental assessment and environmental impact analysis.

The process of land management is predicated upon the subdivision of the land into manageable regions or units. This subdivision can be based on many criteria including rights of use or ownership, planning controls or government policy and requires that we introduce a new entity: the land parcel. A land parcel can be defined simply as a unit of land and the processes of land management are usually concerned with individual land parcels or groups of parcels.

Data and information

There is often considerable confusion about the use of these two terms, and indeed the difference in the meanings that they convey is quite subtle. Let us begin by considering the term 'data'. We can define data simply as 'records of facts' following the pragmatic suggestion of Larner (1996). To take an example, the layout plan of a housing estate is a set of data. Individual data items associated with this example might be the pair of X and Y co-ordinates that define the geographical position of the corner of a building or the name of the owner of the land on which the estate is constructed.

By themselves, data are inert. For example, data may simply be the encoding of records of facts. In order for data to become 'information', we need to assimilate, understand or interpret them in some fashion. To achieve this we may need to classify or organise the data in order to convey meaning. By analysing the data the user extracts the meaning (or an interpretation of the meaning) and obtains information.

Geographical information

We can define geographical information as information that carries some form of geographical or 'spatial' reference allowing us to pinpoint its location in some fashion. Spatial references tend to fall into one of two categories that we will call numeric and symbolic references. A numeric or direct spatial reference uses a co-ordinate system, usually in conjunction with a mathematical model called a map projection, to position objects precisely in space. Examples of such systems include geographical co-ordinates (latitude and longitude) and the Ordnance Survey's (OS) National Grid of Great Britain.

Symbolic or descriptive spatial references use less precise methods to locate the object, but can still provide useful information. Examples, in decreasing order of exactness, might be:

- A postcode such as WC1 1AA.
- A postal address such as '15 Acacia Avenue, Bromley, South London'.
- A written description such as 'next to London Bridge'.

It is often cited that approximately 70 per cent of the data held by local and regional governments and private sector organisations are geographically referenced in some way. By capitalising on this fact and bringing computer analysis techniques to bear on our geographically referenced information we can maximise their usefulness.

Property information

We can define property information as information that describes the characteristics of a defined unit of property such as a land parcel or a building. Examples of property information might be the floor space of an industrial property, the leasing arrangements on a commercial office or the price paid for a house. Usually, a property will have a range of descriptive characteristics that are of interest.

Information systems

It is increasingly recognised that high quality information is a critical resource for the effective strategic functioning of organisations. This recognition is the result of our increased awareness of evolving technology and its potential to revolutionise information handling. The role and significance of information resources has not changed, but what was missing before was the existence of a mature Information and Communication Technology (ICT) that was able to manage and manipulate it. Miller (1956) has shown that humans are able to process between five and nine 'chunks' of information at any one time – recall the difficulties that we experience when trying to remember ten and eleven digit phone numbers or numerous computer passwords. Computers and appropriate software are capable of handling substantially more pieces of information in a fraction of the time. The problem has become one of interaction with the computer in order to retrieve this information quickly and efficiently.

Developments in ICT, much of which barely existed or was prohibitively expensive twenty-five years ago, have been rapid and continuous in the last decade. Since its first business application in the mid-fifties, the use of computer technology has expanded dramatically. The picture confronting us is constantly evolving and few weeks go by without the announcement of new products, major technological advances, improved cost performance ratios and myriad allied services. Fired by parallel developments in marketing techniques, the typical business is being bombarded by a host of technologies, many of which bear the most inscrutable names, mnemonics or just plain numbers that convey little or nothing to the uninitiated. Similarly, the range of activities to which computers and cognate technologies can be applied is vast and there are few aspects of the working and leisure lives of the individual which are not influenced in some measure by ICT.

The 'revolution' in ICT has been the subject of numerous publications and the topic of many conferences and seminars. Assertions about the pervasiveness of this revolution are frequently made in findings of surveys and studies of the economic environment. ICT has become a component part of so many human activities that it is regarded as ubiquitous, impacting upon us throughout our lives. The result of the ICT revolution is that the penetration of digital technology into the fabric of society has reached far and wide and is, by now, almost certainly irreversible.

Information systems and the property sector

The application of ICT to the management and exploration of information is realised in the concept of an 'information system'. We can define an information system as an integrating technology where resources and activities are brought together to support the decision-making process of an organisation. If we consider the property sector, computerised information systems have taken a number of forms. Early examples were property management systems that evolved from accounting software. Relational database technology improved these systems because the hierarchical nature of property interests is well suited to the structure of relational database management systems. Spreadsheet technology allowed the information stored in property databases to be analysed in more sophisticated ways and subsequently bespoke software packages have been developed that are capable of supporting many property-related decisions and automating the more mundane data processing elements of property management.

Over recent years demand for high quality property information has risen, partly due to the increased availability of affordable computerised analysis tools. Advances in IT have led to more analytical applications within the profession; econometric modelling, regression analysis and other statistical modelling techniques have been used. Multiple regression analysis and econometric modelling underpin the house price indicators developed by building societies in the UK and the application of expert systems and neural networks to valuation has received attention from academics.

However, these techniques depend on access to large volumes of high quality data if they are to provide reliable results.

Geographical Information Systems

Geographical Information Systems are a family of information systems that are designed specifically to address the handling, manipulation and visualisation of data that are geographically referenced. By this we mean that the data are linked to a location in space, usually somewhere on the surface of the Earth.[1] Such geographical linkages can be precisely defined using co-ordinate systems, or may use less exact frames of reference like a street address or a postcode. That said, there are many different definitions of what GIS is and what it does, probably because the technology has applications across a wide range of the sub-disciplines of geography. In trying to draw together the many ideas about what GIS is and what it does, we can draw out three common strands. The first is that GIS is associated primarily with the storage and analysis of information about the Earth or that is geographically referenced in some manner. The second is that there is a set of subsystems within a GIS that together define the types of functionality that can be achieved with it. The third is that GIS activity usually takes place within the context of a particular organisation and that the nature of that organisation will support and shape the use of the technology. We will explore these issues in more depth later in this section. Further discussion of GIS definitions can be found in Burrough and McDonnell (1998) and DeMers (2000).

Geographical Information Systems is one of a number of spatial data handling technologies. Others include cartographic drafting and graphic design tools for publishing maps and Computer Aided Design (CAD) systems for engineering and architectural design. Although these other technologies are also useful for handling and visualising geographical data, GIS is unique among them in that it permits the formulation of questions and the integration of results using spatial location and the geographical relationships between objects as the basis for query generation. If we examine the core technology of GIS, we find that most computerised GIS systems consist of a database engine whose records allow for the positioning of objects in space and a set of associated visualisation and analytical components. These permit the user to interrogate data using their geographical characteristics. For the purposes of this book, we include image processing and manipulation procedures for satellite and aerial photographic analysis as a subset of the broad range of GIS application software.

Geographical Information Systems technology has seen widespread application in the fields of natural resource management, planning and development control, socio-economic analysis and the inventory of infrastructure and facilities. The technology has been taken up by a range of different communities of users who deal with information that is referenced to the spatial domain. Examples of such information include postal addresses, roads, drainage basins, socio-economic zones, areas of land, pipeline networks and groups of properties. In all its diverse application areas GIS offers the unique ability to analyse and display information geographically, often revealing trends and patterns that were not apparent when examining the information in a spreadsheet or database. The power of GIS lies not only in its ability to analyse and display data referenced to the spatial domain but also in its function as a database that can bring data together from a wide range of sources. Using a GIS to manage property data offers two advantages: the unified storage of property data using a spatial representation of the real world and the ability to analyse and visualise that data in new ways by bringing together a wide range of source information.

GIS and Land Information Systems (LIS)

Since this book is primarily concerned with the use of GIS for the management of land and property, it is important that we introduce the concept of Land Information Systems at this point. A Land Information System (usually abbreviated to 'LIS') is a specialised application of GIS technology that is concerned with issues of land ownership, land planning and land management. Such systems usually rely on large-scale maps and store information about land ownership and land use. More formally an LIS can be defined as 'a system for capturing, storing, checking, integrating, manipulating, analysing and displaying data about land and its use, ownership, development, etc.' (Department of the Environment [DoE] 1987). LIS are primarily concerned with the storage and interrogation of property data. Within such systems a digital map base provides the foundation on which to build real estate information using land parcels as the basic building block. In urban areas the land parcel may be subdivided into buildings and parts of buildings such as a shop on the ground floor with a flat above. Information that might be contained within a map-based LIS includes:

- land rights and restrictions, including precise delineation of boundaries;
- land values and tax assessments;
- land use (other planning information may also be stored);
- information relating to buildings situated on the land parcel;
- population and census data;
- administrative data such as local government or national park boundaries;
- environmental data such as contaminated land, pollution and other hazards.

Other property data can be linked such as topography, geological and geophysical data, soil type, vegetation, wildlife, hydrology, climate, industry, employment, transport, utilities data (water, sewerage, gas, electricity, telecommunications) and emergency services.

Cadastral LIS maintain information on property ownership, use and value and are common throughout the countries of Europe, with the exception of Britain. In Sweden, for example, the on-line Land Data Bank System integrates the Land Taxation Register, Census Register, Housing Statistics and local authority information systems. Applications include planning, natural resources management, banking and real estate brokerage.

It is important to clarify that the core technology underpinning LIS is essentially the same as that of the more generic GIS and given the list of candidate data types for a LIS described above there would appear to be little difference between the two. For the purposes of this book, we consider that a LIS is simply an application of GIS technology for cadastral and land management purposes and henceforth the term GIS can be considered to encompass LIS.

GIS in the UK

Early implementations of GIS in the UK were project specific. All of the necessary data were collected in whatever format they happened to be available and were geo-referenced and input into the system, often using manual data entry procedures. Significant time and effort were required to convert the data into appropriate formats for query and analysis. The data were analysed, some sort of output (map or report) was produced and the information acted upon. Such GIS projects tended to be time-consuming and expensive in terms of manpower, data, hardware and software.

During the 1980s, GIS methods began to emerge from the specialist laboratories of university departments and government research institutions and the technology experienced a wider uptake by information management professionals across a range of sectors. The uptake of GIS as a viable information management technology with applications in many areas was predicated on several factors. The increasing affordability of computer processing technologies meant that users no longer needed to invest in large, mainframe computer systems to use GIS. At the same time, the availability of high quality digital geographical data was coupled with a growing awareness of the potential uses of geographical information and an increasing number of trained GIS operators. National digital datasets were nearing completion, the cost of computer hardware and software was tumbling and the global Internet was born. In 1987, the UK Department of the Environment published a report called 'Handling Geographic Information'. This was the culmination of two years' work by a Committee of Enquiry led by Lord Chorley into the way in which geographical information was used in the UK. The aim of the report was to map out future directions in this area in the light of advances in IT. In the report it was suggested that GIS was 'as significant to spatial analysis as the invention of the microscope and telescope were to science, the computer to economics and the printing press to information dissemination'.

The 1990s saw a move away from project-specific GIS implementation and a more integrated approach to systems development. In the US, GIS software development mushroomed, fuelled mainly by the affordability of a critical mass of spatial data in the United States and driven by Federal government initiatives like the National Spatial Data Infrastructure (Federal Geographic Data Committee, 1996) and the associated uptake of GIS by US government authorities. An international geographic information industry soon followed, including specialist data suppliers, value-added re-sellers and technology providers.

In the twenty-first century, GIS is an established desktop technology, albeit a computationally intensive one. The standardisation of the personal computing market around the Windows family of operating systems and the fact that a modern PC processing, storage and graphics platform is capable of tackling the intensive computational tasks that previously were only practicable on expensive UNIX workstations have brought GIS into a new marketplace. The availability of cost effective geographical data from government and commercial suppliers in the UK has advanced the situation still further.

The nuts and bolts of GIS technology

In this final section of Chapter 1, we introduce the technology of a GIS in more detail. Our aim is to provide an overview of the main components of a GIS and to describe how those components fit together. We can subdivide a GIS into four key building blocks. These are computer hardware, computer software, information (which forms the core resource of the GIS) and finally the people who build, manage and operate the system. A GIS usually operates as part of a larger organisation such as a local authority, government department or property development company and the people that are associated with its day-to-day operation might range from a small research team to a large corporate decision support department.

Although we have emphasised the use of computer systems in modern GIS activity, it is important to consider that computers are not an essential part of GIS. Many paper-based records management systems utilise maps and involve the storage and retrieval of geographical information, effectively performing the same decision support functions as a computer-based GIS. Having said that, modern GIS is usually heavily reliant on digital technology since the computer introduces new and powerful ways of exploring and managing information that are too time-consuming or expensive to reproduce using a paper-based framework.

Computer hardware

Computer hardware is the machinery that drives a GIS. Hardware includes the computer itself and its internal architecture as well as connected equipment such as keyboards, printers and plotters, disk drives, modems and network cards, tape drives, monitors, pointing devices such as the mouse, digitising

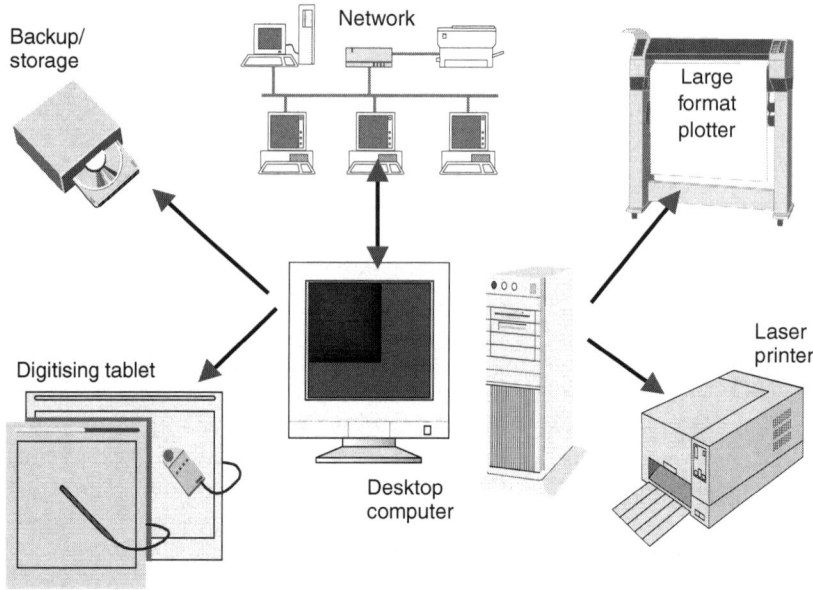

Figure 1.1 An indicative GIS hardware configuration.

tablets, computer projectors and digital cameras (Figure 1.1). Alongside the hardware itself we often find items like diskettes, CD-ROMs, DVDs, magnetic tapes and ZIP diskettes. These smaller items are often classed as 'consumables' rather than hardware and are usually portable or removable components that can be transferred between hardware platforms.

Geographical Information Systems operate on a wide variety of different computer platforms from the desktop personal computer through to high-powered workstations, servers and mainframes. At the present time, the typical GIS hardware platform seems to be a medium to high performance PC for desktop GIS, with support for advanced database activities from net-worked servers or UNIX computers. The increasing power and falling cost of the everyday PC, coupled with the introduction of operating systems like Windows XP® and LINUX mean that high performance GIS packages will run quite happily on a desktop computer. Only the fast processing power that is required by intensive graphics tasks such as complex 3D animation, image processing or digital photogrammetry is likely to require additional hardware support. This was certainly not the case five years ago, when PC GIS functionality languished behind more expensive UNIX solutions.

GIS software

The software programs that control a GIS come in many shapes, sizes and price brackets. However, virtually all GIS software includes the following basic subsystems (Peuquet and Marble, 1990; DeMers, 2000) which together provide a set of core GIS functionality.

1 A data input system that collects and pre-processes data from various sources. Such systems usually include the facility to restructure data and perform editing operations on digital maps, imagery and databases.
2 A data storage and retrieval subsystem that organises the data in an efficient manner and allows for retrieval, updating and editing.
3 A data manipulation and analysis subsystem that performs query and analysis tasks on the data. This might include aggregating and disaggregating data, combining and comparing datasets, statistical analysis or modelling functions.
4 A reporting/presentation subsystem that displays all or part of the database in tabular, graphical or map form and allows for the production of customised hard copies or output files.

Together, these subsystems supply the GIS user with a formidable arsenal of data manipulation, analysis and visualisation functionality. Different GIS software packages have different ways of handling each subsystem and will exhibit varying degrees of sophistication within each, but all four subsystems are present in most of today's commercially available GIS software packages. We will examine the tasks that might be carried out using a GIS in more detail in the next chapter.

Data for GIS

Data are the most expensive component of most GIS projects. At the same time a GIS is of little use without them. Based on the findings of numerous projects, it is generally reckoned that from 60 to 70 per cent of the cost of GIS implementation will be generated during the acquisition and development of an appropriate database (Korte, 1997). Given this, the choice of source data, its formats and structures is of critical importance in establishing an effective solution. The types of data that are used in GIS are very wide ranging, but three broad categories of information can be defined.

Digital maps

A digital map is a computerised representation of traditional paper map information. The features on the map are digitally encoded either using data gathered from a land survey or through a process of digital conversion and stored in a computer database. Chapter 2 explores some of the commonly applied techniques for storing and manipulating digital maps in more detail. For now, we will restrict ourselves to saying that the digital map has two major advantages over its paper equivalent. First, it is effectively seamless – many paper maps can be digitally encoded and combined in a single database eliminating problems of map sheet overlap and allowing the user to select areas of interest when required. Second, the digital map is fully customisable. The user is able to specify exactly what information is displayed and how it is depicted and to select the scale at which this visualisation is produced.[2] Such flexibility comes at a price because in customising the map and choosing an appropriate representational scale the user will be required to make decisions about symbolisation, representation and issues of scaling and data quality that would traditionally be made by a trained cartographer. Some formal training may be necessary before this can be achieved with confidence.

Digital imagery

A digital image is a computerised representation of pictorial data. Examples of such information include traditional photographic images and aerial photographs that have been digitally captured using a scanner or digital camera and images of the Earth taken by orbiting satellites (see Figure 1.2). Digital imagery adds a valuable additional dimension to the GIS database both in terms of increasing the quality of visualisation and in providing new information. For example, some airborne and orbital satellite sensors

Figure 1.2 Cities Revealed® – Digital aerial photograph of Oxford Circus and its environs.

Source: © 1994 The GeoInformation Group Ltd.

collect sufficient data to allow for the semi-automatic classification of land uses. When combined with digital mapping, such imagery can produce some very useful results.

Database information

Today's commercial and public sector databases about every aspect of our society and environment are larger, more comprehensive and more readily available than ever before. As the world becomes more data rich it has been readily pointed out by several authors (e.g. Longley, 1998 or Openshaw, 1998) that our ability to manipulate and extract meaningful patterns from this 'cyberspace' of data appears to be less and less certain. The modern analyst is data rich but information poor. We need new tools to interpret the wealth of data that is now available quickly and easily. GIS is a comparatively new weapon in this quest for interpretation.

Many phenomena exhibit some kind of spatially related behaviour. With GIS, we are able to examine these patterns, in many cases for the first time. One of the most useful aspects of GIS technology is its ability to integrate non-geographical information like house prices, rent values, socio-economic profiles or land use statistics with digital maps. Standard databases can be linked to the features on the digital map such as houses, roads, land parcels or administrative areas to provide an enormous range of ancillary data about them. The GIS can then be used to visualise and interrogate this information. We can begin to ask new questions of our databases – Which shops are best served by the public transport system? Which areas of a city have the most buoyant property prices and how are those prices changing? Which neighbourhoods have the best local facilities for leisure, shopping or healthcare? Such questions require us to bring together our database information by way of an underpinning knowledge of local geography, and using GIS we are able to do this.

Summary

In this chapter we have introduced the ideas that form the foundation of the rest of the book. We have defined what we mean by land, property, data and information, and introduced the concept of an information system for managing and analysing data from different sources. We have discussed land and geographical information systems and the take up of these technologies in the UK. Finally, we introduced the computing platforms that GIS rely upon, and discussed hardware, software and data concepts.

Geographical Information Systems is special because it allows us to bring together information from many different sources using geography as the common framework to link them together. This has been possible using paper maps for many years, but with GIS we can displace paper maps and explore the data in real time without having to spend months redrafting it.

We can combine the information we hold in whatever way we deem to be appropriate, producing customised visualisations of different combinations of data quickly and effectively.

Notes

1 Although the majority of GIS databases are earthbound, the technology has also seen some application in the mapping of other planetary bodies, notably the visualisation of the surface of Mars using satellite imagery.
2 Although a computerised map is 'scale free' in the sense that a user can precisely define the map scale at which information is displayed, the accuracy and precision with which the data were originally collected should inform this decision. Data should not be displayed at scales larger than the accuracy of the original survey would reasonably permit.

References

Burrough, P.A. and McDonnell, R.A., 1998, *Principles of Geographical Information Systems*, Oxford University Press, Oxford.

Dale, P.F. and McLaughlin, J., 1988, *Land Information Management*, Clarendon Press, Oxford.

DeMers, M.N., 2000, *Fundamentals of Geographic Information Systems* (2nd edition), Wiley, New York.

Department of the Environment, 1987, *Handling Geographic Information*, HMSO, London.

Dorling, D. and Fairbairn, D., 1997, *Mapping: Ways of Representing the World*, Longman, Harlow.

Federal Geographic Data Committee, 1996, The National Spatial Data Infrastructure, FGDC Internet publication, URL: http://www.fgdc.gov/nsdi2.html

Korte, G.E., 1997, The GIS Book: *Understanding the Value and Implementation of GIS Systems*, 4th edition, OnWord Press, Santa Fe.

Larner, A.G., 1996, Balancing rights in data – elementary? In Parker, D. (ed.), *Innovations in GIS 3*, Taylor and Francis, London, pp. 25–35.

Larsson, G., 1991, *Land Registration and Cadastral Systems*, Longman, Harlow.

Longley, P.A., 1998, Foundations. In Longley, P.A., Brooks, S.M., McDonnell, R. and MacMillan, B. (eds), *Geocomputation: A Primer*, Wiley, New York, Chapter 1, p. 1–15.

Miller, G., 1956, The magical number seven, plus or minus two: Some limits on our capacity for processing information, *The Psychological Review*, **63**, 81–97.

Openshaw, S., 1998, Building automated geographical analysis and explanation machines. In Longley, P.A., Brooks, S.M., McDonnell, R. and MacMillan, B. (eds), *Geocomputation: A Primer*, Wiley, New York, Chapter 6, p. 95–115.

Peuquet, D.J. and Marble, D.F. (eds), 1990, *Introductory Readings in Geographic Information Systems*, London: Taylor and Francis.

2 Maps and mapping

Introduction

In the same way that the management of land and property has been an essential human activity since the earliest civilisations, the maps and charts that provide key information for land management have been familiar tools throughout history. The map as a means of spatial representation is a diverse and flexible tool and its development has resulted in many different approaches to the visualisation of the world. Although map products are very diverse, there are a number of common ideas that underpin their creation. These concepts have accompanied the map on its journey of development through the ages and have moved with it from the printed page into the digital computer and the GIS. In this chapter we will explore the fundamental concepts behind maps and map making. We will investigate how computerised mapping techniques have enhanced our ability to make maps and explain how these ideas affect the creation and use of maps for property management.

An introduction to spatial referencing

Maps show us where things are in relation to one another. This requires a method of determining position on the surface of the Earth, a process known as spatial referencing or geo-referencing. The method by which geo-referencing is carried out is very important as this will largely determine the accuracy of any map that is produced.

Let us take an example of spatial referencing in practice. We will use this to illustrate the different assumptions and choices made when deciding how to define a geographical position. Consider the case of a British estate agent who wants to map all of the properties that his agency is currently marketing. The agent is at the cutting edge of technology and has purchased a Global Positioning System (GPS) receiver to facilitate his work. The receiver picks up signals from a network of orbiting satellites and uses these transmissions to calculate its own position. It allows the agent to pinpoint his location and obtain a geographical co-ordinate reference at the touch of a button. When he does this, the receiver presents him with two sets of numbers:

0.0328E, 51.5081N
541100, 180800

Both sets of numbers refer to exactly the same location, but express it in different terms. The first is the *longitude and latitude* of the agent's position. The second is the same position expressed as *grid co-ordinates* in metres on the British National Grid. In order to understand the difference between these two ways of expressing the same position, we need to introduce some basic concepts of geographical positioning.

Measuring position using longitude and latitude

First, let us consider how we can define where we are on the surface of the Earth using longitude and latitude and explain the first set of numbers on the GPS receiver readout. Imagine looking at the Earth in cross section from one side, with the North Pole at the top, the South Pole at the bottom and the Equator running horizontally across the centre. Latitude determines our position north or south of the Equator and is the angle measured from the centre of the Earth between our position and the plane of the Equator. Latitude can range from 0 to 90 degrees north or south of the plane of the Equator and thus latitude references are usually given the postscript North or South to indicate whether they are above or below the equatorial plane. Figure 2.1 illustrates the concept.

The second parameter, longitude, specifies our position east or west of the Greenwich Meridian. This is an imaginary circle around the Earth

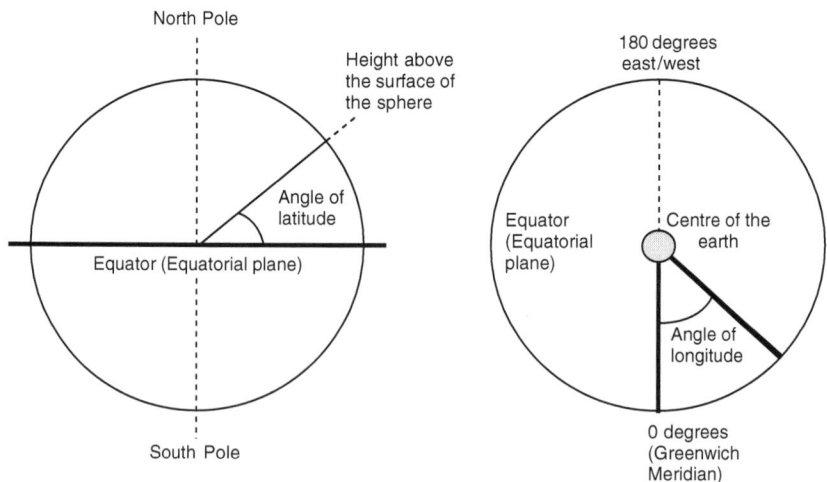

Figure 2.1 Latitude, longitude and height measurements on the sphere.

between the North and South Poles with its centre at the Earth's core and a diameter equal to that of the Earth. It passes through the Greenwich Observatory in East London, which has a longitude of 0 degrees. Figure 2.1 provides an illustration. In the figure, we are looking downwards onto a cross section of the Earth at the Equator. Longitude is the angle measured from the centre of the Earth in the equatorial plane from our position to the Greenwich Meridian and ranges from 0 to 180 degrees east or west.

Now we can return to the GPS receiver readout. The first number, 0.0328E is the agent's longitude, the angle between the Greenwich Meridian, the centre of the Earth and the agent's position. In this case, the estate agent is standing in East London, close to the Meridian, and this angle is very small. The second number, 51.5081N, gives the agent's latitude, the angle between the equator and his position. In this case, the angle is larger as the agent is some way north of the Equator in the United Kingdom.

The sphere, the spheroid and the geoid

So far, we have assumed that the Earth is round. In fact, this is not actually the case. Our planet is not perfectly spherical because of the gravitational effects of its rotation. The term spheroid gives a better approximation of the shape of the Earth than the sphere. An example of a spheroid is depicted in cross section in Figure 2.2.

Using a spheroid model of the shape of the Earth, we can still define positions in terms of latitude and longitude although there are some differences in the way these references are calculated. Rather than take positions based on the centre of the sphere, latitude and longitude are defined as the angle between the equator (in the case of latitude) or the Greenwich Meridian (in the case of longitude) and a spheroid normal. This is a line at right angles to the surface of the spheroid between our location on the surface and the plane of the Equator or of the Greenwich Meridian. Latitude on a spheroid

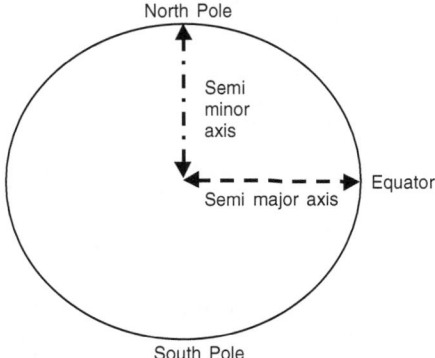

Figure 2.2 Cross section of the Earth as a spheroid.

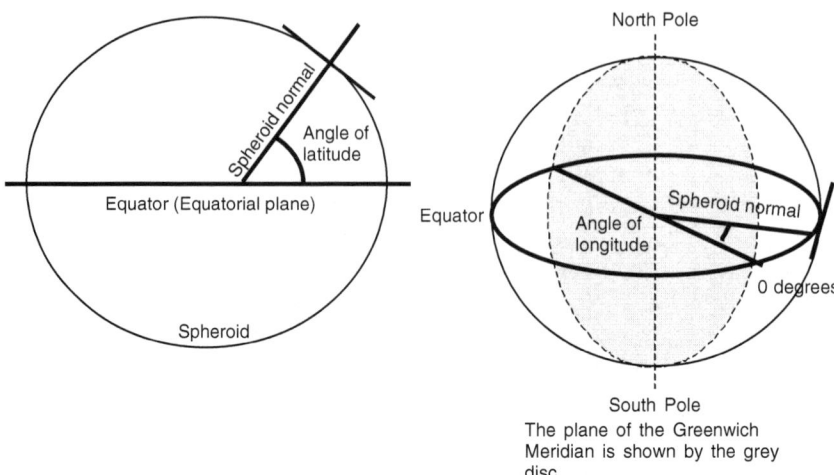

Figure 2.3 Latitude and longitude measured on the spheroid.

is thus defined as the angle between the Equatorial plane of the spheroid and the spheroid normal, while longitude is the angle between the plane of the Greenwich Meridian and the spheroid normal. Examples of latitude and longitude calculation on the spheroid are shown in Figure 2.3. Co-ordinates defined in this way are referred to as geodetic co-ordinates (Iliffe, 2000).

Although the spheroid represents a better model of the shape of the Earth than the sphere, the true surface of the Earth, known as the geoid, is somewhat different again. A more comprehensive discussion of the geoid and its implications for surveying and mapping is outside the scope of this book. The interested reader can find a useful introduction to the topic in Iliffe (2000), Cross *et al.* (1985) or an introduction to the study of geodesy and co-ordinate systems such as NOAA (1985) or OS (1999). For now, it is sufficient to say that the geoid coincides with mean sea level. It represents the surface to which the oceans would move if free to adjust to the combined forces of the mass attraction of the Earth and the centrifugal force of its rotation (NOAA, 1985).

Parallels and meridians

At this point, we must also introduce the concepts of parallels and meridians. A horizontal circle known as a parallel joins points of equal latitude on the spheroid. Parallels decrease in radius as they approach the poles,eventually becoming a single point at each pole. A circle that is equal in circumference to the circumference of the Earth itself is known as a great circle. The Equator is the only parallel that is equal in circumference to the Earth.

Just as parallels join points of equal latitude, points of equal longitude are joined by vertical circles known as meridians. However, all meridians are

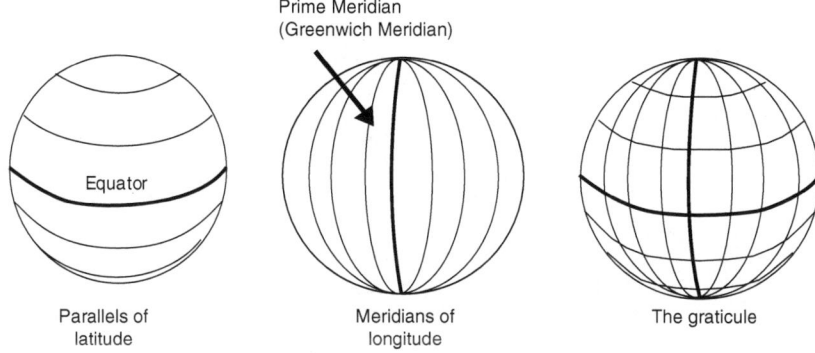

Figure 2.4 Parallels and meridians on the sphere.

great circles since they pass through both North and South Poles, and converge at the poles. We have already introduced the Greenwich Meridian, the great circle around the Earth in the polar plane at 0 and 180 degrees of longitude. The network of parallels and meridians that surround the Earth is known collectively as the graticule. Figure 2.4 provides an illustration of these concepts.

Co-ordinate systems and map projections

So far we have considered positioning on the three-dimensional (3D) surface of the Earth and how we can determine our position using latitude and longitude. We have also seen that we can approximate the true shape of the Earth using a shape called a spheroid. In order to explain the second readout on our estate agent's GPS receiver, giving his location in National Grid co-ordinates, we must introduce some additional concepts.

To draw a map of the spherical Earth on a flat surface we will need to transform the 3D spheroid into two dimensions. We carry out this process using a mathematical model known as a map projection. Map projections manipulate or transform the 3D surface of the spheroid so that it can be represented in two dimensions. Imagine taking an inflatable globe of the Earth and then stretching it – perhaps even cutting it – so that it can be placed flat on a piece of paper. The process of map projection is the mathematical equivalent of this. The surface of the spheroid is flattened mathematically so that it can be represented in two dimensions instead of three.

A very important side effect of this process is that it will always result in distortion. Areas, directions or distances on the surface of the spheroid are stretched in the transformation. Unfortunately, it is impossible to preserve all three of these properties when transforming the 3D spheroid into a two-dimensional (2D) representation. The choice of an appropriate projection is therefore a critical consideration for the mapmaker and the projection used

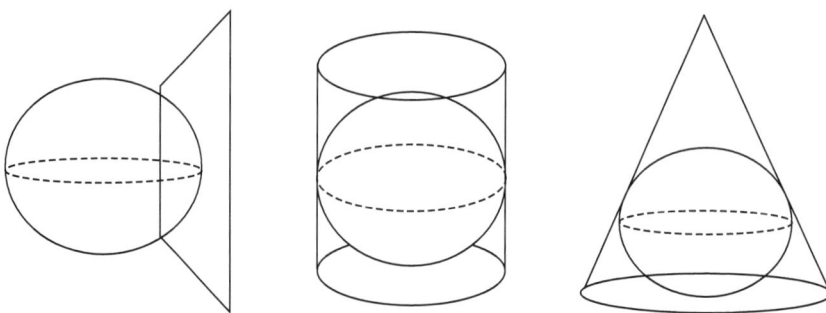

Figure 2.5 Three families of map projections: planar, cylindrical and conic.

will determine which geometric properties of the 3D spheroid can be depicted without distortion.

There are three well-known groups of map projections. Each is based on the use of a different shape for creating a 2D representation from the Earth's surface. These shapes are shown in Figure 2.5. There are numerous individual projections within each group, but the three share a common property, which is that the shape that the spheroid is projected onto can be transformed easily into a flat surface. The planar projection is already flat. The cylindrical and conical projections can both be 'unwrapped' mathematically so that they lie on a flat plane.

Different types of map projection are available that allow the user to preserve particular properties of the spheroid. The first of these types is called a conformal projection. Conformal projections preserve angles around a single point on the sphere, so that, for example, lines of latitude and longitude in the projection are always at 90 degrees to one another. In order to preserve angles, distortions must be introduced into the shapes of objects and therefore areas will not be correctly maintained. It is also very difficult to maintain angles over large areas, and so this type of projection will only work well for small portions of the Earth (DeMers, 2000). If the preservation of area is important we can choose an equal area or equivalent projection. In this case, the areas of objects are preserved, although their shapes may be stretched or otherwise distorted. At the same time, angular measurements will be distorted on an equal area projection.

A third type of projection, the equidistant projection, preserves the measurement of certain distances on the map. Caution is needed when using it because this property of distance preservation only holds true for certain parts of the map. There are two approaches to equidistant projection. The first ensures that distances measured along one or more parallel lines called standard parallels are correct. The second maintains the distance in all directions from either one or two points (DeMers, 2000). This means that distances measured from these points to any other location on the map will be correct, but unfortunately does not preserve accurate distances anywhere else on the map. The choice of the points where distance is preserved is therefore extremely important.

In summary, the process of map projection requires the mapmaker to make two critical choices. First, we must select a suitable spheroid model to represent the shape of the Earth. Next, we must decide on a map projection that preserves the properties we require (areas, distances or directions) and transform the spheroid into two dimensions using this projection.

Determining position on the plane with a grid co-ordinate system

Once we have projected the surface of the Earth onto a 2D plane, we need a method of determining our location on that plane. This is commonly achieved by superimposing a grid co-ordinate system onto the plane. The grid co-ordinate system makes use of linear measurements from a pair of fixed axes to determine location (Maling, 1973). The most commonly used form of grid co-ordinate system for mapping is the regular or rectangular grid. Here, the two axes are at right angles to one another and each is a straight line. The axes extend outward to infinity from an intersection point called the origin. The horizontal axis is called the abscissa or the X-axis and the vertical axis is called the ordinate or Y-axis. Any unit of measurement can be used to determine distance along each axis although both axes should use the same units.

Figure 2.6 shows an example of a rectangular, planar grid co-ordinate system in practice. Numbers on the X-axis to the right of the origin are positive while numbers to the left of the origin are negative. Similarly, numbers above the origin on the Y-axis are positive while those below the origin are negative. The geographical location of a point on the grid is defined using

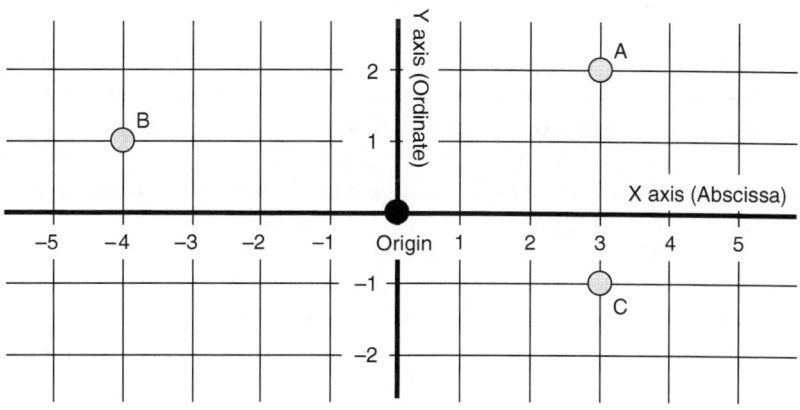

Figure 2.6 The grid co-ordinate system. The origin is the point at which both X and Y are equal to zero.

a pair of co-ordinates given in the form (X, Y) which give the positions of the point in terms of its distance from the origin along the X and Y axes. For example, in Figure 2.6 point A lies at location (3, 2) while point B lies at location (–4, 1) and point C at location (2, –1). In general, the X co-ordinate of a point is known as its Easting and measures distances east or west of the origin while the Y co-ordinate is its Northing and measures distance north or south of the origin.

We can now extend the process of map projection to a three-stage one:

1   Select a spheroid to model the shape of the Earth.
2   Choose an appropriate map projection.
3   Superimpose a grid co-ordinate system onto the projection and use this to measure off distances and obtain positions.

The British National Grid

Now let us return to the readout on the estate agent's GPS receiver. The second set of numbers, 541100 and 180800 are the Easting (X co-ordinate) and Northing (Y co-ordinate) of the agent's position on the British National Grid co-ordinate system.

The British National Grid uses a spheroid model developed by Airy in 1830 to represent the shape of the Earth. It is based on a conformal (angle preserving) cylindrical projection called the Mercator projection. In the example shown in Figure 2.5 the cylindrical projection is centred on the Equator and would give a good representation of the surface of the Earth at that point. In the British case, the cylindrical projection is rotated or transverse. The central meridian, which is the line of latitude that runs through the centre of the projection, runs down the centre of the British mainland. It is the meridian at 2° West of the Greenwich Meridian and the central point of the projection is at 49°N, 2°W, a point about 20 km south-east of St Helier in Jersey (Maling, 1973). This gives good local coverage and minimal distortion within a region of ±3° east and west of the central meridian and means that the projection covers the mainland of Britain well. Figure 2.7 shows how the Transverse Mercator projection for Great Britain is orientated.

A regular grid co-ordinate system is superimposed onto the Transverse Mercator projection to produce the National Grid itself. The Y-axis of the grid is aligned with the central meridian of the map projection at 2° West, so that all vertical grid lines are aligned in this direction which is known as grid north. Because meridian lines converge at the North Pole and therefore are aligned to true north, we will find that as we move east or west away from the central meridian, grid north, which is aligned in the direction of the central meridian, will begin to differ from true north. Figure 2.8 illustrates this effect.

This difference between true north and grid north is referred to as the angle of grid convergence. In general, the graticule is not usually shown on

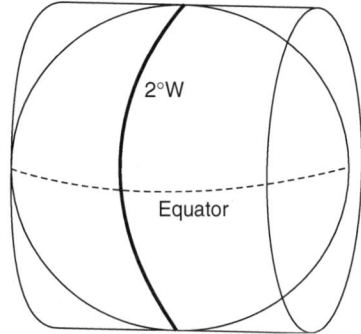

Figure 2.7 Orientation of the Transverse Mercator projection for the British National Grid. The cylindrical projection has been rotated so that its central meridian is at 2°W rather than the Equator. Transverse projections are often used for areas that are not close to the Equator.

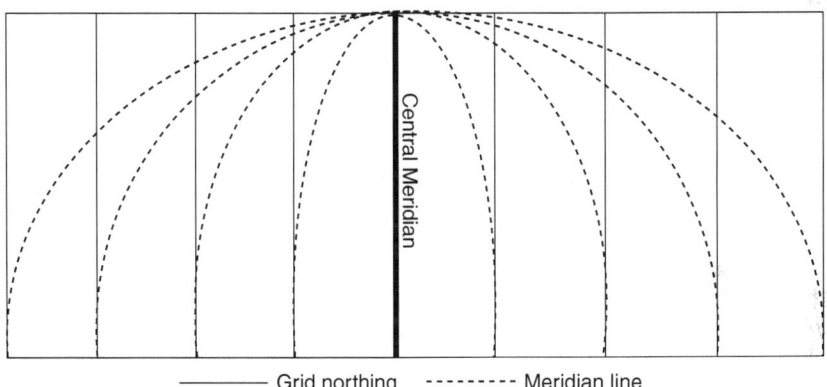

——————— Grid northing   - - - - - - - - Meridian line

Figure 2.8 The relationship between true north, the direction of the meridian lines, and grid north, the direction of the Y-axis of the grid aligned to the central meridian. Note how the divergence between grid north and true north becomes more pronounced as we move further east or west from the central meridian.

a map where a rectangular grid co-ordinate system has been superimposed. Instead of depicting the curved lines of the graticule on the map, and perhaps causing confusion with the grid co-ordinate system, the values and positions of lines of latitude and longitude are often marked along the edge of the main map. For an example of this practice, refer to the British OS's 1:50,000 Land Ranger series maps.

There is one additional complication. Consider Figure 2.9. The origin of the National Grid is at the central point of the Transverse Mercator map

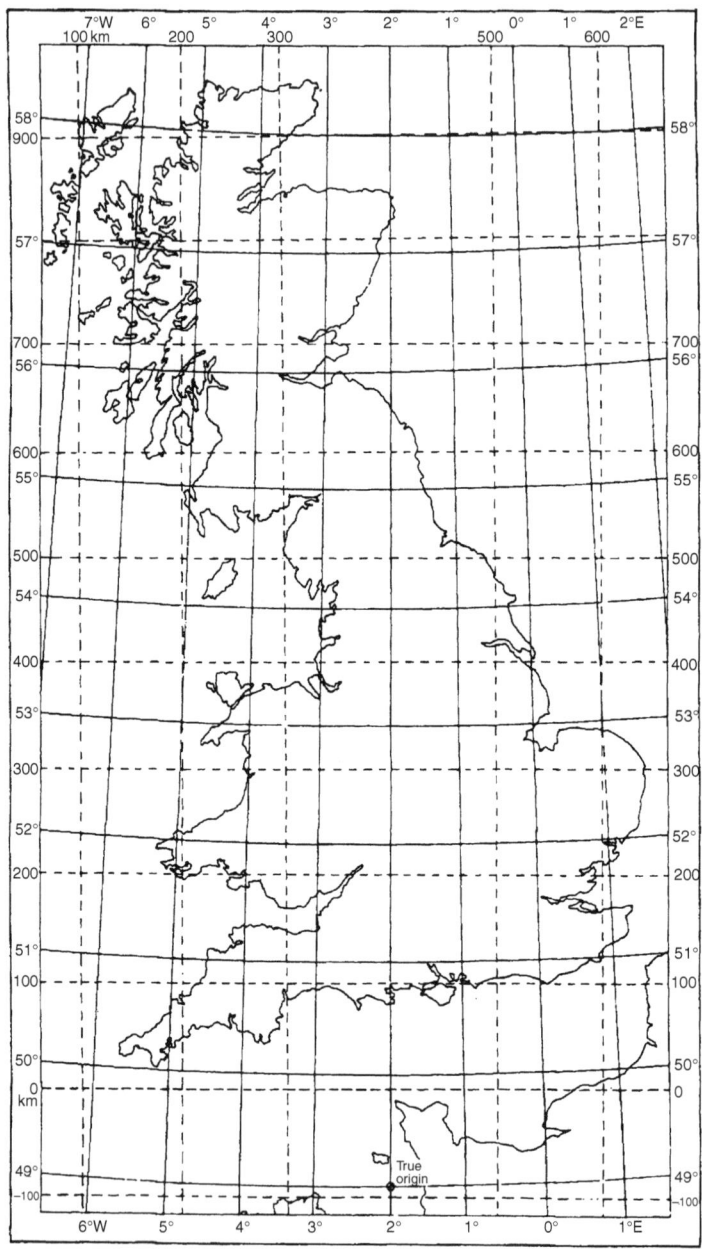

Figure 2.9 The National Grid with the true and false origins shown (Maling, D.H., 1973, Co-ordinate systems and map projections, George Philip & Sons).

projection at 49°N, 2°W. Although the British mainland is north of the X-axis, some parts of Britain are west of the central meridian of the projection and will therefore have negative X grid co-ordinates. While this is perfectly acceptable mathematically, it could be very confusing for users of the National Grid. To overcome the problem, we renumber the grid lines as if the origin of the grid was actually west of the British mainland, creating a false origin. This is located 400 kilometres west and 100 kilometres north of the true origin, and lies about 80 kilometres west of the Scilly Isles (Maling, 1973). The false origin is given a grid reference of X = 0, Y = 0. The National Grid is then numbered from the false origin and so the true origin actually has a grid location of X = 400,000, Y = −100,000. Figure 2.9 illustrates this point.

We are now able, finally, to explain the second readout on our estate agent's GPS receiver. The first number represents the X co-ordinate of the agent's position 5,41,100 metres (541.1 kilometres) east of the false origin in the Scilly Isles. The second figure is the Y co-ordinate of his position, 1,80,800 metres (180.8 kilometres) north of the false origin.

The map: a window on the world

Now that we know how to define a position on a map, we can discuss other aspects of the process of making maps and visualising information about land and property data. Let us begin with a very big problem: the world we live in is infinitely complex. This has far-reaching implications for the mapmaker, since it is impossible for us to create a perfect representation of such complexity. The solution has always been to be selective about what should be included on the map and what can be omitted without losing the information that we really need. The process of making a map is thus one of selection and simplification. Rather than trying to consider everything, we simplify our representation of the world by including only those features that are of immediate interest. A map is thus a representation or model of those aspects of the world that are of interest to us. It follows that different people will probably have different ideas about what is important and what is not important, depending on their objectives and the tasks they need to carry out. The process of map-making requires us to make choices about what to visualise and what to omit and how to go about representing the phenomena of choice effectively in order to convey the information we wish to depict efficiently.

Map scale

The scale of a map governs how much of the surface of the Earth it can depict. A large-scale map represents only a small area, usually in great detail, while a small-scale map covers a larger area but at the expense of detail. The most common method of expressing map scale is the representative fraction. This expresses the scale of the map as the ratio between map units and ground units. The larger the value of the fraction, the larger is the scale of the map.

Let us take some examples. The most detailed large-scale maps published by the OS of Great Britain, which cover a ground area of 500 square metres, have a representative fraction of 1/1,250. This means that 1 unit of measurement on the map represents 1,250 ground units, so 1 centimetre on the map represents 1,250 centimetres = 12.5 metres on the ground. The decimal value of the representative fraction is 0.0008. In the small-scale case, a map about 1 metre across representing the entire planet would need to be drawn at a scale of approximately 1:40,000,000 (Clarke, 1999), giving a representative fraction of 1/40,000,000 or 0.000000025. Notice that the decimal value of the fraction is significantly smaller. In this case, one unit on the map represents 40 million ground units, so 1 centimetre on the map = 40,000,000 centimetres = 400,000 metres = 400 kilometres on the ground.

There are two other common methods of expressing the scale of a map (Robinson *et al.*, 1984). The first is a simple verbal statement such as '1 inch represents 1 mile'. The second is to use a scale bar. This is a line on the map page which is subdivided to show the distance on the ground represented by particular distances on the map. An example is shown below in Figure 2.10.

Representation and symbolisation on maps

Objects are represented on a map by means of a wide range of cartographic symbols whose size and shape will be governed by the scale of the map, the objects that the cartographer wishes to describe and the properties of those objects that are deemed to be important. Over the centuries, the symbolism of cartography has become partially standardised, especially on topographic maps, but the enormous choice of symbols and variety of map types has prevented rigid standardisation (Robinson *et al.*, 1984).

In cartography, geographical phenomena tend to be represented using three categories of point (non-dimensional), line (1D) and area (2D) features

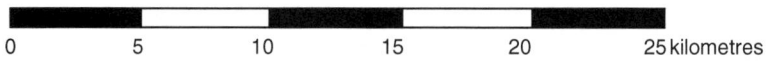

| 0 | 5 | 10 | 15 | 20 | 25 kilometres |

Figure 2.10 Example of a scale bar.

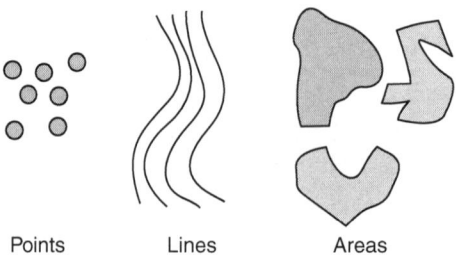

Points   Lines   Areas

Figure 2.11 Point, line and area representations.

(Figure 2.11) (Wright, 1955; Robinson *et al.*, 1984). We will introduce each of these briefly before continuing with our discussion.

Point features   A point is a simple position, a dot on the map. It represents the distinct location of an individual phenomenon. Point features are used for different purposes at different scales. At large scales, features like bus stops, trees, lamp posts or road signs might be represented by points. On smaller scale maps, towns and cities might be represented by points. The key concept behind the use of point data is the idea of existence of the object in question at a single location (Robinson *et al.*, 1984).

Line features   A line is used to represent any feature that is elongated or one-dimensional (1D). A river, a road, a footpath and an electricity transmission line are all examples of ground features that might be represented well using lines. Even though such features always have width as well as length (the breadth of a road or river, for example) they are often represented on maps using single lines of varying degrees of thickness (Robinson *et al.*, 1984).

Area features   Area features are 2D extents. Parks, fields, development sites, supermarkets, catchment areas, census districts, zones of vegetation cover and floodplains are all examples of common geographical phenomena that could be represented by such features.

Each of these representational approaches can be applied to many different geographical phenomena, but all geographical phenomena can be represented by one or more of them (Robinson *et al.*, 1984). It is entirely possible to represent the same phenomenon differently on two maps. For example, the city of London would be represented as an area on a map of southern Britain. On a small-scale map of the world, London would probably be represented with a single dot.

Representing descriptive information

In an effective cartographic representation it is not usually sufficient simply to mark the existence of geographical phenomena on a map. We are interested in conveying descriptive information about those phenomena as well, such as whether a house is detached or terraced, whether a river carries fresh water or salt water or whether the level of employment in an area is high or low. Such descriptive information is an essential component of the map-making process and will usually determine how the map features themselves are symbolised. For example, roads may be shaded in different colours to represent varying traffic carrying capacities. Different types of woodland area might be assigned different symbols to differentiate between evergreen and deciduous forests. In both these cases, specific descriptive properties of the features on the ground have been used to differentiate them from one another and govern the approach to symbolising them on the map.

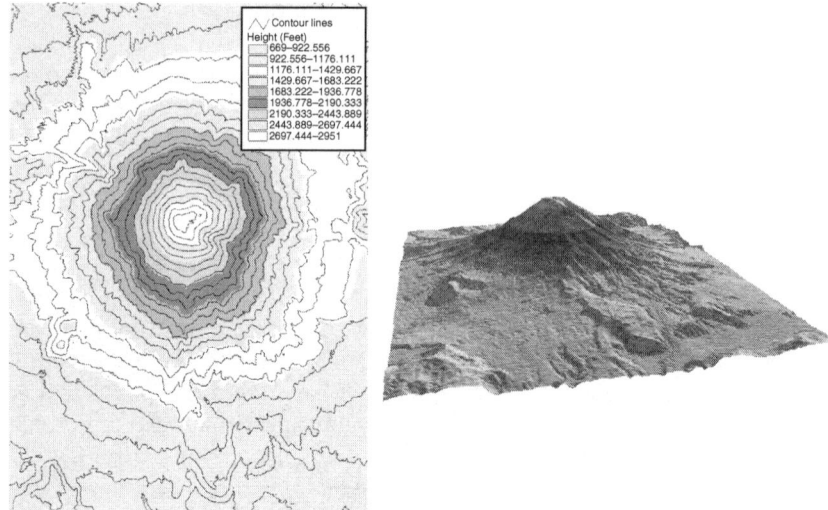

Figure 2.12 An example of surface representation. This is a visualisation of height variation across the Mt Saint Helens volcano in Washington, United States of America. The picture on the left shows a relief shaded map of the volcano, with contour lines superimposed. The same model is shown in three dimensions on the right. This technique can be applied to the visualisation of a wide range of continuous phenomena. Source data were obtained from the US Geological Survey.

The use of attribute data in symbolising geographical phenomena allows us to introduce a fourth method of representing geographical data: the use of surfaces or volumes. It is often useful to consider how a particular attribute of the surface of the Earth such as temperature or height above sea level changes as we move through space. An effective method of visualising this variation is through 3D representation using a surface. Examples of such surface models include the modelling of height above sea level, the density of population, or mean surface temperature. In many cases such attributes are measured according to a base level or datum (as in the case of height discussed above). In each case the variation of the attribute of interest across space is visualised on the map. An example of surface modelling is shown in Figure 2.12.

Discrete and continuous characteristics

We can make a further distinction between attributes that are present at all locations on the surface of the Earth and can potentially be measured anywhere and those that have distinct, clearly defined boundaries. The first type is known as continuous characteristics. Two examples of such continuous geographical characteristics are height above sea level and air temperature at the ground surface. Each could potentially be measured at an infinite number of points across the Earth. In contrast, the second type, discrete characteristics, apply only to individual features like houses, cities or land parcels whose geographical extents are clearly defined.

It is possible to convert discrete characteristics into continuous ones. A common example of this is the measurement of the density of occurrence. For example, although the value of a particular parcel of land is a discrete quantity that applies only to the geographical space occupied by the land, the average value of land per square kilometre is a continuous characteristic that could potentially be measured anywhere on the surface of the Earth.

Real and imaginary map features

Not all of the features on topographic or thematic maps are part of the physical world we live in. Often we may wish to place abstract or imaginary features on a map as well as real ones. We can therefore separate the features on a map into tangible ('real') and intangible ('imaginary') objects. Tangible map features exist as a part of the landscape. Examples are the buildings, rivers and roads we have already discussed. Intangible features are more abstract and are used to visualise particular characteristics of the landscape without actually being a part of the physical world. The extents of administrative areas like counties or districts are good examples of such intangibles – these lines do not exist on the ground physically, but help us to define and visualise information about the world. Similarly, measurements such as population density or average income are abstract concepts that can be used to visualise particular properties of ground features.

Generalisation and visualisation of cartographic data

We have said that the map can be thought of as a model of the world, and that the process of constructing that model is one of selection and representation. One of the most important tasks for the cartographer is to decide how to represent different map features so as to display them clearly and unambiguously for the intended audience. By its very nature, this process is scale dependent and the selection of the map scale will have a profound influence on what can be depicted on the map and the most appropriate way of carrying out that representation.

This brings us to generalisation, a fundamental process in cartographic construction, which is concerned with the effective representation on the map of features of interest on the ground. Consider the task of mapping a small town. At large scales such as 1:1,250, 1:2,500 or 1:10,000 it will be possible to show individual features within the town on the map – the detailed shapes of the buildings, the roads and their kerbs, bus stops, detailed descriptions of parks including tree positions, footpaths and so on. If we reduce the scale considerably, for example to 1:50,000 or 1:1,00,000, these features will begin to blur and mesh together. At small scales it is no longer effective from a representational point of view to depict them on the map individually or to map their boundaries with great precision. Instead we might use a single, shaded area with a greatly simplified boundary to represent urban spaces. At even smaller scales it would probably be most effective to represent the entire town as a single dot.

It is also important to note that symbolisation, especially at smaller scales, is sometimes more effective if the size of map features are exaggerated or reduced. Exact representational accuracy is not always compatible with effective communication. The lines representing features such as major roads or railways are often much wider on small-scale maps than they would be if represented accurately. The reason for this is simple – the accurate lines would be so thin that the map-reader would hardly be able to distinguish them.

The user should have an idea of the accuracy and precision required from a map depending on the application that is envisaged. As an example, OS 1:1,250 scale paper maps are stated to be accurate to approximately 0.4 metres. When we buy a paper map the only way we can 'zoom in' is by peering more closely at the map, perhaps with a magnifying glass. As we do so the lines and other features on the map appear thicker and therefore tell us intuitively that their accuracy cannot be assured at that closer level of inspection. With digital maps that are viewed using a GIS, we can zoom in to whatever scale we choose and (depending on the GIS we use) the lines may well look the same, giving a false sense of precision at scales greater than that which the map may have been surveyed at.

As a way of tackling this problem most GIS software allows the user to 'switch' thematic map information on and off depending on the scale that has been chosen for display purposes. For example if a map comprises two layers of information; one surveyed at 1:10,000 and the other at 1:2,500 scale, the former will be switched off as the user zooms in to a scale of 1:9,999 or more. In this way, cartographic generalisation is built into the GIS.

It is important that the user considers the level of cartographic detail needed at the start of a GIS project. For example, a local authority department might ask for a small-scale boundary map to use in day-to-day work. Later, when it is realised that the exact positioning of the map boundary may determine whether a property receives a grant or not, a much more detailed large-scale map may be needed. Such a request could be met with little difficulty if the original data were surveyed on a large-scale map, but if data collection were tailored to the more generalised small-scale map output, then the data would have to be resurveyed.

Topographic and thematic maps

Maps come in many different shapes and forms, but there are two important families of maps that we will introduce here: topographic and thematic

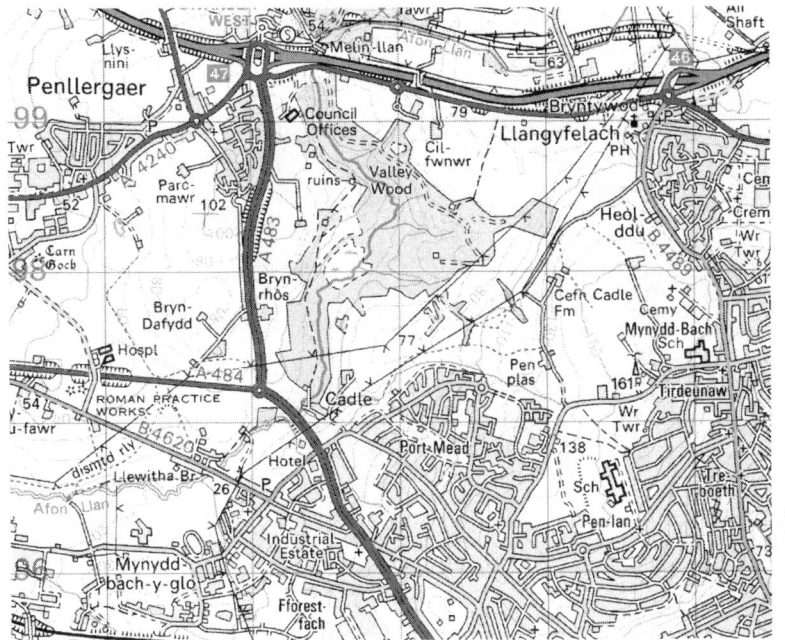

Figure 2.13 Topographic mapping from the OS 1:50,000 Landranger series of maps of Great Britain.

Source: Crown Copyright, 2001, used with permission.

maps. Of these, the topographic map is probably the most familiar to most of us. Topographic maps depict natural and man-made features on the surface of the Earth such as roads, rivers, towns and woodland. An example is shown in Figure 2.13.

Thematic maps are somewhat different. They are drawn to depict specific descriptive information about particular ground features and often greatly simplify or omit other ground features entirely. Here are three examples of thematic maps:

1 A map that uses data from the Census of Population to show levels of unemployment in the counties of England;
2 A geological map showing the extents of soil and rock types;
3 A weather map showing rainfall levels or temperature characteristics across a region.

In each of these cases ground features that are not directly relevant to the theme of the map are usually omitted for clarity. For example, the map of population at county level would probably show only the county boundaries and the counties themselves colour coded according to unemployment levels.

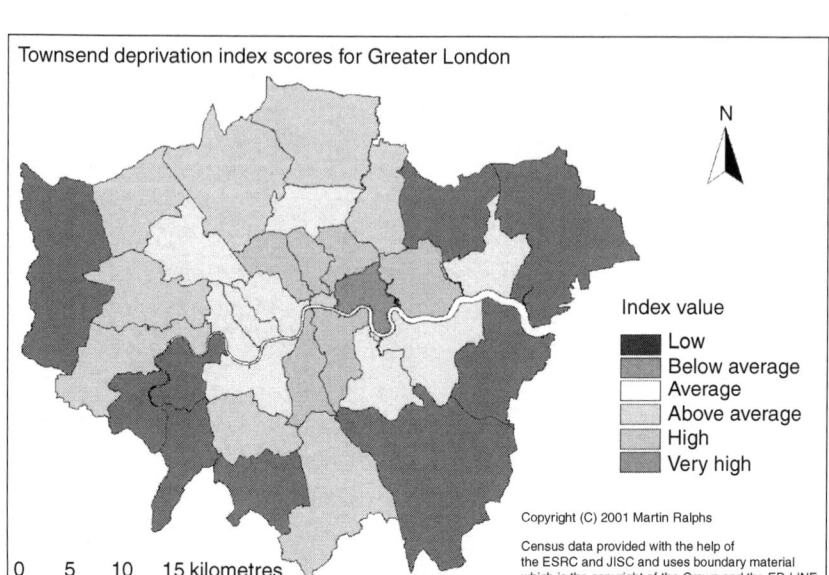

Townsend deprivation index scores for Greater London

N

Index value

■ Low
▨ Below average
□ Average
▨ Above average
▨ High
■ Very high

Copyright (C) 2001 Martin Ralphs

Census data provided with the help of
the ESRC and JISC and uses boundary material
which is the copyright of the Crown and the ED-LINE
consortium.

0   5   10   15 kilometres

Figure 2.14 Choropleth map depicting relative scores on the Townsend Index of
Deprivation for the Boroughs of Greater London. In this example, a set
of reporting zones (the Boroughs) are superimposed onto the landscape
and the characteristics of all the points within those areas are
summarised at Borough level.

We can further subdivide thematic maps into three types that are widely
used in GIS. The first of these is the choropleth map. Choropleth maps
summarise data according to a defined set of reporting zones such as
numbers of votes for political parties in different electoral districts or the
average price of property in the different counties of England. The
socio-economic mapping example discussed above fits into this category.
On such a map, the theme of interest is summarised according to the
amount of occurrence in each of the reporting zones. An example is shown
in Figure 2.14.

The second type of thematic map is the area class map. In this case, it is
variation on the ground that defines the size and shape of the areas. This is
different from the choropleth map, where the sizes and shapes of the areas
are defined first and the variation on the ground is summarised in those
areas. Examples of area class maps are vegetation maps that show the
extent and size of different types of vegetation across a study area and
geological maps that show subsurface geological patterns. An example of
an area class map for vegetation types is shown in Figure 2.15.

The third type is the isopleth map (from the Greek 'iso', meaning 'equal').
Isopleth maps are used to visualise a surface by joining up points of equal
value with lines. Perhaps the most familiar example is the contour map, on

Figure 2.15 An example of an area class map showing soil types. Different soil types are colour coded and each area on the map is shaded according to the underlying soil type. Image classification created using ERDAS imagine and LANDSAT Satellite Imagery.

which the lines join points of equal height, and an example is shown in Figure 2.16.

Another common application is the map of barometric pressure familiar from television weather forecasting bulletins where isobar lines join points of equal pressure. In the case of land and property management an application of this type of map might be contour mapping of property prices across a city or the calculation of drive times ('isochrones') from a development site to other key locations.

Bringing geographical information into the computer

Modern cartographic processing is usually carried out with the aid of a computer and the analytical capabilities of GIS are almost exclusively carried out using computer technology. We will now begin to investigate how the concepts and principles of mapping that we have described in the previous

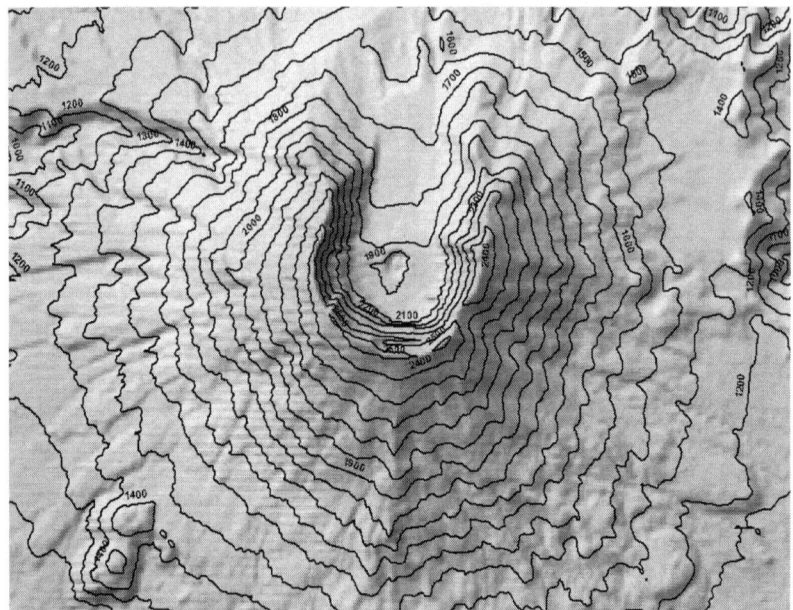

Figure 2.16 Contour mapping of height on Mt Saint Helens.
Source: Data provided by US Geological Survey.

section are applied when we enter the digital environment. Computers allow us to manipulate and visualise geographical information in exciting and powerful new ways. That said, the principles and concepts that determine how effective a map is in representing aspects of the world apply just as strongly to the creation and manipulation of digital maps as they do to our use of traditional paper maps.

Contrasting digital and paper maps

Because of the way that printed maps are made, they have limitations. A printed map is a drawing on a piece of paper or plastic/film media. Points, lines and areas are displayed using various symbols, shadings, colours etc. A key or legend is then used to explain what all the symbols mean. This has several implications.

First, the data on which the map is based must usually be simplified and classified to represent them effectively on the map. This implies that the mapmaker must interpret and filter the information subjectively and decide how to deal with the issues raised in our discussion above. The map must be drawn very accurately and the presentation has to be clear. Once the data are on the map, they are fixed. It is time-consuming and expensive to extract them again and altering the manner in which they are displayed will require a complete redraft of the map.

Second there is the issue of the physical size of the map sheet. A common problem with paper maps is that the area of interest may need to be represented by a number of map sheets, particularly when the map scale is large. This may require the map user to move sheets around or work with more than one map sheet at once which can be cumbersome and time-consuming. Also, the printed map is a snapshot at a given point in time, and may quickly become outdated and the material on which the map is drawn may be unstable, so that its size and shape may change given exposure to different humidity levels or temperatures. Finally, because printed maps are usually paper-based they may easily become damaged. While it is certainly true that computer disks can also be damaged, the same map can be stored lots of times on a computer system with relatively low cost.

By using computer systems, we can overcome a lot of the problems with traditional maps. Large amounts of information can be stored cheaply and accessed quickly. We can use the computer to bring together lots of different pieces of information and synthesise them in an integrated product. The computer can carry out complex and time-consuming cartographic assembly and drafting work quickly and easily. Additionally, because we can store many map sheets simultaneously in the computer we can browse an area without worrying about map boundaries. We can also change scale freely and redraw our map to show different levels of detail. Finally, we can project data into three dimensions and explore and visualise them in ways that are impossible with traditional maps. We can animate the progress of information through time. Most importantly of all, the computer allows us to decide for ourselves what we want to map and how we want to map it. Given all these advantages, a word of warning is also necessary. Although the computer liberates us from many of the restrictions of traditional cartographic drafting, the principles of good cartographic practice still apply and many of the same decisions about generalisation and symbolisation will still have to be made to ensure effective depiction of the required information.

Storing geographical information in a computer

The world is in effect infinitely complex and this presents us with a seemingly intractable problem. How can we represent an infinitely complex world in a computer system with limited processing and data storage capabilities? The answer is that instead of trying to represent everything, we make a computer model that includes the essential information that we need. The computer can be used to store simplified representations of those things that we need to study and explore. A GIS database is therefore a model of the world.

How does this modelling process work in practice? We can define three broad stages of activity:

1   Definition – what phenomena interest us and how are they defined?
2   Classification – how do these phenomena relate to one another?
3   Representation – how can we represent them in the computer?

A GIS contains representations of things in the real world: objects like houses, trees, roads, rivers and lakes. It may also contain representations of artificial objects like contour lines and political boundaries or historical information that no longer exists. It is important to remember that a computer model of the world is NOT the same as the real thing: it is an abstraction. A computer model is limited by the decisions of the information collector, the data that have been used to build the model and the way in which it is represented in the computer.

Making a model of the world

We will take an example by considering the representation of a simple everyday object: a house. The first stage of the modelling process requires us to define what we mean by the term 'house'. This might be more difficult than it sounds. For example, is the definition a generic one so that single storey bungalows, apartments, detached houses and terraced houses all fall within the same grouping, or do we require more detailed discrepancy between different types of dwelling place? By thinking about these issues, we have already introduced the concept of classification – we may have a class of objects 'house' or 'dwelling place' within which there are a number of subclasses.

Representing the house in the computer requires us to make a number of new choices. The first concerns the level of detail that we are interested in. At smaller scales, we may only require the location of the house to be shown, so that a single dot indicating its position will be sufficient. At larger scales, a more detailed plan of the extent of the property may be required. At this stage we might like to revise our definition to take into account the geography of the house – do we include the garden and any outbuildings in our definition, or does this require a new set of concepts to be included in our model of the world? We also need to decide what descriptive information we need to include in the model. For example, we may wish to record the total floor area of the house, the number of rooms it has, the year it was constructed or its current market value.

Let us assume that we decide to record the house, its gardens and outbuildings in the model. We have defined our objects of interest and how they relate to one another. The final stage of the process is to encode this information digitally so that we have a computer record of the house including all of the information we need about it.

Storing information in a GIS

A GIS commonly stores two types of information. The first is geographical or spatial information that is referenced in some way to the surface of the Earth using a co-ordinate system or other framework and defines the geographical extent of the objects of interest. The second is descriptive or attribute data about those objects.

In order to use the GIS to answer questions about the world, we need to organise this information efficiently. To do this, we need to employ one or more data structures. A data structure can be simply defined as a way of organising information inside a computer system. A simple example of a data structure is an ordered list, for example a set of different tree species arranged alphabetically:

- Ash
- Beech
- Cedar
- Chestnut
- Elm
- Oak
- Olive
- Pine
- Walnut.

The way information is organised in a computer will have a profound effect on our ability to retrieve it quickly and to process it in different ways. In the list of trees shown above, the information is organised into ascending alphabetical order and we can use this fact to search efficiently through the list. If the list were not alphabetically ordered we would need to look through each entry individually to find the information we required. If the list were much longer, perhaps including thousands of names, this process would be very lengthy. In a similar manner, the more complex data structures that are used to store geographical information will affect our efficient use of that information. We will introduce data structures for storing geographical information by considering some of the most widely used methods.

There are two commonly used data structures in GIS – vector approaches and raster approaches. In the vector approach, objects on the ground are represented using three simple building blocks: points, lines and areas. Line and area features are composed of linked sets of points and objects like houses or parks are modelled by connecting points together with straight lines and grouping the lines into areas. In the raster approach, the area of interest is divided up into tiles or grid cells of equal size and geographical extent. Cells are then coded according to the properties of the area they cover.

Organising and representing vector data structures

For a vector database to adequately represent information about the real world, it must allow us to define the geographical extent of objects and permit the attachment of pertinent descriptive information to these objects. The software that is used to achieve this goal normally consists of a mapping program attached to a database. Information about the vector structure is stored in the database and used by the mapping program to display information on the screen.

We have seen that the vector structure is based upon three primitive objects: points, lines and areas. How might we represent these features in the GIS? Individual GIS systems have different approaches to this problem, but the example below shows how we might use the simple building blocks of point, line and area to create a vector database and demonstrates the broad principles of the data structure.

Representing points

Point features are quite simple to handle. Each point has an X co-ordinate, a Y co-ordinate and a unique identification number. The identification number allows us to differentiate between the different points. Geographical co-ordinates for each point are stored in a database alongside its identification number. Figure 2.17 provides an example.

Representing lines

In the simplest case, line features can be stored as strings of co-ordinate pairs. Consider Figure 2.18, which consists of five different lines. Each line

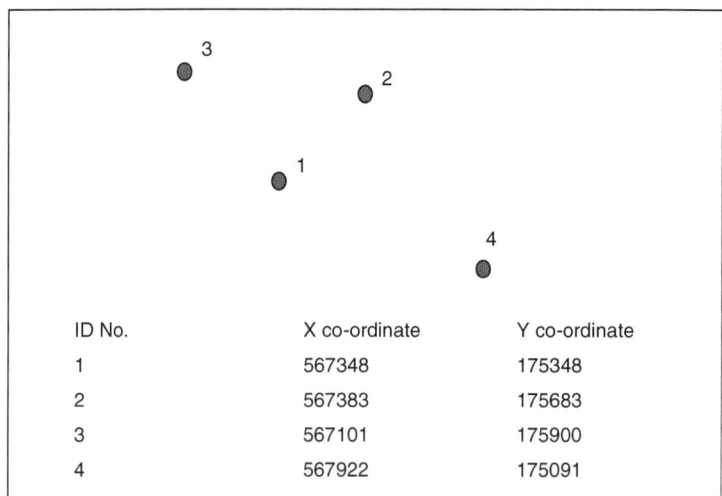

ID No.	X co-ordinate	Y co-ordinate
1	567348	175348
2	567383	175683
3	567101	175900
4	567922	175091

Figure 2.17 Representation of point features. Each point on the map is given a unique identification number in the database. These numbers link to the co-ordinates of the points in a table and can be used to position them on the map.

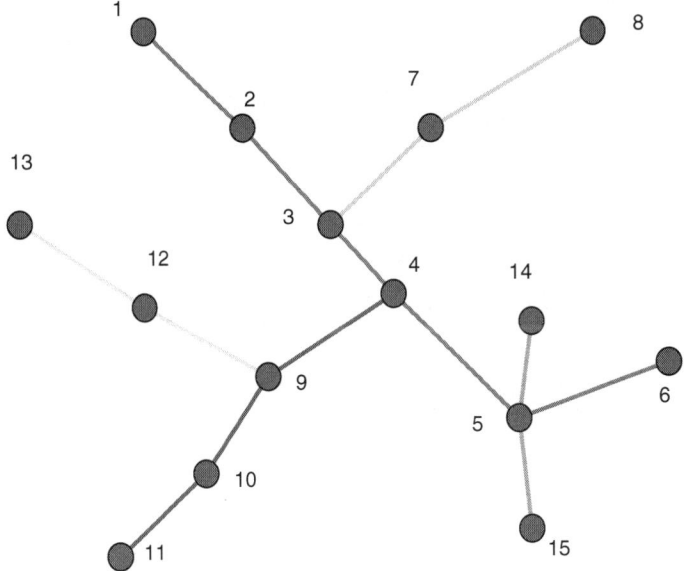

Figure 2.18 Representation of lines in a GIS.

is made up of a series of points, whose co-ordinates are known and stored in a database. We can store a line by making reference to the identification numbers of the points that comprise it. This in turn can be cross-referenced to a table of points that contains the geographical co-ordinates of each input point. For Figure 2.18, that table might look like this:

Line	Points
Orange	1,2,3,4,5,6
Cyan	3,7,8
Blue	4,9,10,11
Purple	9,12,13
Green	14,5,15

The problem with representing line features in this way is that we do not know very much about them. We cannot answer questions such as 'Does the blue line join the orange line?' without carrying out additional calculations to work out which points are at the beginning and end of each line. It would be very helpful to know more about the interconnections between the lines, and so the simple data storage method that has been shown is often extended to include additional information.

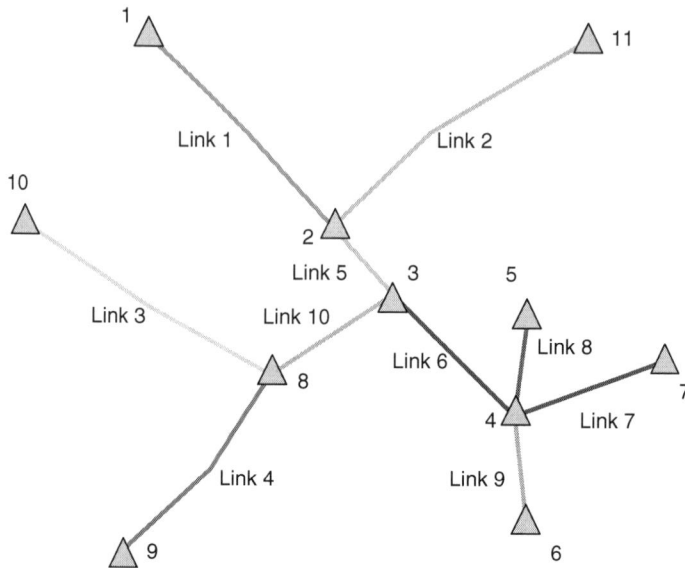

Figure 2.19 Example of a link and node vector data structure based on Figure 2.18.

Link and node structures for line features

A commonly used addition to the simple method of storing lines described above is the link and node approach. Here, a line is defined as having beginning and end points referred to as nodes, while the line and its component points between the nodes become a link. A link can be defined formally as a series of non-intersecting line segments with no connection to another link except at the start or end point. A node is defined as the start or end of a link, can be shared by many links and is only stored once, which is not the case in the simple structure described above. Consider Figure 2.19, which shows the link and node structure for the lines in Figure 2.18. You will see that the original set of five lines has been split into ten lines, each beginning and ending at an intersection between lines or an unconnected end point.

Storing the link and node structure

The link and node data for the structure in Figure 2.19 can be represented using two cross-referenced tables – an example is shown here.

The link and node structure allows us to store information about the connectivity of points and lines. The connectivity of links is expressed in

Links table			Nodes table		
Link	*From node*	*To node*	*ID*	*XY*	*Links*
1	1	2	1	X1, Y1	1
2	2	11	2	X2, Y2	1,2,5
3	8	10	3	X3, Y3	5,6,10
4	9	8	4	X4, Y4	6,7,8
5	2	3	5	X5, Y5	8
6	3	4	6	X6, Y6	9
7	7	4	7	X7, Y7	7
8	5	4	8	X8, Y8	3,4,10
9	4	6	9	X9, Y9	4
10	3	8	10	X10, Y10	3
			11	X11, Y11	2

terms of their 'direction of flow', usually in the direction of data capture, and recorded using a 'from node' and a 'to node'. In the example, link 3 goes from node 8 to node 10. From link 3, we know that we can go to link 10 or link 4, for example. By storing information in terms of links and nodes, we are able to navigate through the line network.

Extending the link and node structure to polygon features

Until now we have only considered point and line features. We also need to consider the storage of area features, ideally in a manner that will allow us to find out information about the relationships between them. Fortunately, the link and node structure can be extended to cope with the storage of information about areas.

Consider the diagram in Figure 2.20 and the accompanying tables. In this case we have constructed four areas (polygons) from a link and node structure. Because we know about the connectivity between nodes we can work out which links are connected and which polygon each set of links encloses. Similarly, we can use the connectivity information in the structure to work out which polygon is next to others.

Using such a structure, we can derive information about the spatial relationships between polygon features. The new table is then cross-referenced to the previously derived link and node structure that defines the positions of nodes and the structure of individual lines. In this structure, there is little duplication and geometrical updates are easy to maintain. However, if geometry is altered, topology will need to be reconstructed.

This type of structure can be easily integrated with a standard relational database architecture. It is predicated on the existence of a unique identification number (or sequence of unique attribute values) that sets each record apart from the remainder of the database. Using this unique number,

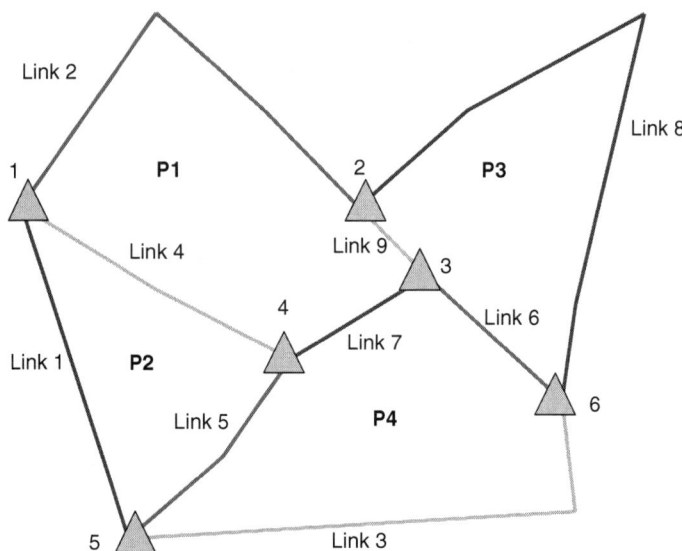

Figure 2.20 Storing polygons using the link and node structure.

Polygon	Links
P1	2,9,7,4
P2	4,5,1
P3	8,6,9
P4	6,3,5,7

Link	Left poly	Right poly	From poly	To node
1	P2	0	1	5
2	0	P1	1	2
3	P4	0	5	6
4	P1	P2	1	4
5	P4	P2	4	5
6	P3	P4	3	6
7	P4	P1	3	4
8	P3	0	6	2
9	P3	P1	2	3

information about a single geographical feature can be held in several data tables and cross-referenced. The vector structure can also be easily linked to a standard database management system, allowing for the storage of a wide variety of descriptive information for each vector feature.

Benefits and limitations of the vector data structure

The precision of the vector structure allows for high levels of cartographic accuracy in the representation of geographical phenomena. It also permits selective sampling, so that the user can choose which objects to record and how much detail is required. At the same time, the structure can be extended to permit the easy encoding of relationships and connectivity between objects. As such, the structure is highly effective for the representation of networks and the retrieval of information using spatial relationships. Additionally, the vector structure can cope with the storage of a wide range of descriptive information about the phenomena of interest, particularly when linked to a database management system. Given these advantages, vector structures are the most commonly used method of storing information about land and property databases. Features like buildings and land parcels can be stored with high levels of accuracy, which is particularly important when recording property boundaries and measuring characteristics such as square footage.

There is another side to these benefits. The vector data structure can be misleading because of its precision. The appropriateness of high precision data storage is dependent on the accuracy of the land survey or other recording method that was used to collect the vector data in the first place. For example, the Ordnance Survey of Great Britain quotes an average accuracy of 0.4 metres for its large-scale Landline data sets. This would allow for 1 decimal place of precision in the representation of National Grid co-ordinates. However, many modern GIS systems store co-ordinate information with a precision of 5 or 6 decimal places. This is much more precise than even the most accurate of land surveys and in the case of OS large scale data would suggest that measurements are accurate to one thousandth of a millimetre!

Boundaries in the vector structure must be clear-cut. This works well when modelling the built environment but is less successful when used to represent natural phenomena that do not exhibit such behaviour. Boundaries between soil types or vegetation zones tend to be graduated and do not lend themselves to precise representation in this manner and a note of caution must therefore be sounded when using the structure for modelling these types of geographical data.

The vector structure is quite complex and geometric operations such as the overlay of different sets of data or the generation of new data sets based on spatial relationships are computationally intensive. Effective vector representation also requires large amounts of preparatory structuring to build topological relationships.

Organising and representing raster data structures

An alternative method of spatial data representation can be found in the raster data structure. Some common examples of raster data sets used in GIS include satellite imagery, aerial photography, scanned imagery (e.g.

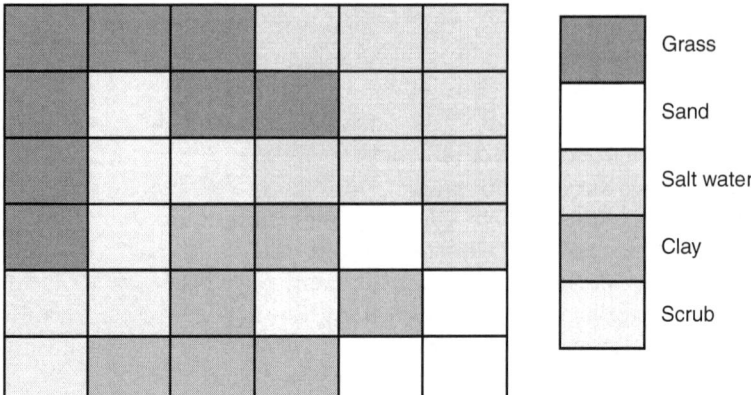

Figure 2.21 The raster data structure. Cells are coded according to the ground cover onto which they are superimposed.

maps that have been digitally scanned) and elevation data. A raster is created by subdividing geographical space into contiguous cells (usually square or rectangular and all the same size). The cells cover the area of interest, forming a mosaic or grid. Figure 2.21 provides an example.

Each grid cell in a raster can store a single attribute value. For example, in the case of a black and white photograph, this value might be a number from 0 to 255 describing the brightness of the cell from black (0) through many levels of grey to white (255). It follows that to store many different pieces of information using the raster method we will need to create many different rasters covering the same area. These are commonly known as raster layers.

Consider the familiar topographic map. Such maps often depict lots of information – county boundaries, road networks, railways, cities and towns, water features and so on. Each of these different themes would normally be stored in a different raster layer. A typical raster database might contain hundreds of layers, each consisting of thousands of grid cells.

Properties of raster layers

There are two important characteristics associated with raster layers. The first of these is the resolution of the raster. We can define resolution as 'the minimum linear dimension of the smallest unit of geographic space for which data are recorded' – in other words, the length and breadth of a single cell in the raster.

The reader will probably be familiar with the terms 'high resolution' and 'low resolution' from the computer graphics industry, where they refer to the amount of detail that is recorded in a graphical display. In the same way, a 'high resolution' raster data set contains lots of detail – this means lots of

 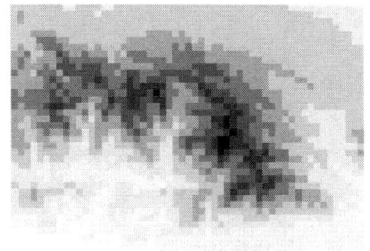

Figure 2.22 An example of high resolution (left) and low resolution (right) raster
data. Notice that the grid cells on the right-hand example are much
larger than those on the left, giving a correspondingly coarser representation of the landscape.

cells, where the geographical area covered by an individual cell is quite
small. 'Low resolution' refers to the opposite case – fewer cells that cover
a larger geographical area and provide correspondingly lower levels of
detail. Figure 2.22 provides an example. The figure shows the same set of
elevation data portrayed with a raster grid cell size of 90 metres (left) and
1,000 metres (right). Notice how the detail levels on the lower resolution
data set are substantially reduced.

The second important characteristic of a raster is its orientation. We can
define this as the angle between true north or grid north and the direction
defined by the columns of the raster. Figure 2.23 shows two examples. The
first grid is oriented in the direction of grid north. The second has an
orientation of approximately 45 degrees from grid north.

Although raster data produced from scanned paper maps will usually
have the same orientation as the map projection of the original map, the
same is unlikely to be true of raster data obtained from other sources such
as orbiting satellites or aerial photography. Before raster data can be combined with other information in a GIS, it must first be correctly aligned in
relation to grid north and the map projection that is being used.

Storing information for raster cells

Raster cells can be assigned a wide range of different data values, including numbers, labels, logical variables and dates. Exactly what can be stored
will be dependent on the capabilities of the GIS software that is used to
manipulate the raster. Each cell in the raster layer is assigned a value.
Multiple layers will be required to store a range of descriptive information.

Locating raster cells

The location of a raster cell is usually defined by an ordered pair of
co-ordinates that identify the location of an individual cell in the raster by

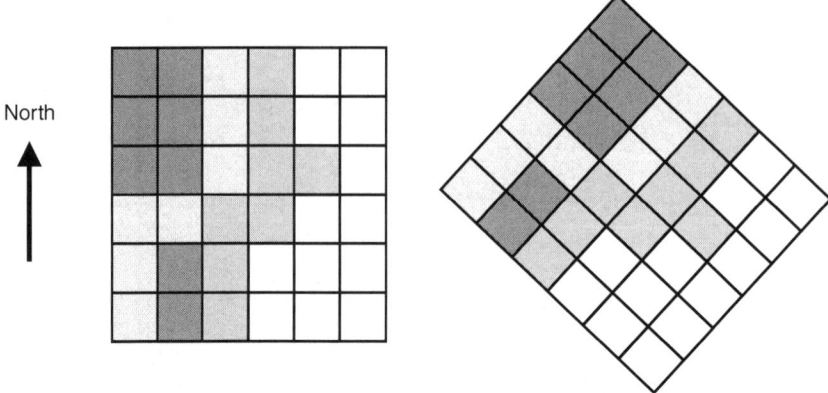

Figure 2.23 Raster orientation in relation to true north. The left raster is oriented in the direction of north. The right raster is oriented to approximately 45 degrees from true north.

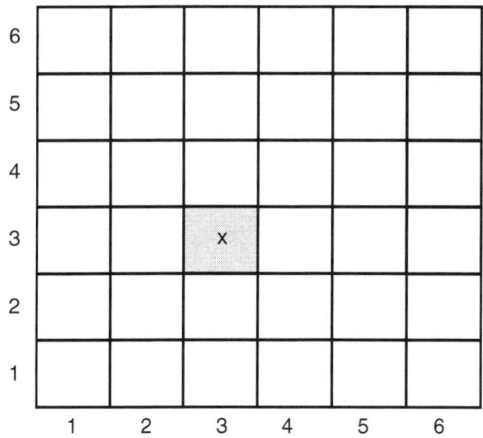

Figure 2.24 Positioning on a raster grid. Cell X has grid location X = 3, Y = 3.

its row and column references. Usually, the true geographic location of one or more corners of the raster is known and assuming that the resolution of the raster is also known it is then easy to derive grid-referenced co-ordinates for individual cells. Let us take an example.

Consider the simple raster data set in Figure 2.24. It consists of six rows and six columns. If we want to describe the location of grid cell X in the figure, we do so using its row and column numbers. Usually, numbering of raster cells starts at the bottom left. If we assume that this is the case in our

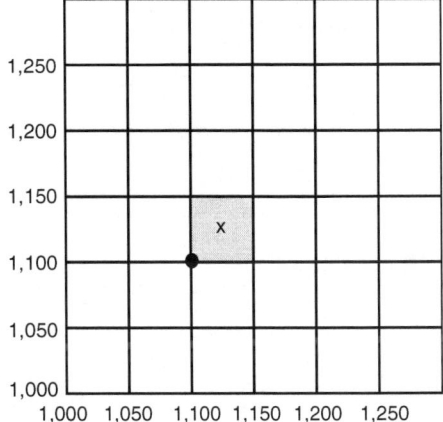

Figure 2.25 Spatial referencing on a raster grid.

example, grid cell X is in row 3 of the raster and column 3 and so its cell location would be (3,3).

Now let us assume that the raster has a cell resolution of 50 metres. Each grid cell is 50 metres across and 50 metres from its northern edge to its southern edge. If we know the location of the bottom left-hand corner of the grid in map co-ordinates, we can calculate the position of cell X in map co-ordinates. Assuming that the bottom left-hand corner of the raster has map co-ordinates (1000,1000) then it is quite straightforward to work out the co-ordinates of other cells in the raster. The bottom left-hand corner of our grid cell X at row 3, column 3 of the raster (marked with a black dot in Figure 2.25) will have map co-ordinates of (1100,1100)* while the centre of cell X will have grid co-ordinates (1125,1125).[1]

Benefits and limitations of raster structures

The key advantage of the raster data structure is its simplicity, and this brings with it some important benefits. Many operations such as the overlay of data layers on top of one another or the definition and selection of groups of cells based on distance from specified points can be carried out very efficiently using raster methods. Similarly, operations requiring the re-coding of grid cells like colour shading or reclassification are very rapid. These techniques are discussed in more detail in the next chapter.

The simplicity of the structure is also its major limitation. It is difficult to represent objects with a sufficient level of cartographic precision using raster methods, since the resolution of the raster grid needs to be prohibitively high to capture an equivalent level of detail. Similarly, the topological relationships that can be encoded into the vector structure are much

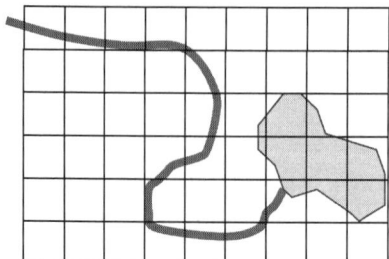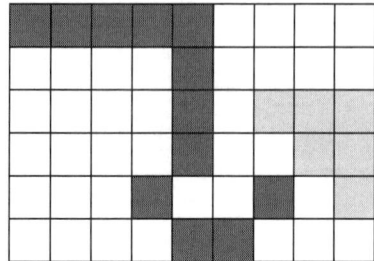

Figure 2.26 Comparison of vector and raster representations. In the vector case on the left, it is possible to model the sinuosity and extent of the stream and lake quite accurately. The raster on the right provides a much cruder representation and cells are classified as 'stream' or 'lake' when only a small component of the cell actually contains these features. Representation could be improved by increasing the resolution of the raster, but this increase would need to be substantial to achieve the accuracy of the vector structure.

harder to represent using a raster. For these reasons, raster data structures tend to be used less than vector structures for handling land and property information. Their most common application is in the representation of aerial photography for overlay by vector data.

A related problem is the assumption that the occurrence of a cell value is geographically consistent throughout a grid cell. This may prove to be inaccurate, as the boundary of two features or categories on the ground might cut across the middle of a raster cell. The possibility that the cell is not internally homogeneous increases as the resolution of the raster is reduced. Figure 2.26 illustrates this using the simple example of a stream flowing into a lake.

Although the layer-based model described above is commonly used it often results in large numbers of raster layers. Some GIS packages have overcome this problem by allowing for the linkage of raster data to database management systems which permits the storage of multiple variables against a single raster layer (DeMers, 2000).

A final problem is that of storage space. The simplicity of the raster structure means that it requires large amounts of disk space to store in a raw, uncompressed form. Although the cost of computer storage today is extremely low, there are related performance issues in the transfer of large volumes of data to and from the computer disk. As a result, extensive use has been made of compression methods and more complex data structures for the representation of rasters. A review of these methods is beyond the scope of this chapter. The interested reader is referred to standard texts on GIS such as Burrough and McDonnell (1998), DeMers (2000), Laurini and Thompson (1992) for a more detailed treatment of the topic and some operational examples.

Summary

In this chapter, we have considered ways in which geographical information can be represented visually using maps and the underlying co-ordinate systems and models of the shape of the Earth that underpin them. We went on to consider how such information can be encapsulated digitally in a computer system, and the data storage methods that are used to hold and manipulate digital geographical data. We explored the use of simple spatial features such as points, lines and areas to characterise particular objects. We also considered ways of using the map to depict the characteristics of those objects with different types of symbolisation and use of colours.

Two families of maps are commonly used within a GIS environment. Topographic maps depict important features of the landscape in symbolic form. Thematic maps allow for the representation of particular ideas about landscape features or the populations that live in them, such as political affiliations or social characteristics, in detail. Both types have uses in the context of property management.

Maps are dependent upon underlying methods of locating phenomena on the surface of the Earth. Concepts of co-ordinate geometry, the use of map projections and national grids are fundamental. These 'behind the scenes' characteristics of geographical data have a profound effect on the shapes and areas of objects on the map and will determine what types of analytical processing can be undertaken with them.

There are numerous methods of representing geographical information digitally. The most commonly used are vector and raster-based methods of representation. In the former case, objects are represented by interconnected points, lines and area features. In the latter, a mosaic of cells is superimposed upon the landscape and coded according to the information that lies within each. These structures can be used to store information about geographical features, including data about connectivity and linkages between features.

It is important to emphasise that the production of a map is a selective procedure. It is impossible to represent fully the infinite complexity of the world we live in (or even a small part of it) on a 2D piece of paper. We restrict the content of the map to features of particular interest. The choices we make about which features are important and the map projections and co-ordinate systems we use to position them are critical.

The choices of the mapmaker about the use of colour and the sizes and shapes of symbols on a map can radically affect the way in which the information on the map is perceived. The goal of the mapmaker is therefore to produce a map that shows the information that is to be portrayed as clearly and unambiguously as possible and minimise the possibility of misleading or misinforming the map reader. Monmonier (1996) provides an excellent treatise on the various ways in which maps have been used to confuse or mislead (deliberately or accidentally), and his book is heartily recommended as an introduction to the many pitfalls of cartographic endeavour.

These issues of selectivity and personal choice extend to the digital representation of geographical phenomena. The digital mapmaker is bound by the same rules of cartographic good sense as the traditional cartographic draughtsman, but has additional problems to contend with. The ways in which geographical data are represented digitally and the nature of the linkages between them will have far-reaching consequences for the types of analysis that can be performed upon them. The GIS user is simultaneously a cartographer and an information analyst and must treat both roles with appropriate gravity.

Note

1 We can calculate these values because we know the co-ordinates of the bottom left corner of the raster and the cell size in metres. Cell (3,3) has a bottom left-hand corner that is 100 metres east of the origin and 100 metres north of it, hence it is located at (1000 + 100), (1000 + 100). The centre of the cell is a further 25 metres east and north of the bottom left corner, i.e. at location (1000 + 125), (1000 + 125).

References

Burrough, P.A. and McDonnell, R., 1998, *Principles of Geographic Information Systems* (2nd edition), Wiley, Oxford.

Clarke, K.C., 1999, *Getting started with Geographic Information Systems* (2nd edition), Prentice Hall, Englewood Cliffs, NJ.

Cross, P.A., Hollwey, J.R. and Small, L.G., 1985, Geodetic appreciation, Working Paper No. 2, Department of Land Surveying, University of East London, ISSN 0260–9142, School of Engineering, University of East London.

DeMers, M.N., 2000, *Fundamentals of Geographic Information Systems* (2nd edition), Wiley, New York.

Iliffe, J.C., 2000, Datums and map projections for remote sensing, GIS and Surveying, Whittles.

Laurini, R. and Thompson, D., 1992, Fundamentals of Spatial Information Systems, APIC Series No. 37, Academic Press, London.

Maling, D.H., 1973, *Coordinate Systems and Map Projections*, Philip, London.

Monmonier, M.S., 1996, How to Lie with Maps (2nd edition), Chicago.

National Oceanic and Atmospheric Administration (NOAA), 1985, Geodesy for the Layman (5th edition), US Department of Commerce.

Ordnance Survey (OS), 1999, A guide to co-ordinate systems in Great Britain, Ordnance Survey.

Robinson, A.H., Sale, R.D., Morrison, J.L. and Muehrcke, P.C., 1984, Elements of Cartography (5th edition), Wiley, New York.

Wright, J.D., 1955, Crossbreeeding geographical quantities, *Geographical Review*, **45**, 52–65.

Further reading

Egenhofer, M.J. and Frank, A.U., 1989, Object oriented modelling in GIS: Inheritance and propagations, Proceedings of the 9th International Symposium on Computer Assisted Cartography, ACSM/ASPRS, United States, pp. 588–599.

Environmental Systems Research Institute, 1994, Map Projections – Georeferencing spatial data, Environmental Systems Research Institute, Redlands, California.

Meyer, B., 1988, *Object-oriented Software Construction*, Prentice-Hall, New York.

Worboys, M.F., Hearnshaw, H.M. and Maguire, D.J., 1990, Object oriented data modelling for spatial database, *International Journal of Geographical Information Systems*, 4(4), 369–383.

Zhan, F. and Mark, D.M., 1992, Object oriented spatial knowledge representation and processing, Proceedings of the 5th International Symposium on Spatial Data Handling, International Geographical Union, pp. 662–671.

3 Mapping and analysis using GIS

Introduction

Chapter 2 introduced the basic concepts that underpin the technology of GIS and discussed how GIS systems store and manipulate geographical information. In this chapter we will move on to examine the use of GIS as a tool for exploring, interrogating and visualising data. We introduce some of the common analytical capabilities of GIS and show how standard techniques that are available in most commercial GIS packages might be applied to the analysis of land and property data.

In the first part of the chapter, we consider some different approaches to mapping and presenting information with GIS. We go on to discuss the use of GIS for exploring spatial relationships and combining data from different sources. Finally, we consider some of the possibilities that are available for data visualisation and manipulation using GIS.

Many of the topics introduced here are quite complex and the objectives of the book preclude a detailed treatment of all of them. Other authors discuss the application of particular methods in more depth, and the annotated bibliography at the end of the chapter is designed to point the reader to other references that complement the material covered here and provides more detail about the application of specific techniques.

Mapping and presenting information with GIS

All GIS packages have the ability to produce high quality thematic or topographic maps from the feature data stored within the system. Datasets such as topographic maps, aerial photographs, satellite imagery and complex databases can be combined and visualised concurrently. Map annotation, scale bars and legends can all be generated and placed on the map by the user. In many cases, insets showing more detailed portions of the main map can also be produced. More advanced systems also allow the user to define their own symbols to fit in with house styles and existing practice. Because the GIS commonly incorporates a database of thematic data, extra analytical information such as graphs and charts or summaries of tabular data can often be included alongside the maps.

Geographical Information Systems brings a high level of flexibility to the processes of topographic and thematic mapping. Whether or not features are included and how they are symbolised on the map, the map scale and ancillary textual information are decided by the user. Complex drafting procedures are completed automatically by the computer from the geographical data structures discussed in Chapter 2. This high level of cartographic power brings with it some potential pitfalls. The user needs to be aware of cartographic good practice such as the choice of appropriate symbolisation, an appropriate number of categories and colour schemes for effective visualisation of the information that is to be mapped and the limitations of scale and generalisation caused by source data. We introduced these topics in Chapter 2. More in depth coverage is available from standard cartographic reference works such as Robinson *et al.* (1987) and the reader who is unfamiliar with cartographic principles is advised to investigate further.

Mitchell (1999) has produced a useful typology of common mapping operations that can be carried out using standard GIS techniques:

- Mapping a single feature type
- Mapping by category
- Mapping using quantities and ratios.

We will discuss each of these approaches briefly.

Mapping a single feature type

It is often necessary to produce a map showing the geographical distribution of a single feature type such as houses, roads, railways or towns. In this case, all of the features are usually assigned a single symbol. For example, in the case of roads this might be a red line, or, in the case of towns, a black dot. It is unusual to produce a finished map showing a single feature type. Instead, some background information is often included to provide a context for the information that is presented.

Consider Figure 3.1. It shows the distribution of electricity generating power stations in Great Britain. Each facility is represented by a grey dot. Context is provided by an underlying map of administrative areas that allows us to locate the power stations.

Mapping by category

It is a common requirement to depict different categories of the same feature on a map. In the power station example we might be interested in the generation methods used, for example coal fired, oil fired, gas powered or nuclear facilities. We may require information about different classes of road from local routes to national motorways or different types of land use zoning in a town centre.

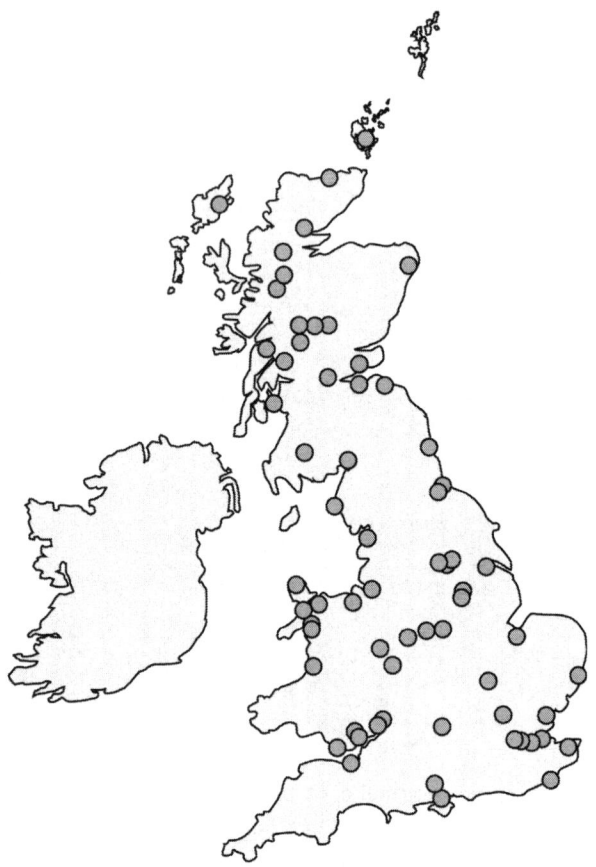

Figure 3.1 An example of single feature mapping showing the location of power generating facilities in Great Britain.

Figure 3.2 shows the same power stations categorised by type into six groups and shaded accordingly.

Mapping using quantities and ratios

More elaborate possibilities become available when we introduce numerical information into a mapping project. The most common methods of visualising such information use graduated colours or different shading patterns to describe the values of variables in particular areas while variations in symbol size and colour can be used to depict different levels of magnitude for a numerical value at point locations or along lines. Consider Figures 3.3 and 3.4. The first shows the power stations from the previous examples using proportional symbols to map an artificial variable whose value ranges from 1 to 10 at individual stations. The second uses local education authority data

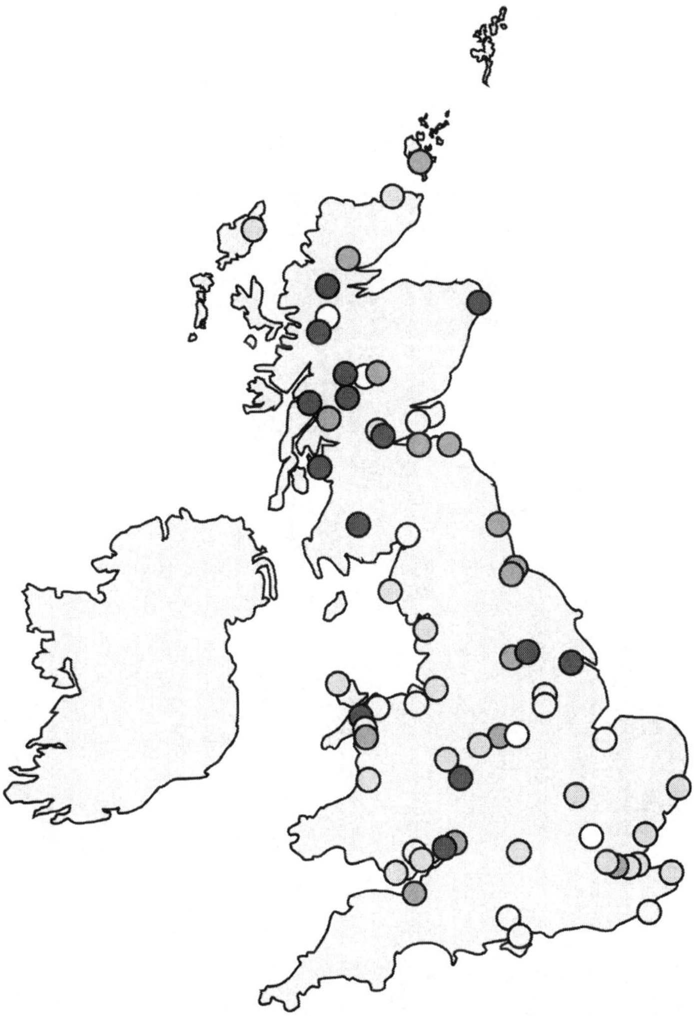

Figure 3.2 Shading of a single feature type according to category.

for choropleth mapping to visualise General Certificate of Secondary Education (GCSE) performance in the boroughs of East London. The choropleth map is combined with a second GIS data layer showing the actual location of the schools from which the data were collected. A proportional symbol map could have been created just as easily at the individual school level, with large symbols representing higher levels of exam success than small ones.

When mapping numerical data it is common practice to introduce ratio measurements to visualise averages, proportions or densities, particularly when working with data that are aggregated to a particular area level

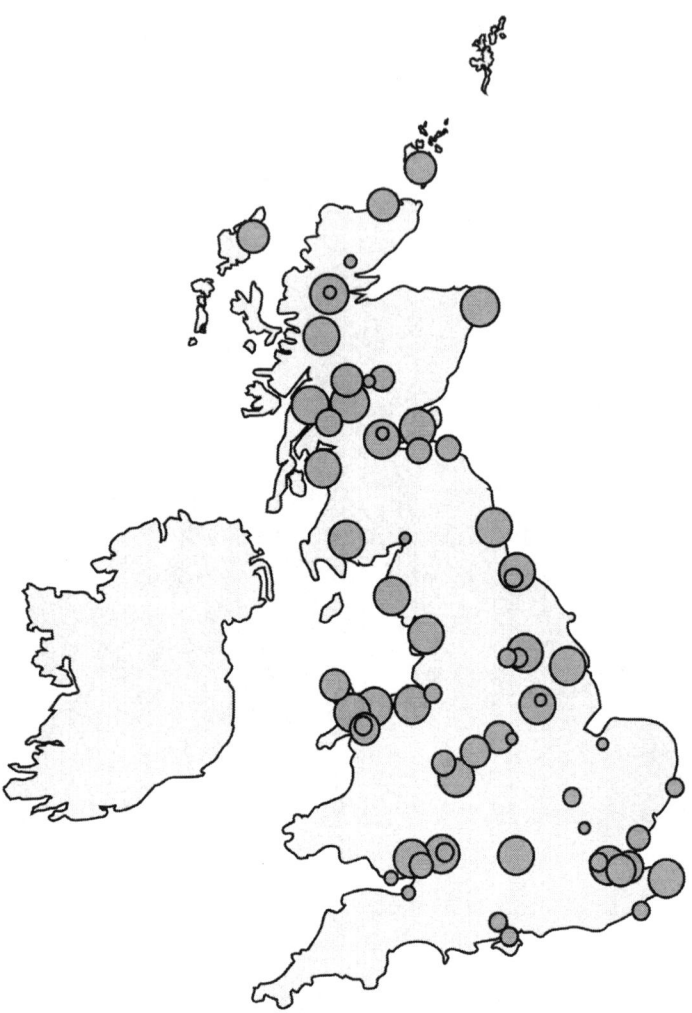

Figure 3.3 Proportional symbolisation of power generation facilities.

(Mitchell, 1999). We produce such ratios by dividing the variable of interest by a second measurement for the same area. Percentage measurements are commonly used to summarise ratio variables, and the exam success example shown in Figure 3.4 provides an illustration of this approach. In each borough, the total number of students passing at grades A to C has been divided by the total number of pupils and the result multiplied by 100 to give a percentage pass rate between 0 and 100.

An average value for an area can be created by dividing the variable of interest by a second variable to give us a rate per unit of the second variable. For example, if we know the total value of all of the residential properties

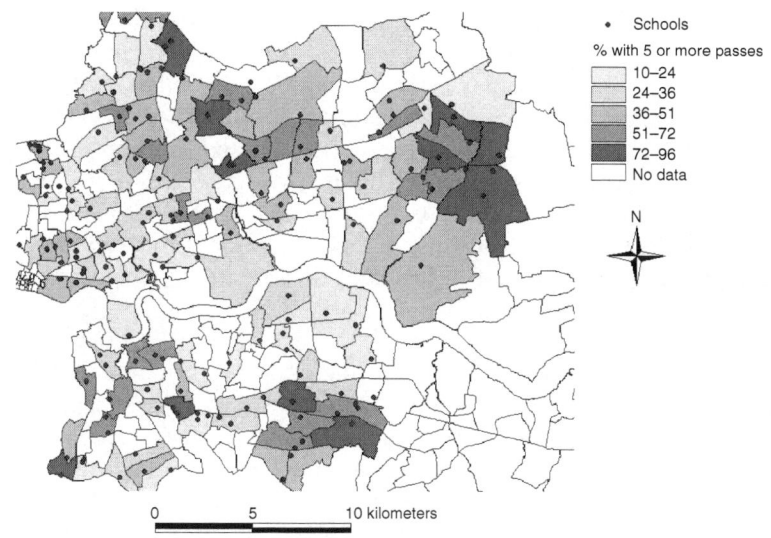

Figure 3.4 Choropleth mapping of student GCSE performance in the Boroughs of East London. Percentage of students attaining 5 or more GCSE passes at grades A–C, 1999.

Source: London East Training and Enterprise Council, 1999.

within an area, we can calculate the average residential property value for that area by dividing the total value by the total number of residential properties present in the area. Similarly, if we know the total amount of retail floor space that is available in a town centre and the number of retail premises in that centre, we can calculate the average amount of floor space per shop in the town centre. Such information is very useful for visualising the differences between areas as we can quickly see which ones score more highly than others do.

Dividing the variable of interest by a second variable of which it is a subset produces a proportional measurement. For example, to visualise the proportion of vacant properties in an area we would divide the number of vacant properties by the total number of properties in the area. Percentages are particularly useful for visualising and comparing proportional measurements.

Density measurements provide a value of incidence per unit area (the area units of interest might be square metres, square miles, square kilometres, hectares or any other valid area measurement). They are very useful for describing the differences between areas, since they take account of the relative sizes of the areas. Densities are calculated by dividing the variable of interest by the size of the reporting region given in the area measurement units of interest. To calculate the average property value per square kilometre for a region, we would divide the total value of all the properties in that region by its area in square kilometres. Once again, these measurements are very useful for differentiating between areas.

Choosing classes and colours on a map

In general, it is advisable to limit the number of categories shown on a map to at most six or seven different types for a specific geographical feature type (Mitchell, 1999; Robinson *et al.*, 1987), particularly when using colour to distinguish between map features. Robinson *et al.* (1987) note that the ability to recollect hues and retain a visual impression is quite restrictive and that where colour is used to distinguish one feature type from another the colours should be as different as possible. Mitchell points out that too few categories may not reveal sufficient information about the geographical patterns present in the study area, while too many categories may hide important patterns.

In general, if the number of categories to be mapped is less than seven, it is acceptable to use different hues of a single colour for mapping. Where a distribution of values fits clearly into two groups then it is useful to employ one colour for the first group and a second for the other. If the data to be mapped varied from -50 to $+50$, for example, we might choose shades of blue for values below zero and shades of red for those above zero, with zero itself represented using white. This rule of thumb can be extended where multiple groups occur, with the restriction that the more colours and shades we use, the more difficult it will become to interpret the map.

Observers will tend to associate darker hues with higher values, so it is important to take account of this when producing maps. Additionally, the human eye is more attracted to some hues than others and is able to distinguish different shades of some colours better than others. Red, green and yellow elicit more response than blue and purple (Robinson *et al.*, 1987) but it is easier to distinguish between shades of purple and blue than some other colours.

There may be important cultural or social significance attached to particular colours. This could influence the visual impact of the map for different observers. There are also cartographic precedents such as the use of 'standard' colour gradients for relief or bathymetric shading that the conscientious mapmaker will need to take account of.

When the number of categories to be displayed and the colour scheme to illustrate those categories have been decided, there remains the question of how to apportion the individual map features to different classes – where should the break points between classes lie?

Choosing break points between classes

A critical choice for the mapmaker is where to define the boundaries between classes that are used when mapping values of a particular variable. There is considerable cartographic literature on this topic. Generally, it is useful to look at the distribution of the variable under consideration using exploratory statistical procedures and graphical tools like histograms or cumulative frequency graphs to gain some understanding of how variation occurs and use this to determine an appropriate method of classification.

Most GIS systems are able to calculate an appropriate set of class intervals using standard classification procedures given a user-defined number of classes for the map and will apportion individual features to classes appropriately. Many also have the functionality to explore the characteristics of the data using graphs and summary statistics before determining classes.

There are several common techniques employed for determining breaks:

1   The 'natural breaks' method developed by George Jenks (1977) using a refinement of Fisher's earlier algorithm for finding class boundaries (Fisher, 1958). This method searches for large jumps in the data values and fixes the class boundaries where these jumps occur, reflecting the natural groupings of the data. It is quite effective at showing the differences between high and low values where there are marked discontinuities in the data, but less useful if the distribution of data varies smoothly from low values to high ones.

2   Equal interval classification, which simply divides the data into *n* classes of equal size. While simple in concept, this approach should be treated with caution as it imposes an arbitrary scheme onto the data without any consideration of how the data themselves behave. The method is least successful where there is bunching of values or sharp breaks in the data and will usually under emphasise patterns if discontinuities do not occur at equal interval class boundaries.

3   Quantile classification, which partitions the data so that there are an equal number of members in each of the *n* categories in the classification scheme. This method is useful for balancing class membership and should emphasise high and low values quite well, although extreme values will be grouped with lower scores.

4   Classification using mean and standard deviation, whereby classes are defined according to the number of standard deviations from the mean into which they fall. This method is very useful for defining high and low extremes in the data, but will only work well if the data are evenly distributed around the mean and do not exhibit skewness. Plotting the source data using a histogram should quickly show whether or not this is the case and whether the method is effective.

In order to show the differences between the methods four maps were created. The Townsend Index of Deprivation was used as the input variable. This measures social deprivation, with a negative score indicating low levels of deprivation and positive scores indicating progressively higher levels of social deprivation. All of the maps had five classes at census ward area level. The only change that was made was in the classification method used to determine class boundaries. The results are presented in Figures 3.5 to 3.8. Consider the differences between the resulting maps. The class boundaries that result from each classification technique are shown in Table 3.1.

Figure 3.5 Classification of deprivation using the natural breaks method.

Figure 3.6 Classification of deprivation using the equal intervals method.

Figure 3.7 Classification of deprivation using the standard deviation method.

Figure 3.8 Classification of deprivation using the quantiles method.

Table 3.1 Upper boundaries of classes for different classification procedures

Classification	Class 1	Class 2	Class 3	Class 4	Class 5
2.6 Nat. Breaks	−1.08	1.49	4.16	6.89	9.71
Eq. Intervals	−1.3	2.06	5.43	8.80	12.17
Quantiles	−0.65	1.77	3.87	6.10	8.73
Std. Deviation	−2.36	1.87	6.1	10.3	14.5

Using geographical relationships in spatial analysis

A key advantage of GIS technology lies in its ability to manipulate and retrieve information based on geographical patterns as well as the more traditional selection by descriptive criteria that is available in less specialised database query and analysis. We can use GIS not only to select features of interest, but also to see where they are and how they are related to one another using digital maps. This allows us to ask questions that are difficult to formulate using a standard database. Examples might be:

1   Identify all of the vacant one- or two-bedroom residential properties in the London Borough of Camden that are within 2 kilometres of an underground or main line rail station, are less than 1.5 kilometres from the nearest superstore and have a market value of between £150,000 and £250,000.
2   Select all of the office buildings in a property portfolio that have broadband internet access, from three to five vacant office units of at least 1,000 square feet, existing information industry clients, short lease terms available and are within 2 hours drive or 1 hour by train or tube from central London.
3   Locate all of the vacant warehouse retail premises within 15 minutes drive of the M25 motorway that are connected to a slip road that is accessible by heavy lorries, and which are already part of an established retail park and can offer at least 1 square kilometre of customer parking space.

In this section we will examine the capabilities of GIS for retrieving information using geographical relationships. We begin with basic operations like measuring distances and proceed to consider more complex geographical queries.

Simple geographical measurements

Using a digital map, we can carry out simple tasks like distance measurement quickly and easily. In Figure 3.9 we have measured the straight-line distance from a house to the nearest railway station.

In a similar way, measuring the length of features like fences, roads or railways is straightforward and we can also obtain the perimeter or area of

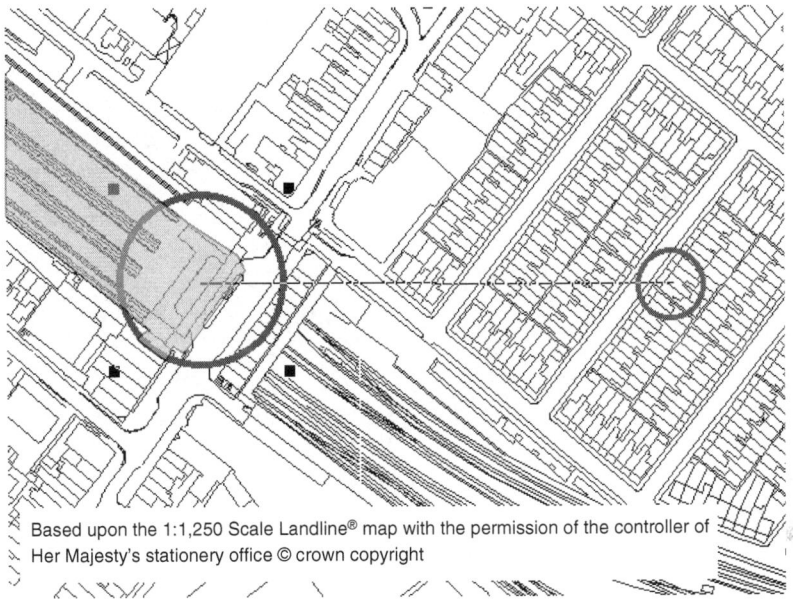

Based upon the 1:1,250 Scale Landline® map with the permission of the controller of
Her Majesty's stationery office © crown copyright

Figure 3.9 Measurement of straight line distance.

geographical features like shops or vacant sites in units of measurement that
we deem to be appropriate.

What is where and where is what?

When we link a digital map with a database, two new methods of geographical query become available. We can find out the characteristics of
features on the map by selecting them interactively, usually with a mouse.
In Figure 3.10 we have selected a land parcel from the map and are able to
interrogate the database record that pertains to that parcel.

The reverse – displaying objects on the map as a result of a database
request – is also possible, assuming that the database contains the information that we need. In Figure 3.11 we have stipulated that we wish to find
all the land parcels that are used for residential purposes, have a value
between £1,20,000 and £2,00,000, have three or more bedrooms and were
built after 1925. This is a simple database query, but the results can be displayed on the map. In this case those properties that fulfil the criteria of the
GIS query are highlighted.

What is next to what?

In Chapter 2, we introduced the concept of GIS data structures. These
methods for storing geographical data provide the user with the ability to

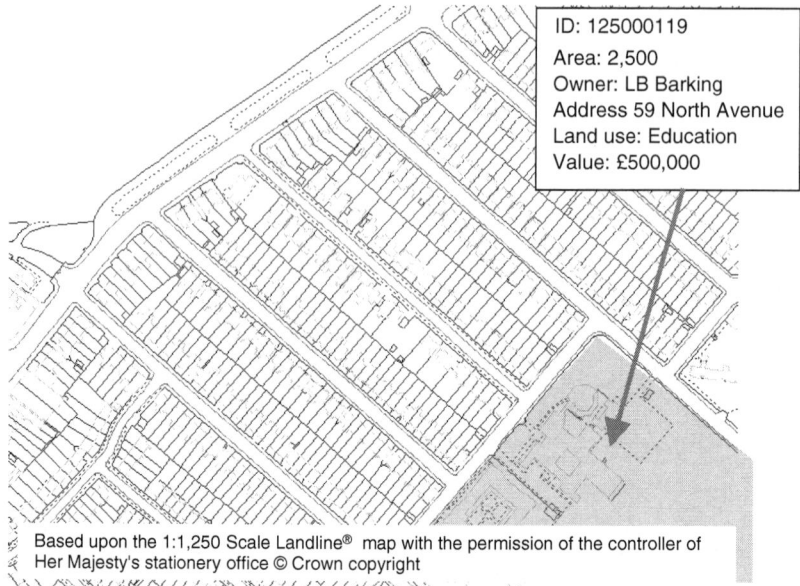

ID: 125000119

Area: 2,500

Owner: LB Barking

Address 59 North Avenue

Land use: Education

Value: £500,000

Based upon the 1:1,250 Scale Landline® map with the permission of the controller of Her Majesty's stationery office © Crown copyright

Figure 3.10 A simple geographical query, using the digital map to identify a feature and then retrieving its characteristics from the database.

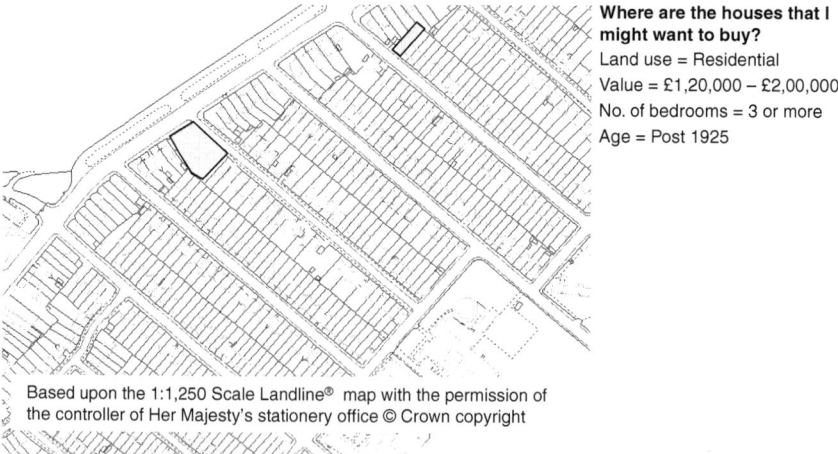

Where are the houses that I might want to buy?

Land use = Residential

Value = £1,20,000 – £2,00,000

No. of bedrooms = 3 or more

Age = Post 1925

Based upon the 1:1,250 Scale Landline® map with the permission of the controller of Her Majesty's stationery office © Crown copyright

Figure 3.11 Visualising the results of a database query on the map.

interrogate a GIS using the relationships between geographical features as the basis for queries. New modes of query become available to us through this facility. We can discover which features are near to or linked to other features, or how different groups of features are related to one another in space.

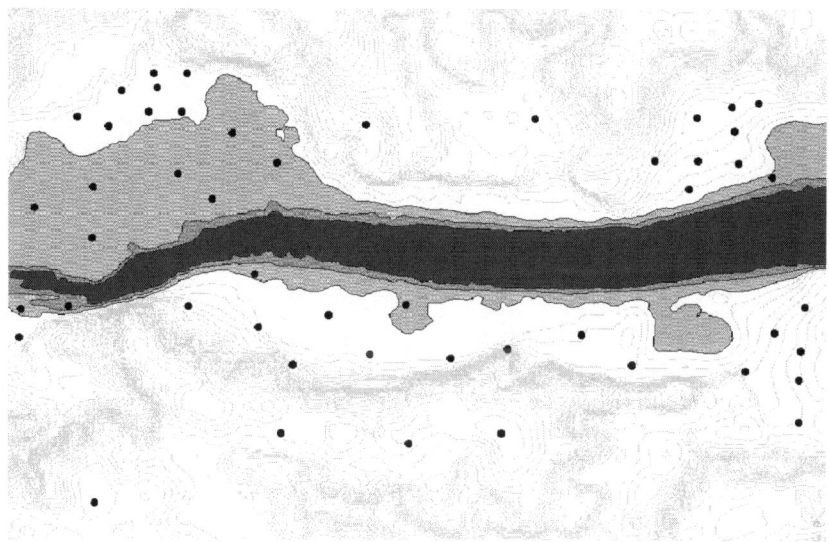

Figure 3.12 Flood risk in a river valley.

In order to show how this might work, let us take an example. Consider the case of an insurance company that specialises in flood risk insurance. Now consider a hypothetical river valley. The valley is susceptible to flooding, and major floods occur on a rough cycle of one every one hundred years with a more minor flood likely to occur once every ten years or so. Using a GIS and a suitable set of input data (heights above sea level measured around our river valley), we can construct a surface model that allows us to map the geographical extent of the flooding during the one-hundred-year and ten-year cycles. In Figure 3.12, we have done just that. The zone around the river represents the maximum extent of flood inundation every one hundred years. The zone close to the water line shows the area that is susceptible to inundation every ten years.

On the same map, we have plotted the locations of houses that lie inside the river valley. As an insurer, we are very interested to know which houses lie within the flood hazard zones. We are even more interested to know whether they insured against flooding. Equally, if we do insure them, should we revise their premiums?

Using the geographical query capabilities of the GIS, we can find all of the houses that lie within the ten-year and hundred-year flood hazard zones. If we also store the postal addresses of the houses in the GIS database and marketing information about local residents we can use this information to target those households that are most likely to take out a new insurance policy or switch their existing policy to a different product range.

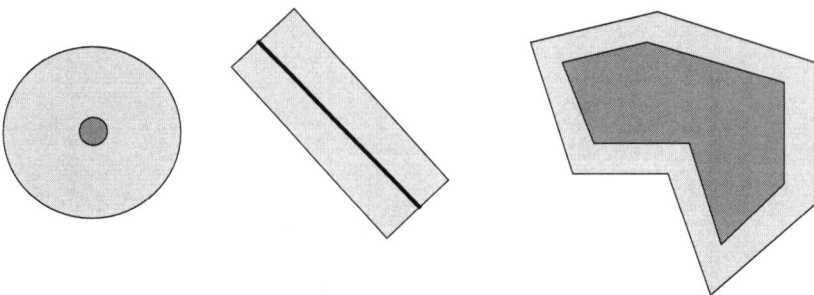

Figure 3.13 Circular, rectangular and polygonal buffer zones around point, line and
area features.

Selection using spatial relationships: the buffer zone

The insurance example introduces one of the most useful GIS analysis
operations, known as buffering. A buffer is a new map feature created at a
specified distance from one or more points, lines or areas on the map.
Consider the examples in Figure 3.13.

Buffer zones are very useful tools for determining spatial proximity and
whether or not features fall within critical distances of one another. They
can be used to model many diverse problems from zones of equal distance
or travel time from a starting point to the catchment areas of superstores or
town centres to the transmission ranges of mobile communication masts. In
Figure 3.13 each zone is of a fixed radius but in many GIS software packages it is also possible to vary the extent of the buffer zones according to
the characteristics of the objects around which they are drawn, or to create
multiple buffer zones around map features. The examples show how this
type of function can be used when deploying GIS methods for decision support. It is with this type of functionality that the analytical potential of GIS
begins to become more powerful. We can use the computer record of where
things are in relation to other things to ask new questions.

Combining different sets of data geographically

Among the most powerful operations that a modern GIS can perform is the
ability to superimpose one set of geographical data on top of another and
to combine the characteristics of both into a new set of information. In its
simplest form, this is a case of visualising two or more sets of cartographic
or aerial photographic information simultaneously, a procedure that has
been used by geographers and cartographers for many years. What the
computer brings to the overlay operation is to examine the similarities and
differences between layers in much more detail than was previously possible
and to carry out such overlays interactively so that exploratory investigation

is possible. Combinations of geographical data that would take days or weeks of drafting without computer technology or would simply be too difficult to visualise can be accomplished quickly and easily using GIS methods.

We can now measure the similarity, overlap or difference between two spatial datasets very precisely. Equally important, we can explore the database attributes of the different features that are overlaid at the same time as their spatial comparison and investigate how those attributes relate to one another across multiple geographical themes in the database. DeMers (2000) suggests that, although there is still a lot to learn about the causal relationships among spatial phenomena, the ability of GIS to perform exploratory spatial analysis may result in the development of new hypotheses, new theories or even new laws about spatial patterns.

The overlay process

Geographical Information Systems overlay involves the combination of different data layers. Practically, this can involve a number of different processes depending on the inputs to the analysis:

1 Point in Polygon Overlay
2 Line in Polygon Overlay
3 Polygon on Polygon Overlay
4 Line on Line Overlay
5 Point on Point Overlay
6 Point on Line Overlay.

Of these, the most commonly used for analytical purposes are Point in Polygon, Line in Polygon and Polygon on Polygon overlays.

Before we look at GIS overlay in more detail, let us consider the mathematical concepts that underpin the overlay process. The overlay operation is a combinatorial process. Different geographical data layers are superimposed upon one another and combined to form a new layer according to a user-defined rule or set of rules. Each input data layer can be thought of as representing a mathematical set where every object in that layer is a member of the set. The result of the overlay is a combination of the sets, depending on the rules that are used. When two sets, A and B, are overlaid, we can use a Venn diagram to illustrate the set concepts of intersection and union. Consider Figure 3.14. In the figure there are two oval shapes, A and B, which overlap each other.

The diagram allows us to think about the different combinations that are possible when we overlay two objects onto one another. The first combination possibility is intersection. The intersection of two sets is the area common to both of them. In the diagram, this would be area C, where A and B overlap. Area C is the only place on the diagram where we are inside both

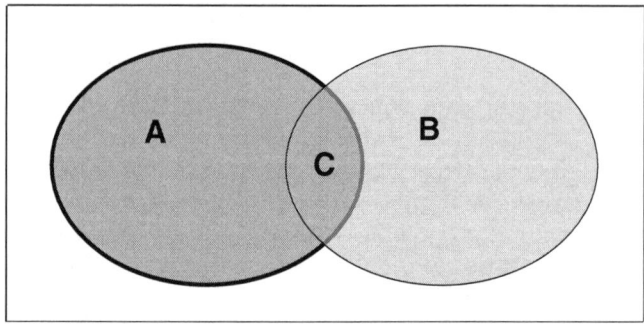

Figure 3.14 An example to illustrate some simple set concepts.

A and B and is the zone of intersection between the two areas. The zone of intersection is the region of the map where both A and B exist. It is common practice when formulating GIS database queries to think of the results of such an overlay as being true (successful) or false (unsuccessful). This characteristic can then be represented easily using a binary value of 0 or 1, where 1 represents 'true' (intersection) and 0 is 'false' (no intersection). In these terms, area C is the only part of the diagram where the intersection of A and B is 'true'. In all other areas, there is no intersection between A and B and we would obtain a value of 0 ('false').

A second common combination is the union of two sets. We can think of the union operation as a combination of all the parts of the input objects. On the diagram, this would be anywhere that is inside area A, area B or both A and B. To obtain a value of 1 ('true') in this case, we must either be inside A or inside B or inside both shapes. Anywhere outside A, B or both of them is not part of the union of the two shapes and would have a value of 0 ('false'). Here are some common overlay combinations and the areas on the diagram that they represent:

Relationship	Area on the diagram
A and B	The area common to both A and B (area C).
A or B	Either inside A or inside B or inside both – A, B or C.
(not A) and B	Outside A and inside B – includes all of B except area C.
(not A) and (not B)	Outside A and outside B – anywhere that is not inside either shape.
A and (not B)	Inside A and outside B – includes all of A except area C.
not (A or B)	Not inside A or inside B – the area outside both shapes.

There are many possible combinations of these relationships and all can be extended to include more layers if necessary. Using them it is possible to create very sophisticated overlay procedures. The sample questions that we posed at the beginning of this chapter are all examples where this type of procedure might be applied. In each case, there are several different geographical datasets under consideration and we need to find places where a series of conditions are fulfilled across one or more of them. We will take a simple example of overlay analysis with a land and property theme to illustrate how these concepts can be applied in a land and property management context. If such an exercise were to be undertaken in reality we would require considerable additional information. The present, simplified case is designed to demonstrate how concepts of GIS overlay might be applied in this type of scenario.

Consider the hypothetical case of a newsagent that wishes to open a new branch selling newspapers, magazines and confectionery in a local suburban centre of London. Let us assume that for a vacant shop to be suitable for this purpose a number of conditions must be met:

1   The vacant shop must be within 100 metres of the nearest railway or tube station.
2   The shop must be within 50 metres of a bus stop.
3   The retail floor must be at least 100 square metres in size, with a frontage of at least 5 metres.
4   The premises must also include a storage area and sufficient space for a small office – at least another 100 square metres of available space.
5   The monthly rent for the premises must be less than £2,000.
6   There must be no competitor premises within 25 metres.

In order to find suitable premises several tasks must be undertaken. First, we would need to review what information is required to perform the analysis. To carry out the query successfully, we require information about the following items:

1   The location of public transport infrastructure within our study area.
2   The location of competitor premises within the area.
3   The location of vacant premises and details about the rent, available retail, storage and office space that they can offer.

Figure 3.15 is a map of the shopping area under consideration. Vacant shops are shown in mid grey, while competitor premises are shown in dark grey. Other shops are shown in light grey and bus stops are shown as squares.

Next, we need to construct a series of queries that find those shops that fulfil the six conditions specified. To do this, we construct a series of GIS overlays. First we define all those areas within 50 metres of a bus stop by defining a buffer zone around each of the bus stop points. These zones are then overlaid onto the map and we can select those vacant premises that are within 50 metres of them. Figure 3.16 shows the result of this process showing the 50 metre zones around the bus stops and suitable shops in light grey.

Next, we select only those shops from the bus stop set that are within 100 metres of the railway station by generating a second buffer zone of

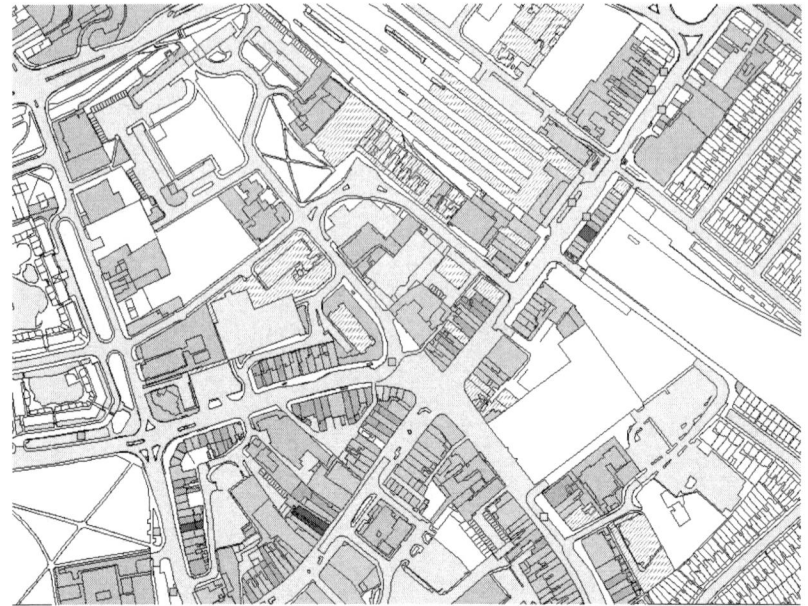

Figure 3.15 Siting a new newsagent in a suburban centre using buffer zones and overlay techniques (1 of 4).

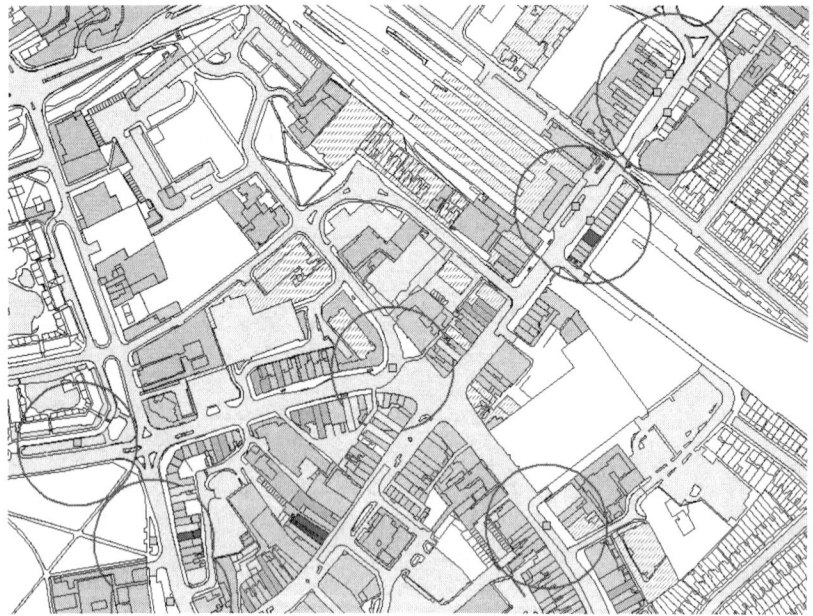

Figure 3.16 Overlay and buffer generation (2 of 4).

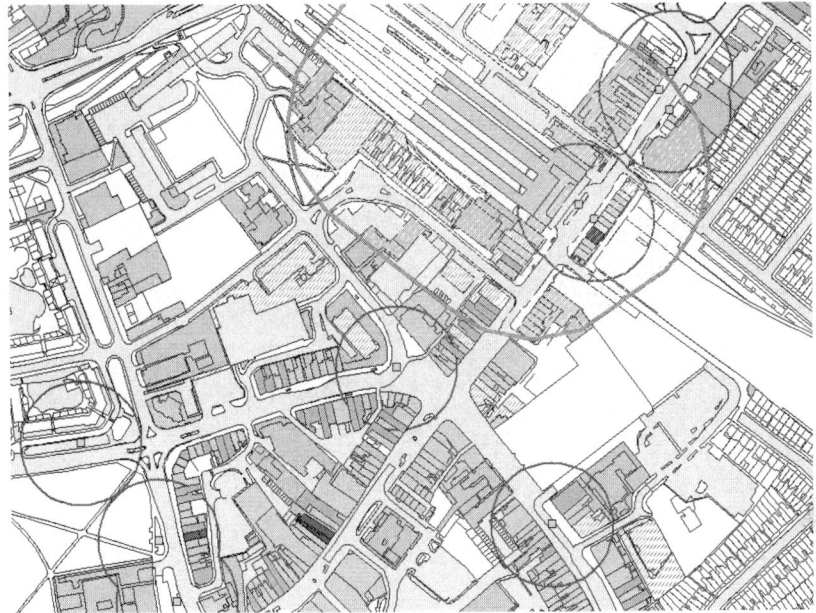

Figure 3.17 Overlay and buffer generation (3 of 4).

100 metre radius around the station and overlaying it onto the shops. The results are shown in Figure 3.17.

We have now restricted our choice to four vacant premises around the railway station. These four are within the required range of the bus stops and the required range of the railway station.

For our final overlay, we find only those of the remaining four shops that are not within 25 metres of a competitor. We can do this by creating a third and final set of buffer zones around the competitor newsagents, shown as small circles on the map. Figure 3.18 shows the results. The hatched grey area in the centre is unsuitable, since it is within 25 metres of competitor premises.

We have narrowed down the choice of premises to just two vacant shops. The final stage of the process is to perform a database query on the remaining candidates. We select only those shops with a retail floor that is at least 100 square metres in size, a frontage of at least 5 metres, a storage area and sufficient space for a small office and a monthly rent for the premises that is less than £2,000.

Combining features through geographical overlay

In the example of vacant shop selection described above, the overlay process was used simply to select out those features on different input layers that met

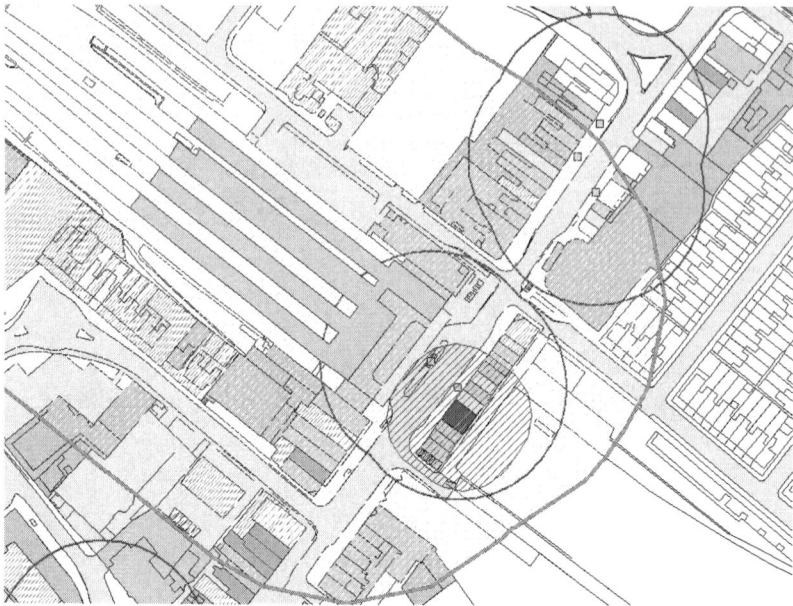

Figure 3.18 Overlay and buffer generation (4 of 4).

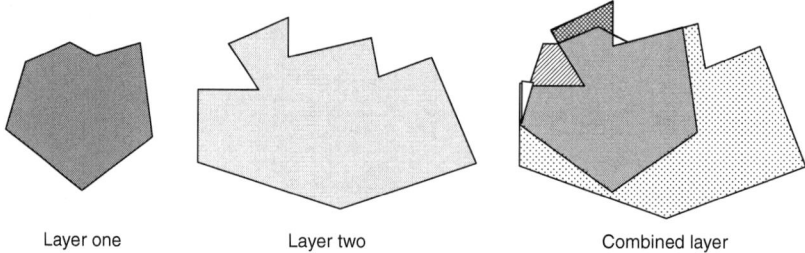

Layer one   Layer two   Combined layer

Figure 3.19 An example of combinatory geographical overlay. Layers one and two
       are combined to form six new objects. Those which are formed by the
       overlap of layers one and two will require some attribute combination.

our criteria for possible sites. It is also possible to extend the overlay process
so that map features on input layers are combined together to create new
geographical objects. Consider the example in Figure 3.19.

Some examples of the use of combinatory overlays

Let us first return to the insurance example depicted in Figure 3.12 to illustrate how combinatory overlay procedures might work. In the example, the
point locations of individual properties are overlaid onto the flood zones on

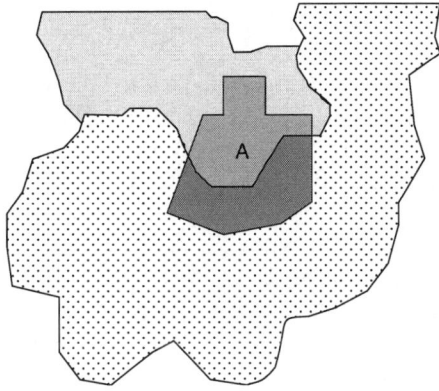

Figure 3.20 Combinatory overlay requiring attribute integration.

the map. This 'Point in Polygon' overlay procedure will result in a set of composite point objects. Each composite point will include all of the descriptive information about the property itself, as well as the information associated with the flood zone that it lies inside.

For a more complex example of combinatory overlay, consider Figure 3.20. In this example, polygon A lies on the intersection of two other polygons. How should the attributes of the two other features be apportioned to A?

This problem requires us to address the question of how to combine attribute information from different source objects in a composite object. How this is done will depend upon the information that we are trying to obtain and the nature of our input data. Chrisman (1997) presents a useful introduction to this topic and suggests that the following general methods of attribute combination can be considered:

1 The DOMINANCE rule. One value is selected from the possible set of attribute values and assigned to the output object, all others are erased. Examples: Take the highest value, and take the lowest value.
2 The CONTRIBUTORY rule. Here, the attribute values of all the input features contribute to the result. Examples include taking the mean of all the input values, taking the most popular (modal) value, taking a weighted value for each input, taking the sum of the input values.
3 The INTERACTION rule. As well as recognising that each attribute score has something to contribute to the result, this method also looks at the interactions between individual attribute scores and other attributes of the input features. It is used where more than one attribute needs to be considered at the same time. See Chrisman (1997) for an expanded discussion and some examples.

Data structures and combinatory geographical overlays

The way in which geographical data are stored will affect the complexity and results of the overlay process. Most overlay and query work in the land and property sector involves geographical features from the built environment. In general these are represented most effectively in a GIS using the vector data structure and stored as point, line or area objects.

Broadly, vector overlay requires the following stages to be carried out:

1   Find all of the intersections between the input themes and create a new link and node topological structure from the results.
2   Identify all of the new polygons formed by the overlay.
3   Find which of the original polygons that the new polygon overlaps with and assign its attributes accordingly.
4   Usually a new attribute table is constructed that consists of the combined old attributes, or new attributes formed by logical or mathematical operations on the old ones.

Each of these operations is quite complex. Of the four, the last one poses the most problems for the analyst because a decision must be taken about how to combine the attributes from the source layer into the new layer where overlaps between features occur.

It is also possible to carry out GIS overlay operations and standard analytical functions like buffer zone generation using raster datasets. In fact, most of the pioneering works on GIS overlay were carried out using raster-based systems because the complex processing requirements of vector overlay were significantly more time-consuming than their raster equivalents. This has become progressively less important as the power of desktop computing systems has increased.

Tomlin (1990) provides a comprehensive account of raster overlay processing. Since each raster grid layer usually has the same cell resolution and spatial extent, the processes of combinatory overlay and functions like buffer generation are generally less time-consuming than their vector counterparts. Vector layers rarely contain objects with exactly similar spatial extents and the most complicated part of the vector overlay process is usually the process of working out partial overlaps between features and creating new features from them.

Analysis using a raster GIS: an example using simple datasets

Let us assume that we have a raster database comprising three different layers of information at the same cell resolution for the same area. Each layer comprises a $5\,km \times 5\,km$ grid, with each cell covering one square kilometre. We wish to locate those raster cells where there is the highest potential for successful residential development. This requires us to generate a composite layer derived from the three input layers:

- Layer One: development potential, a composite score from 1 to 10 that considers soil type, planning restrictions, planning history and local land prices.
- Layer Two: accessibility, a composite score from 1 to 5 that measures access to local facilities, public transport links and the road network.
- Layer Three: flood hazard risk, a score from 0 (no risk) to 5 (high risk).

We have defined a set of rules to define suitable areas for development. An area is suitable if it satisfies the following criteria:

- The development potential score must be at least six out of ten.
- The accessibility score must be at least three out of five.
- The flood hazard risk must be less than two.

The three input layers are shown as follows:

6	7	4	2	2
6	5	3	3	1
4	5	4	2	1
5	6	5	3	1
8	6	4	3	2

Raster layer 1 – Development potential

1–3 = Low potential
4–5 = Intermediate potential
6–8 = High potential
9–10 = Very high potential

3	5	3	2	2
2	3	2	2	2
3	4	3	1	2
3	3	4	1	1
5	4	2	1	1

Raster layer 2 – Accessibility

1 = Very poor accessibility
2 = Poor accessibility
3 = Intermediate accessibility
4 = Good accessibility
5 = Very good accessibility

0	0	2	2	2
0	0	1	1	2
0	0	1	1	2
0	1	1	1	2
1	1	1	2	2

Raster layer 3 – Flood risk

0 = No risk
1 = Very low risk
2 = Low risk
3 = Intermediate risk
4 = High risk
5 = Very high risk

There are many ways of generating a solution to this problem, but all involve two basic raster operations: reclassification and combination of data. Here, we consider a two-stage procedure. First, each layer is reclassified so that cells are coded either suitable (1) or unsuitable (0). Finally, we multiply together the three layers. All cells in the output layer with a value of 1 fulfil all the necessary criteria for development sites. Cells with a value of zero do not fulfil at least one of the criteria. These steps in more detail are:

1   Recode the development potential layer so that cells with a value less than six score 0 and all other cells score 1.
2   Recode the accessibility layer so that cells with a value of 0, 1 or 2 are recoded 0 and all other cells are coded 1.
3   Recode the flood hazard layer so that cells with a value of 2 or more are recoded 0 and those with a score of 0 or 1 are coded 1.

1	1	1	0	0
0	0	0	0	0
0	0	0	0	0
0	1	0	0	0
1	1	0	0	0

Layer 4 – Development potential

1 = Suitable
0 = Unsuitable

1	1	1	0	0
0	1	0	0	0
1	1	1	0	0
1	1	1	0	0
1	1	0	0	0

Layer 5 – Accessibility
1 = Suitable
0 = Unsuitable

1	1	0	0	0
1	1	1	1	0
1	1	1	1	0
1	1	1	1	0
1	1	1	0	0

Layer 6 – Flood risk
1 = Low risk
0 = High risk

The final stage of analysis is to overlay the results of our reclassification to find the answer to our question. We generate a final layer, where cells satisfying all three conditions are set to 1, and cells that do not are set to 0. The answer, in layer 7, would be:

1	1	0	0	0
0	0	0	0	0
0	0	0	0	0
0	1	0	0	0
1	1	0	0	0

Layer 7 – Areas suitable for development

1 = Suitable
0 = Unsuitable

This is an example of map algebra (Tomlin, 1990). Map algebra is the process of performing a sequence of numerical operations on every cell in a grid layer simultaneously. It often involves the combination of grids using mathematical operations (in this case multiplication of cell values) and using a series of input grids with exactly the same resolutions and geographical extents at the same time.

Visualisation techniques

We have introduced many of the standard methods of map customisation and data combination that GIS can offer the mapmaker. However, by bringing data into the digital environment a range of new visualisation options become available. When coupled with the graphical capabilities of the current generation of desktop computer systems, GIS data processing presents new opportunities for the visualisation and interactive exploration of land and property datasets. The user of such desktop systems now has access to data analysis capabilities that were once the preserve of expensive minicomputers and dedicated graphics workstations. In this short section, we demonstrate two approaches to data visualisation in three dimensions that are available in GIS and show how they might be applied. These approaches are not exhaustive, and once again we provide some references for the reader who wishes to learn more about alternative methods of visualisation.

Visualising thematic map information in three dimensions

The use of 3D visualisation often complements 2D representations of geographical data. Patterns may be easier to identify and differences between map features may be made more obvious. The examples in Figures 3.21 and 3.22 contrast a 2D choropleth map showing density measurements for computing sector businesses with a 3D representation of the same information. Marked differences between density levels can be quickly ascertained from the 3D representation.

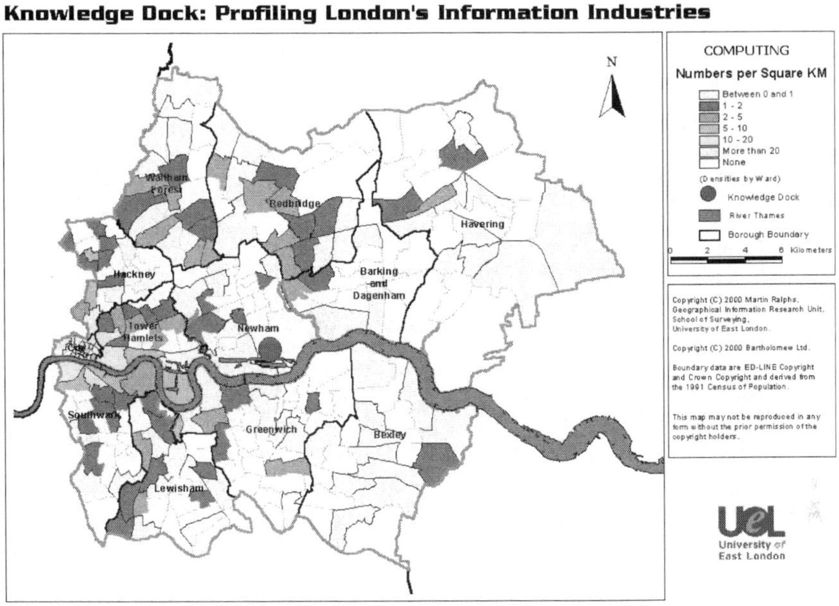

Knowledge Dock: Profiling London's Information Industries

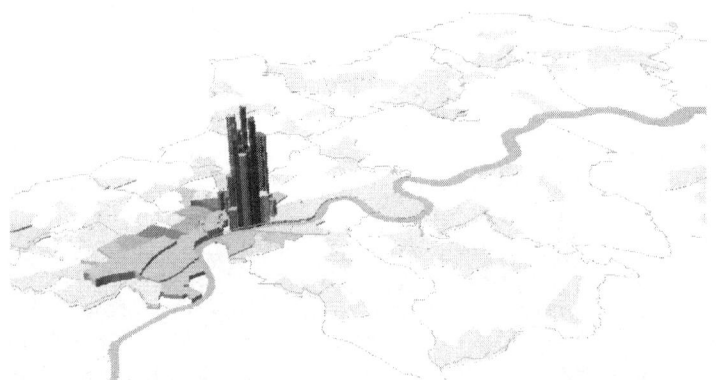

Figure 3.21 Mapping business densities in East London – a traditional cartographic example.

Source: Reproduced by permission of HarperCollins publishers. www.bartholomewmaps.com © Bartholomew Ltd 2002.

Figure 3.22 The same business density data set viewed in three dimensions.

Visualising the built environment

The use of CAD tools for 3D representation of buildings is widespread. Some GIS software now includes the capability to use CAD visualisation methods in combination with the environmental modelling tools of GIS surface modelling to produce realistic visualisations of the built environment. Truly 'immersive' virtual environments (whose application in the land and

Figure 3.23 3D visualisation of the built environment. Spot height and aerial
photography data were provided by The GeoInformation Group
Ltd. The 3D visualisation was generated by the authors.

property sector is discussed in more detail later in this volume) are still the
preserve of university research facilities and dedicated commercial laboratories. They require computing tools that are prohibitively expensive for the
desktop user. However, it is possible to produce effective 3D visualisations
of the built environment with standard computer hardware. These extend
traditional CAD functionality by allowing the user to explore and query the
GIS database in tandem with 3D visualisation procedures.

The example in Figure 3.23 is a still taken from a computer model
created using GIS to show the location of a development site in inner
London. The model was generated from spot height information and building extents and heights obtained from aerial photography and digital aerial
photographs of the study region. It includes a simple block visualisation
of buildings. The landscape is brought to life by draping the aerial photograph of the study region over the 3D elevation model created from the
spot height data. The development site itself is outlined. Realistic shadows
and skyline textures are generated automatically by the visualisation
software.

Such models allow the user to explore the local context of development
sites. Additional information, such as the distance to key facilities can
be superimposed onto the visualisation. Specific features can be highlighted
or emphasised and the whole explored interactively through dynamic
animation in real time. Although it is unlikely that such methods will

replace the 2D map, which still provides a wonderfully concise way of representing geographical variation flexibly and efficiently, they do provide alternative views of information and can bring the built environment dramatically to life.

References

Chrisman, N., 1997, *Exploring Geographic Information Systems*, Chapter 3 on representation discusses different methods of GIS data structuring.

DeMers, M.N., 2000, *Fundamentals of Geographic Information Systems* (2nd edition), John Wiley & Sons, New York.

Fisher, W.D., 1958, On grouping for maximum homogeneity, *Journal of the American Statistical Association*, 53, 789–798.

Jenks, G., 1977, Optimal data classification for choropleth maps, Occasional Paper no. 2, Department of Geography, University of Kansas.

Mitchell, A., 1999, *The ESRI Guide to GIS Analysis, Volume 1: Geographic Patterns and Relationships*, ESRI Press, Redlands California.

Robinson A.N., Sale R.D., Morrison, J.L. and Muehrcke, P., 1987, *Elements of Cartography*, Wiley, New York.

Tomlin, C.D., 1990, *Geographic Information Systems and Cartographic Modeling*, Prentice Hall, New York.

Further reading

Burrough, P.A., 1986, *Principles of Geographical Information Systems for Land Resources Assessment*, Oxford University Press, Oxford.

Burrough, P.A. and McDonnell, R., 1998, *Principles of Geographic Information Systems* (2nd edition), Operations on attributes of multiple entities in space, pp. 177–179.

Chrisman, N., 1997, *Exploring Geographic Information Systems*, Chapter 5: Overlay – Integration of disparate sources, pp. 105–140, John Wiley & Sons, New York.

DeMers, M.N., 1997, *Fundamentals of Geographic Information Systems*, Chapter 12: Comparing Variables Among Coverages, pp. 319–349, John Wiley & Sons, New York.

DeMers, M., 1997, *Fundamentals of Geographic Information Systems*, Chapter 4 on Data Structures, John Wiley & Sons, New York.

Laurini, R. and Thompson, D., 1992, Fundamentals of spatial information systems. Review material on the raster data structure and raster compaction methods, APIC Series No. 37, Academic Press, New York.

Samet, H., 1989, *The Design and Analysis of Spatial Data Structures*. Contains a comprehensive review of the quadtree data structure and its construction, Addison Wesley, Reading, Massachusetts.

Samet, H., 1989, *Applications of Spatial Data Structures*. Contains an extensive review of the applications of quadtrees, Addison Wesley, Reading, Massachusetts.

Part II

GIS applications in land and property management

4 Mapping, land information systems and conveyancing

Introduction

The use of GIS in mapping has been the starting point for many GIS implementations in local government, utilities and the private sector. As far as the user is concerned mapping represents a use of GIS that is easy to understand, provides tangible benefits across an organisation (a corporate rather than a departmental application) and provides a springboard from which to launch more sophisticated applications. There has been acceleration in the development of GIS applications in land and property management and this is due in part to more data becoming available at less cost. Initiatives such as the National Land and Property Gazetteer (NLPG) and the National Land Information Service (NLIS) have also been important catalysts and these are discussed in this chapter.

Many users of GIS do so primarily in order to produce maps. For example the London Borough of Hackney used a web-based GIS to disseminate maps to council personnel and Kingston London Borough Council use GIS to enable the general public to access the authority's geographic data. The display of geographic data using a GIS may require little in the way of analytical capability but, if the data are to be of use to a wide spectrum of users, the display must be as accurate and as up-to-date as possible. This is particularly so for maps illustrating property features because they need to be large scale and require high resolution. Accurate, up-to-date and comprehensive attribute data are also required at the property level. A common misconception is that the introduction of GIS for mapping will be straightforward. Often it is at this stage that an organisation realises that its data are not fit for the purpose intended and an expensive data cleaning exercise is a necessary precursor to GIS-based map production.

This chapter begins by describing mapping applications that are specific to the property industry. It then switches emphasis to land and property information systems that incorporate some form of mapping. Land Information Systems (LIS) are more concerned with data accuracy than analytical capability and their development relies on government policy and action. The main problems that have affected developments in this area are outlined, such as map and attribute data quality and access. The national initiatives that have been implemented to solve these problems are then described, such as the NLPG. This lays the foundation for the introduction of the NLIS, focusing at this stage on the way in which this broad initiative might be used to improve access to and availability of land and property information. The final section of the chapter describes the leading application of NLIS – a service designed to support the property conveyancing process.

Mapping property information

Maps are indispensable tools for understanding and interpreting geographical patterns and relationships between features of the built environment such as shops, residential neighbourhoods and road infrastructure. For example there may be proximity relationships and locational patterns between these features that may not be obvious from text, tables and charts. An historic example of the way in which maps are able to reveal patterns in geographical data is the research undertaken by Dr John Snow, a medical doctor in nineteenth-century London, who was trying to understand what was causing a cholera epidemic in the Soho District of the city. Dr Snow documented the geographical location of every death in London. His map, shown in Figure 4.1, indicated that the outbreak had occurred within a 250-yard radius of the Broad Street water pump. Bars adjacent to the location along each street where a fatality occurred were placed on maps of London along with streets, water pumps and other municipal facilities. The maps revealed

Figure 4.1 John Snow's map of the Broad Street pump cholera outbreak, 1854.

at a glance that the deaths were clustered around the water pumps, indicating that the water supply was a potential carrier of the virus. With further research, Dr Snow explained anomalies and pointed to the water as the main source of the disease. He had the water pump removed and his theory was proved correct within three days when the epidemic abated. With this first-known example of epidemiological research, Dr Snow became the 'father of modern epidemiology' (www.jsi.com).

Map production

The UK benefits from a national, large-scale (1:1,250 in urban areas, 1:2,500 in rural areas and 1:10,000 on mountain and moorland) map base that covers the whole of the British Isles. It is known as the National Topographic Database and has been held in digital form by the Ordnance Survey (OS) since 1995. This is by far the largest and most comprehensive national digitisation programme to be completed in the world at comparable scales, in terms of data volume and map units, if not land area. At the large scale – the scale of most relevance to property decisions – this digital map product is known as Landline. Digital Landline map 'tiles' are available for purchase and are increasingly used in GIS for property-related functions such as planning and infrastructure management. Use of Landline data was spearheaded by organisations directly responsible for undertaking statutory functions such as local authorities and utility companies. Figure 4.2 illustrates how large-scale digital mapping from the OS defines building outlines.

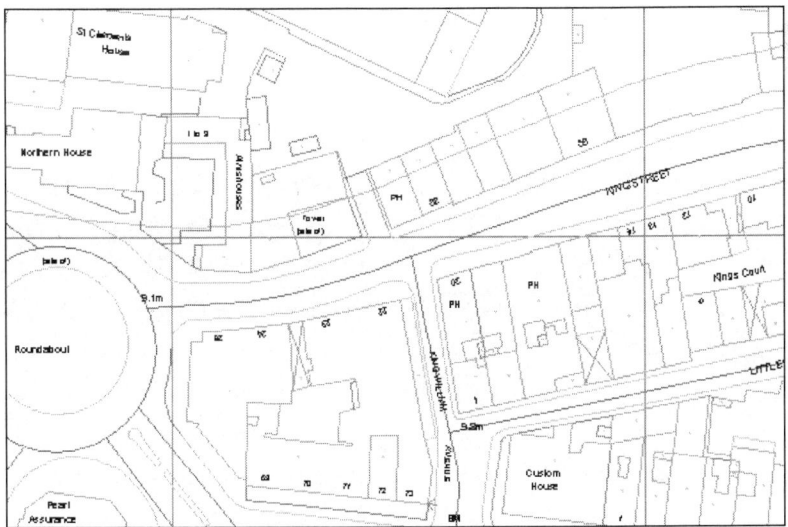

Figure 4.2 OS Landline map data.

The problem with Landline, as far as users of GIS were concerned, was that it was just a digital replication of the paper-based maps that it replaced. In other words, a Landline map tile, once loaded into a GIS was just a collection of lines and points on the computer screen. To really harness the power of GIS, a topological data model for large-scale map features needs to be present so that geographical relationships such as adjacency, connectivity and containment can be represented. Recent development of the National Topographic Database has led to the introduction of a new digital map product called Mastermap. Mastermap is the OS's new topographic view of the real world where nine themes or map 'layers' represent groups of features such as buildings, roads, land and water. Mastermap, unlike its predecessor Landline, is based on a topological data model and uses points, lines and polygons to represent real world features, which are described using metadata. Features that were represented using lines and points in Landline have been reformatted so that, where relevant, polygons represent features such as houses, lakes and parks. Unique identifiers allow attribute data to be linked to map features. Figure 4.3 illustrates, using shading, how Mastermap represents the real world as a collection of polygons rather than linework.

Mastermap will mean that users can link their data to OS map features directly rather than using OS Landline as a backdrop to other land and property data. Some of the enhancements that Mastermap has compared to Landline mapping are particularly useful for property applications. First, buildings, land parcels, fields, open spaces and other real world features are

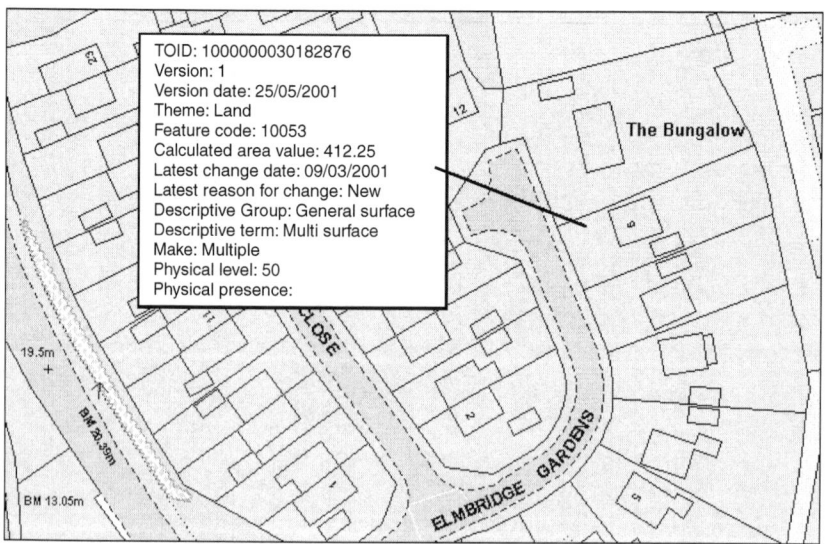

Figure 4.3 OS Mastermap.

represented by polygons with unique identifiers. So a building, for example, can be classified or shaded depending on its attributes. Second, Mastermap themes will be available in a seamless format. Users will be able to purchase data by theme rather than by tile. Indeed, the first Mastermap theme to be made available is the National Buildings Data Set. This data set comprises over 40 million building polygons together with a unique identifier for each. The National Buildings Data Set has been used by housing associations, utilities, insurers and property managers for statistical analysis and to assist asset management, pollution analysis, flood-risk assessment and planning. For example, the OS, the Property Advisors to the Civil Estate (PACE), the Land Registry and English Heritage developed an application that identifies a property, conducts a legal search using a common identifier and returns the legal extent of the property (which may contain several land parcels). It is then possible to conduct a search for, say, listed buildings within these land parcels. The OS is also working with developers to structure Computer Aided Design (CAD) design data from development plans so that they may be more rapidly incorporated into Mastermap.

Integrating maps and property data

Conventional 2D maps display geographical data and are vital for revealing spatial relationships and patterns. But the real world is 3D and two dimensions can fail to satisfactorily display land and property data at the scale of the individual legal interest in a property. For example, the existence of several property interests on successive floors of a multi-storey building, illustrated in Figure 4.4, are not easily depicted using a 2D map.

So although maps provide an intuitive interface for accessing land and property data they are not necessarily the best means of managing such

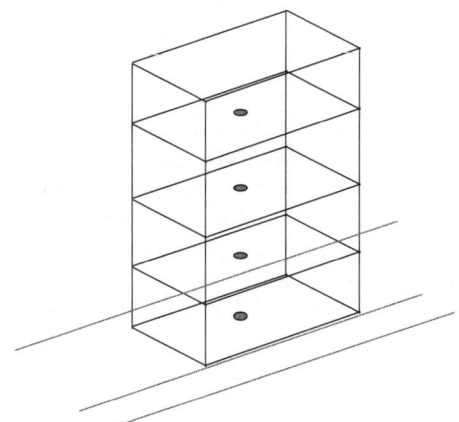

Figure 4.4 Seed points for property interests on each floor of a building.

data. As an example, many local authority services require address-based data sets (e.g. electoral register, environmental health, land charges, housing, council tax and non-domestics rates collection, social services) but only a small number require data sets that have other forms of geographical reference, such as national grid co-ordinates. This has led to the realisation of the importance and subsequent development of a single address database for the UK – the National Land and Property Gazetteer (NLPG). The NLPG is a definitive national index of land and property and is facilitated by adherence to BS7666, the British Standard for describing and locating land and property. The NLPG is therefore more than a map or 'mosaic' of land and property interests – it is a referencing system for objects that occupy 3D space. In this way, it could be said that Mastermap property identifiers are 2D while NLPG address identifiers are 3D, with a relational link between the two. This is illustrated in Figure 4.5. A key objective of the NLPG is to integrate land and property data sets.

BS7666 specifies a consistent format for holding details on every property and street. The standard defines a land parcel as a Basic Land and Property Unit (BLPU). A BLPU is defined in BS7666 Part 2 as a contiguous area of land in uniform property rights or, in the absence of such ownership evidence or where required for administration purposes, inferred from physical features, occupation or use. Each BLPU has a Unique Property Reference Number (UPRN) – illustrated in Figure 4.6, a spatial reference (national grid co-ordinates) and one or more Land and Property Identifiers (LPI). The LPI is basically the address of the BLPU in a standard format that uniquely identifies the BLPU in relation to a street as defined and held in the National Street Gazetteer. The recording of the extents or boundaries of land parcels, rather than just seed points, enables more sophisticated spatial queries using GIS. So, whereas the previous version of BS7666 described only point data, with each BLPU given a single set of national grid co-ordinates, the revised version stipulates that each BLPU can also have one or more extents held within the NLPG. An extent is defined as the spatial boundary of the BLPU, represented by a polygon and/or a closed set of grid co-ordinates.

The NLPG is itself the aggregation of Local Land and Property Gazetteers (LLPGs). Local authorities are the definitive source of addresses due to their statutory responsibility for street naming, numbering, planning and development control. Local authorities are therefore crucial to the effective creation and maintenance of LLPGs and, in turn, the NLPG. For example the Bristol LPG was constructed by integrating Valuation Office (VO), Land Registry, OS and internal data sets to form BLPUs with UPRNs and is BS7666 compliant. The main problems encountered in its creation were a lack of information about flats, commercial properties and industrial estates and differences in the way addresses were treated. Matching non-domestic properties was harder than matching domestic properties. Also, matching in rural areas was less successful than in urban areas due to missing property descriptions, the

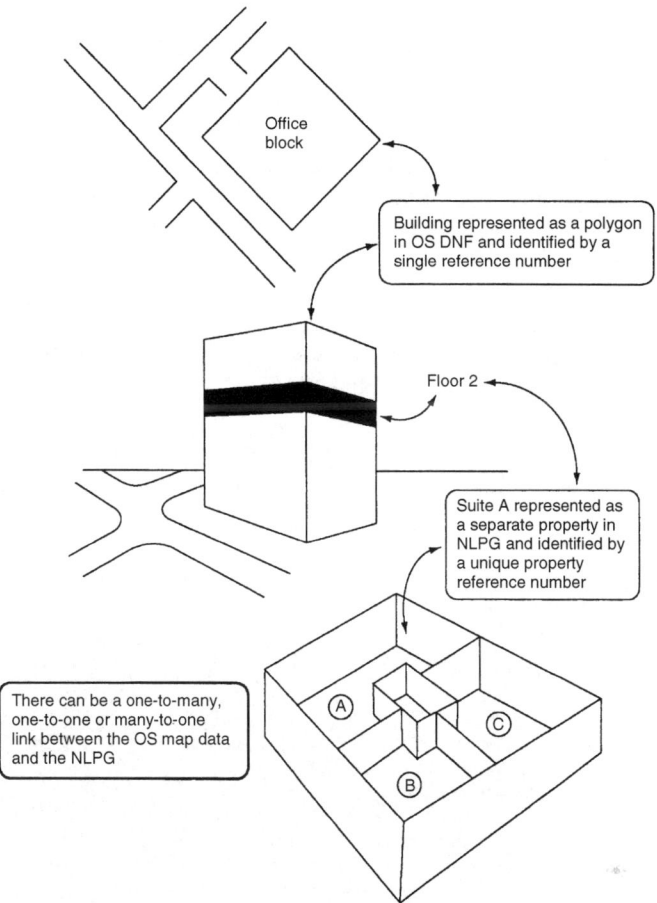

Office
block

Building represented as a polygon
in OS DNF and identified by a
single reference number

Floor 2

Suite A represented as
a separate property in
NLPG and identified by
a unique property
reference number

There can be a one-to-many,
one-to-one or many-to-one
link between the OS map data
and the NLPG

Ⓐ

Ⓑ

Ⓒ

Figure 4.5 The relationship between OS map data and the NLPG.

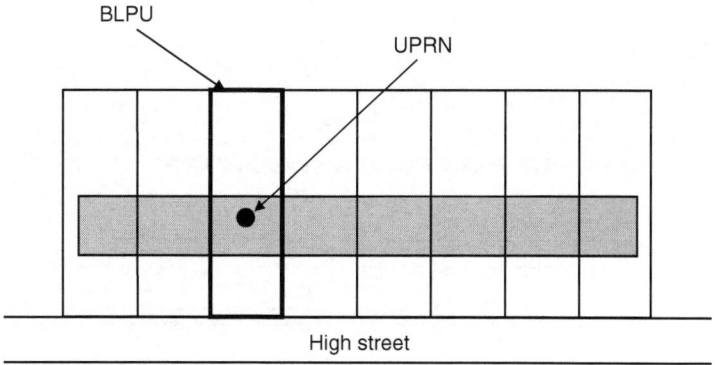

BLPU

UPRN

High street

Figure 4.6 BLPU and UPRN.

absence of definitive street names or presence of alternative street names and differences in settlements in which properties were placed (Smith, 1994). For the NLPG to work, maintenance of LLPGs must be given high priority by local authorities. Successful maintenance will involve the integration of local authority address data sets such as the electoral register, council tax, planning and street numbering/naming functions. Once a LLPG has been created the local authority can use a BS7666 compliant gazetteer application (an example of which is shown in Figure 4.7) to store and manipulate the data.

Development of the 'embryo' NLPG is under way. A company called Intelligent Addressing is constructing the gazetteer from existing national address data sets such as the Council Tax Lists, the Ordnance Survey Address-Point data set and Intelligent Addressing's existing commercial property gazetteer which includes the Non-Domestic Rating List. When this first stage is completed, the results will be matched by Intelligent Addressing to address data sets held by each local authority, including the

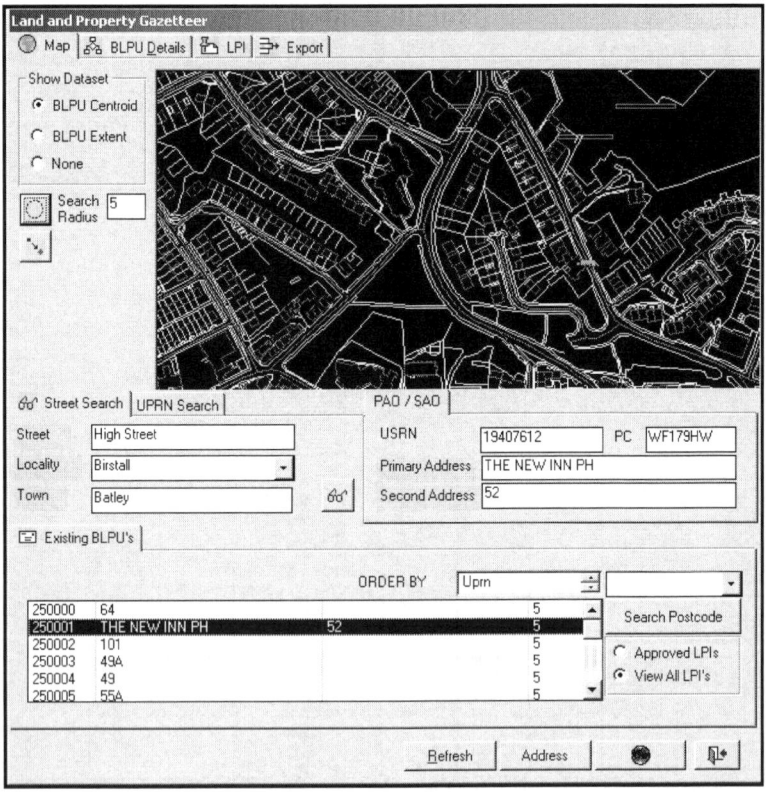

Figure 4.7 BS7666 gazetteer application.

Source: Used with permission of Innogistic Software plc.

Electoral Registers, Address-Point, Council Tax and locally held Non-Domestic Rating Lists. This second stage process will clean and synchronise the national data set with those of each local authority. As each authority's data sets are matched, they will be returned as a BS7666 compliant Gazetteer complete with an official UPRN. A national 'mosaic' of BS7666 compliant LLPGs will thus form the constituent parts of the NLPG. Once complete the NLPG will be a referencing system that is available via a simple tariff to help pay for the compilation of the LLPGs by local authorities. The NLPG will be the master address index to which organisations, public and private, can link their data. Initiatives such as the National Land Use Database (NLUD), NLIS, searches of local land charges, sellers' packs and local authority property asset management will ultimately all rely on the NLPG.

Mapping property data is not a function confined to the public sector. For example, in order to map the location of property-specific data, property consultants Jones Lang Lasalle (JLL) conducted a pilot project at their Leeds office to develop and evaluate a property-related GIS using in-house data. The aim was to develop a property-related application for the Central Business District (CBD) of Leeds, to act as a precursor for other applications across the company using GIS and OS large-scale digital mapping. Currently, a number of internal databases are used to record various aspects of property information within the firm and there are inconsistencies not only between databases in how property data are recorded but also within each database too. This is a problem encountered by many users of GIS – the introduction of an information system that requires data to be integrated in a geographically consistent manner from various sources begins with a data cleaning phase. The pilot project is concentrating on JLL's agency database. A point theme has been created that, to date, includes around 500 properties that have been geographically referenced and plotted on large-scale maps. This theme overlays a raster map of the urban area to create a 'digital atlas' of the Leeds CBD. The GIS provides the ability to query the database on various characteristics such as rents, take-up, availability and occupier type. Also, photographs can be incorporated. The use of points to illustrate the location of properties did not prove popular with surveyors, who preferred building extents to be shown. Building polygons can be shaded according to attributes recorded in the agency database such as rent, number of floors, age, ownership, availability, vacancy and take-up rates and lease details, together with digital photographs.

Developments such as these will remain internal to individual organisations until the NLPG establishes itself and reaches critical mass because one of the major barriers to the development of map-based property data services has been the lack of a consistent standard for referencing property data. Private sector data suppliers and users need the security of a referencing standard in order to ensure that their investment in data is long term.

Land registration and land information systems

The previous section described how digital mapping at the property level and a unified property referencing system allow simple map production to be undertaken and property-based data sets to be mapped. This section will illustrate how map-based property referencing can be used to develop land and property registration and information systems at the national level.

Land registration

Successful market economies depend on national systems for secure ownership of land and property in which all who live or work in the country can have confidence. The cash value of land transacted in the UK exceeds that of the Stock Exchange on a daily basis and a 'land registration system is the cornerstone of a modern land and mortgage market based on guaranteed private rights in land' (Manthorpe, 2001). A Land Register records ownership details of land parcels. In many countries it is not possible to actually 'own' the land, rather it is the 'right to use' land that can be owned, known as land tenure, and a land register is a legal document that records the entitlements or 'titles' to these land rights. The recording of land and property rights, ownership and value is important because effective property management requires the rights of owners to be secure and the value and use to be identifiable. There are two main methods of registering land rights: registration of deeds and registration of title.

Registration of deeds: A copy of the deed of transfer (the legal conveyance contract) is deposited in a public place and provides a priority claim to ownership. However, this is a method of registering legal documents associated with a conveyance rather than title to land. Countries that have a deed registration system have reported errors in the description of the property concerned. These errors can be minimised by improving records management, adopting more flexible survey standards and procedures, making registration compulsory (as in the UK) and computerising access to indices and abstracts of title (as in the UK).

Registration of title: A specimen register of title is shown in Figure 4.8. Here the basic unit for registration is the land parcel rather than the deed. The land parcel is a contiguous area of land within which unique, homogeneous rights or interests are recognised. In a legal cadastre these rights are legally recognised. Each land parcel is identified on a map that is cross-referenced to the register of the owner's details (proprietorship register), nature of tenure and ancillary information. Most registers of title guarantee the information so compensation is payable for incorrect information if loss is incurred. In the UK, registration of title began in the 1860s. Once registered the entry will show:

- Property Register: date, address and Title Plan
- Proprietorship Register: owner's details and nature of title, e.g. freehold
- Charges Register: any adverse interests affecting the title, e.g. mortgage, easement, restrictive covenant.

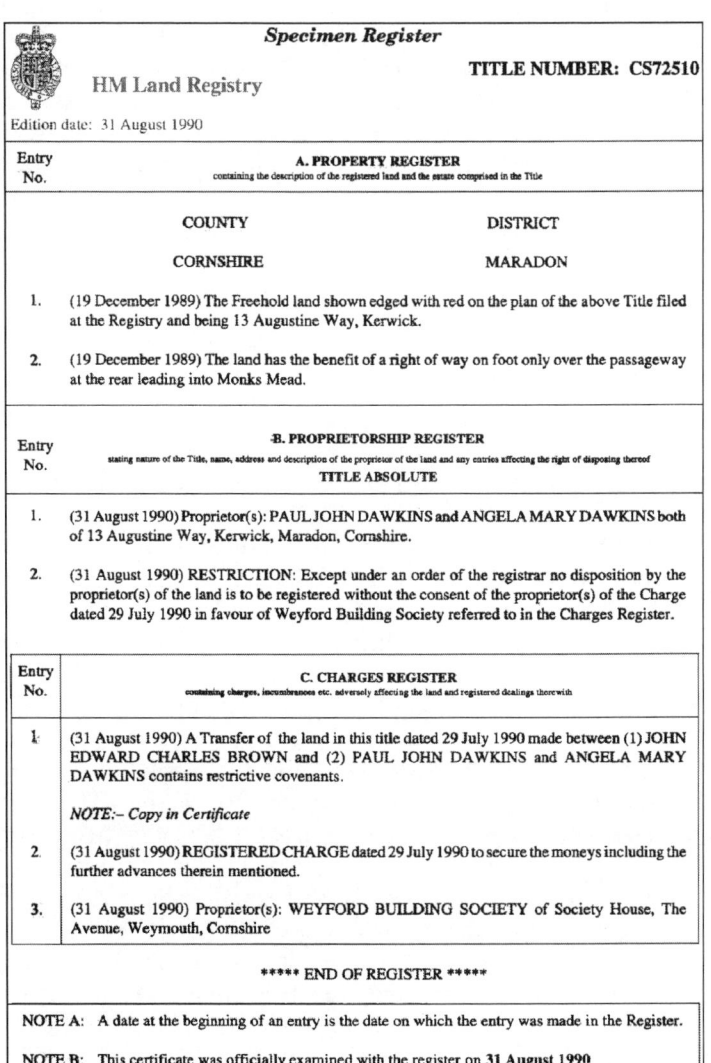

Figure 4.8 Register of Title.

Source: Reproduced with permission of Her Majesty's Land Registry, © Crown Copyright.

In England and Wales Her Majesty's Land Registry (HMLR) is responsible for the registration of title to land in England and Wales. There are currently 17.4 million registered titles (out of approximately 25 million titles in total – many of the unregistered titles are rural) and the records are held at twenty-four District Land Registries. Completion of the Register's geographical coverage is expected by 2010. Further planned improvements include the registration of all leases of three years or more (rather than the present 21-year limit) and the inclusion of details regarding owners, leases, mortgages and other interests from 2003. To date, 98 per cent of the Registers of Title have been computerised and are held online. The 1988 Land Registration Act meant that Land Registry data was opened to public inspection in 1990. Prior to this only the owner or a person who had the owner's consent could inspect the register. The Land Registry now provides a service that allows subscribers to access registers and scanned title plans online.

A Title Plan (an example of which is shown in Figure 4.9) is created for every registered land parcel. The plans are based on OS large-scale mapping and show, as accurately as possible, the extent of the land in a registered title. In addition a Title Plan may also show other plan references which identify any parts of the land or adjoining land affected by entries in the register, such as easements, covenants or areas of land removed from the title. The Title Plan should be read in conjunction with the Register of Title because the plan is part of the certificate of title if it is referred to in the register, which it normally is under a statement such as 'particularly delineated edged in red on the attached plan'.

The extent of the registered land on the Title Plan is normally shown in red edging drawn along the inner edge of the lines of the physical boundaries or the plotted lines of undefined boundaries surrounding the property. Where a registration includes only part of a building, for example, a room over a passageway, a plan reference will be added and an explanatory note given in the Property Register. Complex floor levels will sometimes be shown on a supplementary plan, which will be prepared at a larger scale and attached to the Title Plan. Where the extent of the property is complex (e.g. multiple floor layouts or common parts) a Title Plan is sometimes supplemented by a deed which is referred to in the Property Register.

All of the Title Plans held by HMLR have been digitally scanned and can be viewed as raster images online. However, the scanned Title Plans can be difficult to interpret and do not facilitate spatial searching. The task of digitising the Title Plans so that they are held in vector form is clearly a substantial one. A digital mapping pilot project took place in Peterborough in the 1980s using a Unix computer system but the cost–benefit analysis identified significant costs, particularly in terms of digital OS map data. Subsequently a PC-based system was implemented at the Weymouth

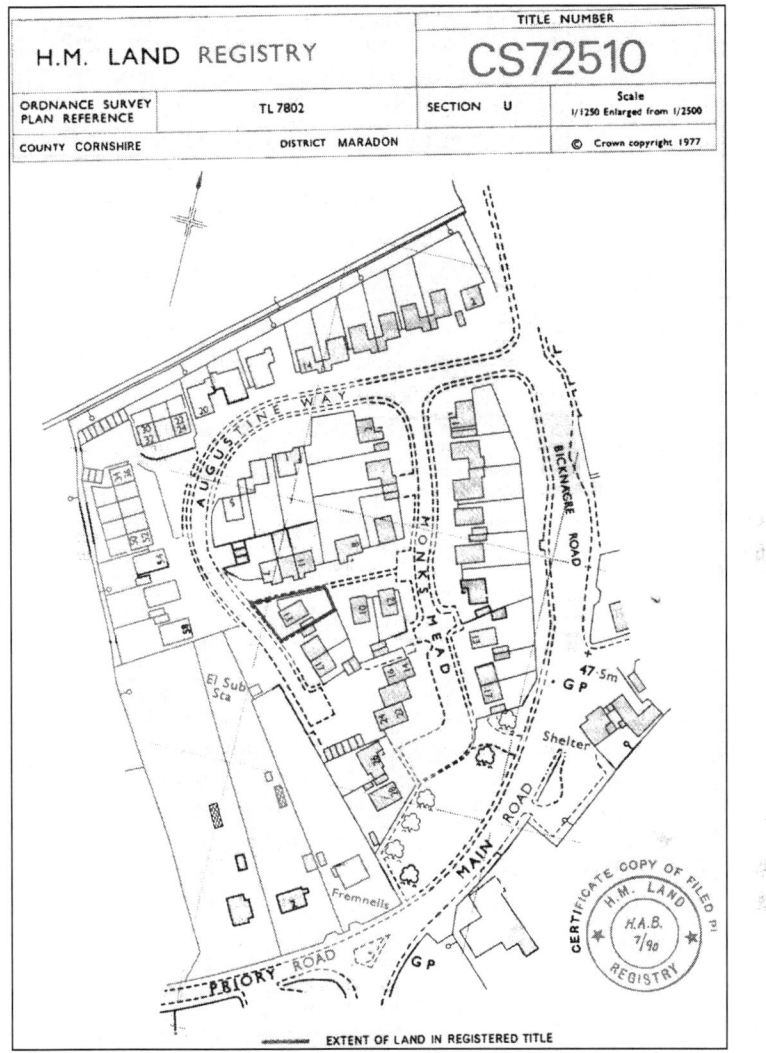

Figure 4.9 Title Plan.

Source: Reproduced with permission of Her Majesty's Land Registry, © Crown Copyright.

District Land Registry. This system provides a geographical interface for the procedures associated with the creation of a Title Plan. The system has now been extended to all twenty-four District offices. Consequently, a hybrid Title Plan system is in place; existing plans are stored in a scanned image format while new registrations are entered digitally.

Every year 500,000 new Title applications are received and these are digitised using a GIS interface. The process of adding a new Title Plan begins by identifying a registered title on the GIS using a map reference, title number, locality or address with an address-check (using the NLPG ultimately). The subject property can then be viewed and all vectorised titles are displayed in the area of the new registration and any overlaps are notified. Scanned and vectorised registered extents can be viewed overlaying current and historic OS map data (the latter aids dispute resolution and registration). The historical mapping can be moved and warped to coincide with current mapping or any other historic mapping epoch. The subject polygon on the OS map is selected and edged prior to 'settling' the extent using a standard template. A benefit of vectorised title plans will be the variety of possible uses of the digital data.

A particular problem in terms of producing a unified map base for all Title Plans is that there is no provision under the Land Registration Act for the OS mapping (on which registration information is recorded) to be updated. The mapping used at the time of the registration becomes a legal snapshot in time. The proprietor of the registered title receives a copy of the Title Plan at the time of the registration and it is that which is legally binding. Therefore, the ideal of a current OS map base on which Title Plans appear as a mosaic is not achievable. Furthermore, the 'General Boundaries Rule' (Rule 278, Land Registration Rules, 1925) states that '...the exact line of the boundary will be left undetermined – as, for instance, whether it includes a hedge or wall and ditch, or runs along the centre of a wall or fence, or its inner or outer face, or how far it runs within or beyond it; or whether or not the land registered includes the whole or any portion of an adjoining road or stream'. This means that the precise position of the legal boundary is often unclear.

The Land Registry also produces an Index Map. This is a collective name for a series of large-scale OS maps covering the whole of England and Wales on which is shown the position and extent of land for every registered title together with the title number and class of title registered. For rural areas the Index map consists of whole OS map sheets on which the registered extents are outlined. For urban areas extracts of the sheets are used for purposes of clarity. Currently the Index Map is held in paper form and consists of approximately 400,000 separate documents held at the District Land Registries. An Official Search of the Index Map is undertaken to determine whether or not a piece of land is registered and, if so, what the title number is and whether the title is freehold or leasehold. The result will also show any pending application for first registration and any caution against first registration, priority notice, manorial interests, souvenir land declaration or rent-charge affecting the land. The Land Registry receives an average of 14 million requests for searches of the register each year.

The Land Registry has embarked upon a programme of digitisation of the paper-based Index Map so that it represents the (non-legal) extents of registered parcels. The web-based vectorised Index Map will act as an interface to the Title Plans and Registers of Title and it is due to be completed by 2004. Because the Index Map is based on current OS mapping, OS Mastermap data will be used so that there is consistency between polygon extents for the estimated 20 million land parcels.

Land Information Systems

A Land Information System (LIS) is 'a system for capturing, storing, checking, integrating, manipulating, analysing and displaying data about land and its use, ownership, development, etc.' (Department of the Environment [DoE], 1987). Pressure on land and property resources has never been greater and the development of LIS as a tool for managing them has been paralleled by unprecedented developments in data management and communications infrastructures. These developments, including the rise of the Internet as an information marketplace and the 'globalisation' of information, have far reaching implications for the effective management of land and property resources.

Land information systems are desirable, particularly in a populous country like the UK, because the task of managing land resources for environmental needs, planning, housing and infrastructure provision is growing in complexity as a result of the need to match wide-ranging requirements to a limited supply of land. LIS, which permit the integration and management of a wide range of land-related data, can provide the detailed and comprehensive information on which to base the effective allocation and management of land resources.

The basic spatial unit of a LIS is the land parcel which has seven attributes (Dale, 1989):

1 an essential component of planning and development control;
2 foundation of the land market;
3 spatial unit for documenting human affairs;
4 basis for property tax assessment;
5 has the ability to be aggregated (postcode areas) and disaggregated (floor levels);
6 can be subdivided or merged according to changes in ownership;
7 collectively they form a contiguous mosaic of legal interests.

Attribute data such as ownership, taxation and occupancy are stored with reference to these basic spatial units in the system (Larsson, 1991) and integrated analysis of many different data sets is possible. The usefulness of a LIS will depend on the currency, accuracy, completeness and accessibility of the information that it contains.

A 'cadastre' is a particular type of LIS that is an inventory of land and property information based on a survey of land parcel boundaries, sometimes referred to as a 'parcel-based LIS'. The UK does not have a cadastre because the Land Registration Act of 1925 did not require the boundaries of land ownership to be surveyed and mapped precisely. Instead we have a 'general boundary' system, defined in the section on 'Land registration'. The principal function of a cadastre is to provide information on land ownership, use and value. Cadastral records consist of maps and text and each land parcel has a unique identifier. There are three common types of cadastre:

Legal cadastre: A legal cadastre is legal record of land tenure that consists of a written register containing owner and land parcel information and a map or plan which is cross-referenced to the written register. A legal cadastre provides certainty of ownership, security of tenure, improved conveyancing in terms of cost and speed, added security for loans that are secured against property and support for a national system of land use planning and property taxation. It also reduces the number of land disputes and helps monitor property market activity.

Fiscal cadastre: A fiscal cadastre is a record of land parcels that includes, primarily, information necessary to determine property value for taxation purposes. Because it is often the owner that pays the tax, fiscal and legal cadastres can be linked together. In order to base a tax on property it is necessary to assess differences in the utility of land parcels or properties situated thereon. Utility can be measured by assessing how much someone would be willing to pay for an interest in a particular land parcel. This process is known as valuation. Thus tax on property is based on the estimated taxable revenue of each land parcel, the amount depending on the use of the land.

The fiscal cadastre, like the legal cadastre, is well suited to computerisation. With regard to market data collected in support of the valuation function, a database would allow the valuer to concentrate on the analysis of these data rather than their collection and assimilation. Computerisation would lead to further benefits in terms of uniformity of tax assessment for property – equity of assessment being a primary concern for any taxation system.

Overseas case study

At the National Land Survey in Sweden, GIS is used to analyse changes in boundaries of valuation areas (areas where the market is assumed to perform in a similar way and market values are uniform) for taxation purposes (Sundquist, 1995). For one- and two-bedroom family houses the country is divided into more than 7,300 areas. Each residence is allocated to a valuation area. The boundaries of these areas change due to the development of new houses and changes in infrastructure. A background topographic raster map of 1:15,000 scale is supplemented by large-scale city maps and valuation area boundaries are overlaid together with information about properties. These attribute data include sales information and are obtained from a central database at the Central Board for Real Estate Data.

Multi-purpose cadastre: A multi-purpose cadastre combines fiscal and legal cadastres and provides links to other land parcel information. For example, it could include data on ownership and legal interests, use, development control, planning history, values and prices and other property-related data such as social, economic and environmental information. It is thus capable of supporting land registration, land taxation and other land administration functions.

Existing LIS and other land and property data sets may be linked together via unique property identifiers (such as UPRNs as defined in BS7666 above) to form a multi-purpose cadastre. A GIS then allows geographical linkages to be created between various property data, as shown in Figure 4.10. In this way a digital map base provides the link between property data sets using land parcels and their unique identifiers as the basic land and property unit. In urban areas the land parcel may be subdivided into buildings and parts of buildings such as a shop on the ground floor with a flat above. Such detail is difficult to represent on a 2D map but can be recorded in a relational attribute database such as the NLPG data structure.

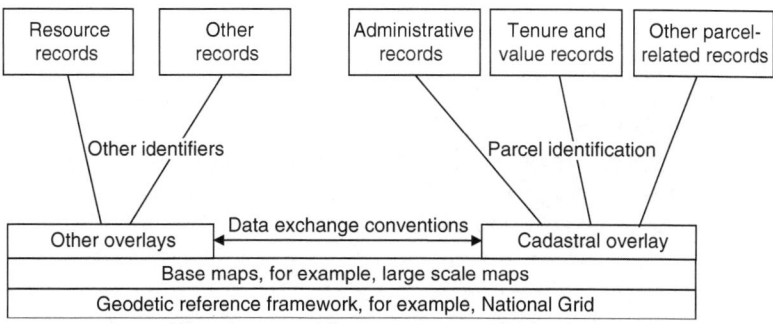

Figure 4.10 Multi-purpose cadastre. Reprinted from 'Land information management: an introduction with special reference to cadastral problems in third world countries' by Peter F. Dale and John D. McLaughlin (1998) by permission of Oxford University Press.

Source: © Peter F. Dale and John D. McLoughlin, 1988.

LIS provide the comprehensive information on which to base decisions relating to the effective allocation and management of land resources. Ralphs and Wyatt (1998) discussed progress in realising the goal of a multi-purpose LIS by examining existing moves towards very large LIS implementations at the regional, state and national level. The discussion drew upon experiences in Europe, the US Australia and the Developing World in addressing a series of key human, organisational and technical issues, including:

- the assurance of privacy of information relating to individuals;
- the need for confidentiality of commercially sensitive information;
- the availability of copyright to protect innovation and original thought;
- the adequacy of legal statutes that enforce the above;
- the creation, implementation and adherence of standards for referencing and exchanging real estate information;
- problems of and prospects for combining diverse information resources and the cost of generating digital spatial information.

We next discuss the development of a National Land Information System (NLIS) for the UK.

A National Land Information System (NLIS) for the UK

The NLIS initiative aims to provide seamless access to integrated land and property data. The service will co-ordinate and distribute comprehensive, accurate, up-to-date and regularly maintained information on land and property ownership, use and value for every land parcel in Britain with links to planning, socio-economic, demographic and environmental information. The service will integrate, via the Internet, all of the various sources of information that are currently held at disparate and unconnected locations. This will give easier access to a wide variety of land and property information on a one-stop basis and in a reliable and secure environment. The goal is to improve the management of one of Britain's most important and valuable resource – land and property. The NLIS will help to speed up the process of gathering land and property related information, while simultaneously creating new markets for this information.

Some of the catalysts for creating an NLIS include:

- a requirement to know more about the land and property assets, for example, local authorities need to record their property assets and the European Union is proposing a register of land holdings for each member state;
- better information about land and property will encourage more effective property management, environmental protection, taxation, planning and marketing;

- many public sector organisations are involved in the collection of property data and are computerising their records. With standardisation of referencing procedures these data can be integrated, thus reducing duplication of effort and adding value through shared use of data;
- developments in information technology provide a foundation on which to build an NLIS;
- many countries, both developed and developing, have taken the plunge in the creation of LIS and the experiences of these countries prove that the cost of development can be recovered provided a critical mass of data is released;
- the establishment of a NLPG which allocates a UPRN to each property will enable unambiguous identification of the land parcels.

An NLIS offers the ability to integrate property data collected by public and private sector organisations. It also provides the opportunity to reuse data that may have been collected for a specific purpose and hence add value to those data and reduce the net cost of collection. Unlike many other countries Britain benefits from a national large-scale digital map base. The UK NLIS will geographically reference land and property information using this map base.

The absence of an NLIS in the UK has long been recognised and there have been many calls for wider access to property data collected by various organisations and for these data to be integrated using a common property-referencing standard. The idea for a national network of land and property data was propounded in a project called Domesday 2000 (Dale, 1991) which put forward the case for an NLIS that would integrate data sources relating to land and property ownership, use and value. A market potential survey was undertaken to gauge demand for computerised access to land information (Capital and Counties, 1992). This survey revealed a substantial market for integrated access to property information sources. Research continued under the auspices of the Domesday Research Group (Fovargue *et al.*, 1992) and a demonstrator system was developed with funding from the Royal Institution of Chartered Surveyors (RICS) to raise awareness of the commercial and social value of this form of property data access. As part of a Citizen's Charter Initiative, in 1992 the government supported the development of a pilot conveyancing application of the NLIS for Bristol (Smith, 1998). This public sector initiative involved collaboration between the Land Registry, the OS, local government and the VO. It successfully used the property-referencing standard, BS7666, to integrate data from various government departments in support of the enquiries made prior to a conveyance of a property. The pilot project also investigated ways of including non-addressable property in order to gain 100 per cent coverage for the study area. Bristol City Council (BCC) now uses the system for the administration of local land charges and the Land Registry plan to develop an on-line conveyancing application of the NLIS pilot. The Bristol pilot was an important first step that was necessary to ensure that all data providers describe properties in the same way (Smith, 1998). It also revealed a significant difference in the way that solicitors and surveyors access property information. Solicitors pay fixed fees for the information that they gather to support their advice on a conveyance. Typical data suppliers include the local authority, the Land Registry and the utilities. Surveyors, on the other hand, have a more informal network of contacts from which they gather information in support of property advice. This has important implications for a more formal means of integrating property information using BS7666 and disseminating this information using an NLIS.

Overseas case study

In Scotland ScotLIS is an initiative to develop a one-stop-shop with easy and affordable access to a wide range of computer-based information about land and property from both private and public sectors. The ultimate aim of the ScotLIS project is that of providing an integrated data set where the user obtains information from a range of providers by means of a single search enquiry. This will be facilitated by means of a gazetteer through which enquiries will be routed to the information providers. A pilot funded by the Government's Invest to Save initiative will cover the information pertinent to the Glasgow City area. The pilot comprises Registers of Scotland, Glasgow City Council, British Geological Survey (BGS), and The Coal Authority. Public and private organisations will be encouraged to participate in the project. Applications include conveyancing, environmental assessments, planning, building control, property searches, estate management, development appraisal, insurance assessment, marketing and land monitoring.

Overseas case study

The Singapore NLIS is an online service provided by the Singapore Land Authority. Land information (including ownership, transaction history, land tenure, encumbrances, land use and transport information, information on surrounding amenities such as shops, schools and roads) and cadastral and locality maps (land parcel boundaries and areas), is made available to the public for a fee. Land information that was previously only available from separate government agencies is now accessible from one web site.

A feasibility study for NLIS, carried out by KPMG Management Consultants in 1997, identified those market sectors for NLIS services that have the greatest commercial value, with many more likely to develop over time. These services are listed in Table 4.1. It is not difficult to envisage how

Table 4.1 Potential applications of NLIS (Local Government Management Board, 1997)

Opportunity	Sectors	Processes
Conveyancing system	Commercial Non-commercial	All aspects of conveyancing
Environmental assessment system	Horizontal market, multi-sector	Site sensitivity, contaminative uses
Geo-marketing system	Marketing	Property enquiries, demographic/trend analysis
Collateral risk analysis	Financial services	Property enquiries, financial/ legal, conveyancing, enviromental and risk analysis
Insurance location risk analysis	Property underwriters	Property enquiries, environmental and risk analysis
Property enquiry service	Horizontal market, multi-sector	Access to owner, value and use of land and property
Policy spatial support system	Central government	Trend analyses, policy monitoring

the applications highlighted by the KPMG study may grow. For example, policy spatial support is probably of equal value to the local as well as the central government. Participants in the feasibility study highlighted the following as the major benefits of NLIS:

- saving of time and money
- ability to provide clients with a better standard of service
- more effective targeting of clients.

From the KPMG study key applications for land and property management would appear to be collateral risk analysis, insurance location risk analysis and a property enquiry service. The first two applications are perhaps better known in the property professions as property valuation for lending and insurance purposes respectively. Indeed, a survey of members of the Royal Institution of Chartered Surveyors (RICS) undertaken in 1997 revealed that the NLIS data of most interest to surveyors were property values and prices, planning and ownership details.

A number of potential service providers were also consulted. They represented a mix of Private Finance Initiative (PFI) operators, system developers and value added re-sellers in the land and property sector. As with the potential users, the service providers demonstrated a high degree of enthusiasm for NLIS. Many are already planning their involvement. The market research clearly identified that many users would pay a higher price for value-added information or services by passing on these charges to clients through better quality of service, time savings or through new business generation.

The NLIS Feasibility Study concluded that the NLIS is commercially feasible and identified considerable enthusiasm in the market place. All of the critical elements for success have converged: open government policies, availability of data and standards, partnership business models, information and communications technology, commercial enthusiasm and an increasing market demand. In January 1999 the Treasury awarded £2.3 million to the Ordnance Survey, the Land Registry, Registers of Scotland and the Local Government Improvement and Development Agency to help implement a live version of the NLIS conveyancing application. It was to be a leading example of 'joined up government' and would help the government pursue its manifesto pledge to speed up and simplify the house-buying process. The application covered the whole of Bristol and enabled the geographical identification and on-line search of twelve data providers as part of the enquiries before purchase of a dwelling.

Now that government support and commercial backing for an NLIS is in place, the development of NLIS applications is under way. Users of the NLIS can access data over the Internet and receive information in a form that is not dependent on the data source. Rather than a central database of land and property data, the NLIS acts as a hub or gateway through which data providers can supply data and users can access data and associated services, as shown in Figure 4.11. Maintenance and accuracy of data sets remain the responsibility of the suppliers but are managed by a hub administrator. The price and quality of each contributor's data will be determined by market forces; if the price is too high or the quality is too low then customers will not request those data. Income received will be apportioned amongst service and data suppliers by a hub manager. The advantages of a hub network are that information may be stored and updated by those organisations skilled in the handling of particular types of land and property information while access, analysis and display may be performed remotely. Also, from a technical viewpoint, a network of distributed

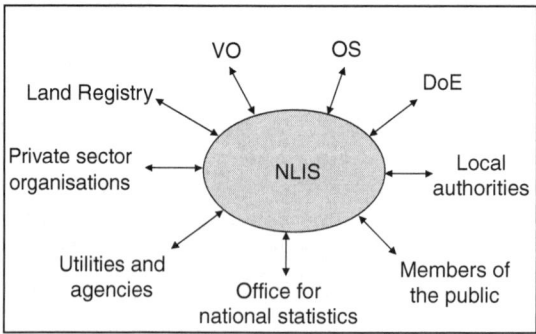

Figure 4.11 NLIS hub.

databases is preferable to a single large database in terms of storage efficiency and access time.

The NLIS infrastructure allows data providers to supply data to the central hub via a local area network, similar to systems in operation in other countries such as Sweden and Australia. The NLIS hub licence has now been let to MacDonald Dettweiler and Associates. With the hub in place 'channel' service providers offer customised access to the data via the Internet. Channels have been licensed to provide NLIS services that will link to the hub and the hub operator will interface with data providers and pass this information to the NLIS channels. Channel licensees will buy data through the hub and sell it to property professionals and the public. Providing land and property information through NLIS will simplify service delivery by creating a single interface for accepting requests, delivering results, billing and payment. In addition data and services will be actively marketed by the NLIS channel operators.

As more than 2.5 million homes are bought and sold in England and Wales each year, initial NLIS channels are being developed to speed up the property searches necessary as part of the conveyancing process. These services will provide access to property information held by local authorities, the Land Registry and others and can then be expanded to help simplify the buying and selling process. This will be achieved by allowing properties to be searched via the web, including faster and more accurate property identification via the NLPG and online maps. The conveyancer will be able to retrieve ownership information from the Land Registry and local land charge information from the local authority as well as searching for other information relating to the property such as environmental or geological data.

In the future services that have been tailored for other property professionals such as surveyors, property agents, lenders, developers, insurers, local authorities and environmental managers will appear. These users may require similar data but packaged in different ways. Additional data may also be required such as data collected from local authorities by English Partnerships on previously developed land, which is being used to create the National Land Use Database (NLUD). The potential for an NLIS is far-reaching and as services appear the user-base is expected to expand from those involved in conveyancing to a much wider ranging community that can benefit from easier access to land and property information, particularly that held by central and local government.

Yet NLIS will only operate effectively with appropriate access to data that are comprehensive, accurate, up-to-date, regularly maintained and reasonably priced. The private sector has traditionally not pooled data resources, preferring to rely on data generated in-house. This has led to a considerable duplication of effort and no single comprehensive or national source. A change in attitude towards data collection, dissemination and use is required for the NLIS to work effectively in the private sector. This is now evident. Furthermore, an NLIS requires the support of government in order to release key property data sets held by local authorities, the VO and other government departments. This is also happening. A comprehensive policy of data release is needed if a 'critical mass' of data necessary for the NLIS to work is to be achieved.

The creation of an NLIS will offer the property professions several advantages. First, the duplication of effort that exists in the collection of property information by individual firms will be minimised. Comprehensive access to national information on property use, ownership and value, with links to many other property-related data sets, will allow firms to concentrate on the collection of more specialised data and develop analysis techniques relevant to their particular area of expertise. Second, small firms, unable to invest in comprehensive data individually, will have access to a 'pooled' data resource to which they can add specialist local knowledge and market experience. Third, the time spent on data collection will be reduced and a quicker response time to client instructions may be possible. Finally, by implementing a GIS-based NLIS, new geographical analysis techniques will be possible and, given the importance attached to location as a factor in property development, occupation and investment decisions, these techniques should be explored by the property professions. Two-thirds of respondents to a survey of property firms in 1998 said that NLIS would replace existing methods of gathering information and 95 per cent of them believed that it would save them time, increase their use of information and open up new sources. In the future land and property managers may generate new business opportunities by being able to interpret and add value to information that is widely available.

Overseas case study

In Saskatchewan, Canada, the city of Regina has added GIS functionality to its LIS that is used for property valuation for taxation purposes (Figueroa, 1998). Many of the variables that are input into the system are geographical, such as proximity to the commercial district. The GIS achieves several objectives:

- construction and maintenance of ownership parcels;
- measurement of proximity to key roads, railways and commercial districts;
- production of maps of comparable evidence for valuation purposes and
- production of valuation neighbourhood maps to aid delineation of boundaries.

Other departments within the city council also use the system for the notification of planning applications to neighbours, route planning for police and fire services, tree and pest control and traffic accident analysis. The city's assessment division revalues properties using

comparable sales. Assessors depreciate or appreciate building replacement costs using market adjustment factors based on comparable
sales by property class and neighbourhood. A computer-assisted mass
appraisal system is used which requires spatial inputs (area, traffic
data and distance from commercial properties). A GIS is used to construct an ownership layer and other data are also input as layers. The
GIS also helps assign value adjustments to residential properties based
on geographical criteria such as contiguity of railway tracks, high traffic roads and commercial properties. Using overlay analysis, traffic
counts from the road network coverage can be assigned to each property. Maps can be output for customers showing comparable sales and
their proximity to the subject property. Map and analysis output from
the GIS has been used as evidence in appeals against property tax
assessments. Ownership data are used to inform residents when a
development is proposed within 75 metres of their property. Engineers
have used GIS to determine flood impact and maintenance division
use it to evaluate garbage collection routes.

Property marketing and conveyancing

It is clear from the previous section that one of the first channel services of
the UK NLIS will be conveyancing. This is because it is widely regarded as
having a sound business case and a service that speeds up and simplifies
house buying in the UK and has political support. This section puts the
NLIS initiative to create a conveyancing application into the context of
related GIS developments in property marketing and agency.

Marketing

In the UK there are many listing services on the Internet for residential
property that use mapping as a window on the data. For example,
www.propertynow.co.uk uses interactive mapping to allow users to search
for property and receive a map showing locations of relevant properties
plus details such as asking prices. Other sites include www.proper-
tyfinder.co.uk, www.propertylive.co.uk, www.housenet.co.uk and
www.homesonline. At the present time none of these listing services has
emerged as a market leader. It is useful therefore to look at the Realtors
Information Network in the US to see where developments might lead.

Overseas case study

In the US the National Association of Realtors (the largest trade association of residential property agents in the country) has established
the Realtors Information Network (RIN) at www.realtor.com. This

Internet-based home search service was established in 1993 to counter the threat of private 'homes for sale' notices appearing on the Internet. The RIN is accessible to all and there are currently 1.4 million properties listed. Potential purchasers and agents access the web site and conduct an initial map-based search followed by a submission of details of desirable attributes (including price range) so the search narrows to a selection of suitable properties within a defined geographical area. Images are included as well as details of the relevant agent.

A search for a property begins by entering the state or province on a map (Figure 4.12), selecting an area within the state (Illinois is shown in Figure 4.13 and northern Illinois in Figure 4.14) and choosing a county (Chicago) that is of interest. An area can then be selected within the county (Figure 4.15 shows Chicago City South) and a specific neighbourhood is chosen (Figure 4.16 shows Hyde Park). Finally the user selects the type of home (single family for example) and inputs a price range and physical characteristics. Matched homes are listed, photographs (such as the one illustrated in Figure 4.17) are included along with physical details and a map.

The RIN has now reached critical mass and is the place on the Internet in the US for properties to be advertised nationally. Critical mass was achieved by making the RIN a complete online service for property brokers, including property advertisements, email, bulletin boards, secure Internet access and web browser for subscribers. It also includes trade journals and newspapers online, membership directory, Multiple Listing Services and an online market for Realtor products. Realtors pay an initial fee followed by payments based on properties advertised and products downloaded. One such product is GeoData, a simple GIS interface that allows Realtors to search for properties geographically, display internal and external images, list all the features and produce reports on a variety of demographic and environmental data sets held on the RIN such as:

- state/county/zip/tract/place geography and demographics, pollution levels, voter registration, income tax rate, unemployment rate, sales tax rate, crime rate, consumer price index, weather data;
- assessor data: owner, assessed value, date purchased, property details;
- details of sales;
- details of foreclosure proceedings in progress;
- school test scores, student/teacher ratios, facilities, enrolment by grade, demographics, equipment, budget, administrators and school district data;

- full telephone directory listings and mailing addresses, identified as business or residential;
- full street coverage with names, address ranges and ZIP codes
- highways, state roads and other major arterial routes;
- landmarks: parks, airports, military bases, golf courses, recreational facilities, retail centres, churches, museums, etc. and
- railways.

These data are updated annually and the aim is to include all conveyancing forms in electronic format too. GeoData means that Realtors can take advantage of GIS capabilities collectively rather than having to invest in data capture and technology independently.

Overseas case study

In Japan a similar system has been developed called the Real Estate Information Management System. The GIS-based system is designed to assist in the analysis of various locational requirements of purchasers and thus improve property matching, share property and purchaser information more effectively and produce 'property statements' that include maps. Properties that meet the requirements of purchasers are searched using the GIS, based on area, desired use, period and price. The system stores data on more than 8,000 properties and is used daily by more than 100 Realtors.

Overseas case study

In Ireland 'My Home' (www.myhome.ie) uses mapping to access a gazetteer of properties for sale. Users can zoom to areas of interest and view details of properties, identify the locations of nearby schools, shops, parks, hospitals and other amenities and a network tool offers directions to these amenities. More than eighty estate agents in Ireland use the site which provides a web-based facility for agents to geo-code new properties for sale interactively as they come on to the market.

There are also commercial property brokerage services available via the Internet such as the EGPropertyLink service from Estates Gazette Interactive (www.egi.co.uk) and Primepitch.com (www.primepitch.com) which use mapping as an aid to property search and selection routines. Many residential and commercial listing services use Multimap for property

Figure 4.12 Selecting a state (1 of 6).

Source: Selected graphic images from Realtov.com and used herein with permission. Copyright © 1995–2002 National Association of Realtors(r) and Homestore. com.Inc. All rights reserved.

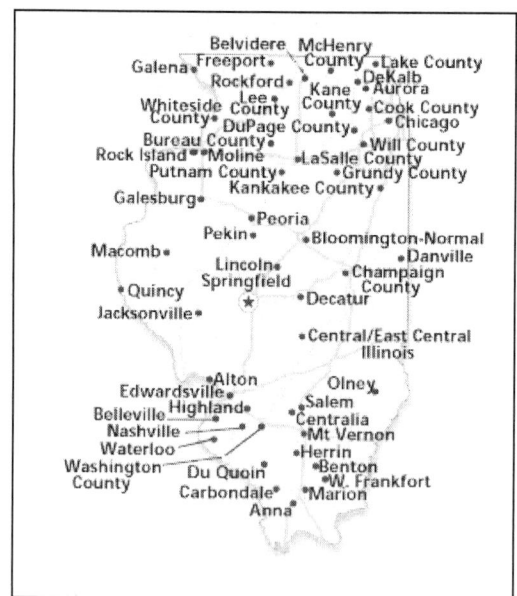

Figure 4.13 Selecting an area within a state (2 of 6).

Source: Selected graphic images from Realtov.com and used herein with permission. Copyright © 1995–2002 National Association of Realtors(r) and Homestore. com.Inc. All rights reserved.

Figure 4.14 Choosing a county (3 of 6).

Source: Selected graphic images from Realtov.com and used herein with permission. Copyright © 1995–2002 National Association of Realtors(r) and Homestore. com.Inc. All rights reserved.

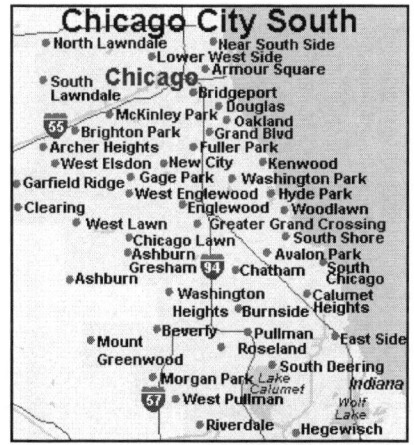

Figure 4.15 Selecting an area within a county (4 of 6).

Source: Selected graphic images from Realtov.com and used herein with permission. Copyright © 1995–2002 National Association of Realtors(r) and Homestore. com.Inc. All rights reserved.

location finding. www.multimap.com provides access to street maps and travel directions for the UK and Europe. It also includes aerial photos of London and other cities in Britain and provides a range of mapping services for businesses.

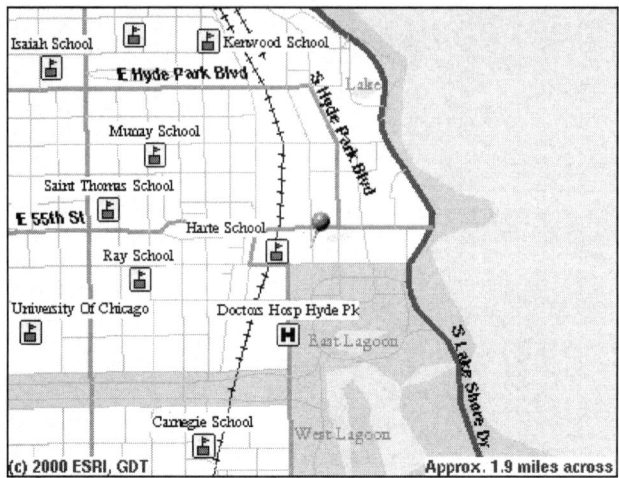

Figure 4.16 Choosing a specific neighbourhood (5 of 6).

Source: Selected graphic images from Realtov.com and used herein with permission. Copyright © 1995–2002 National Association of Realtors(r) and Homestore. com.Inc. All rights reserved.

Figure 4.17 Viewing a picture of a selected property (6 of 6).

Source: Selected graphic images from Realtov.com and used herein with permission. Copyright © 1995–2002 National Association of Realtors(r) and Homestore. com.Inc. All rights reserved.

Most commercial property marketing sites require the user to select a broad geographical region from which properties may be selected. However, this is neither user-friendly nor efficient. As more properties are added to a property listing site a more sophisticated method of refining the search using geographical parameters will be necessary, similar to those that exist on residential web-listing services. The use of such broad geographical regions is also confusing and requires a certain degree of knowledge that the user may not have. For example, is Bournemouth in the south or south-west of England?

The use of large-scale digital map data (perhaps with small scale mapping as a front-end) would mean that a database of available commercial property could benefit from geographical analysis. Potential owners and occupiers could search by location, ask questions about the surroundings, and examine the nature of neighbouring development, for example. In the UK, Heywood *et al.* (1998) created a house-hunting decision support system that allocated weights and scores to a series of criteria including insurance cost, proximity to schools, railways and roads and urban areas. A GIS was used to create thematic layers for each criterion, which are then combined according to their weighting to perform a search for a suitable site. In other words the layer with the highest weight has the most influence on the outcome. Locations of houses for sale were then plotted on top of the map of suitable sites and ranked according to the criteria that they meet. Details of individual houses can then be accessed and maps and reports produced.

Location is a key factor for the success of many commercial businesses so a geographical perspective on property searches would be very useful. Linked to these data could be planning and other relevant information maintained by local authorities. Also, rateable value information from the VO could be linked so that potential owners/occupiers have an idea of their rates liability. FOCUS information could also be linked (FOCUS – www.focusnet.co.uk – provides information on property transactions, rents and prices achieved, landlord and tenant details and the main terms of the conveyance). Once a web-based property enquiry service is operational, it may be possible to provide an incentive for agents to submit transaction details, subsequent to the property being advertised on the site. This would not only provide useful information about the effectiveness of the web site as a marketing medium but would also provide property take-up statistics. Many market commentators have suggested that surveyors should take a lead in the collection, management and dissemination of vacancy rates and take-up statistics. A property enquiry service that collects follow-up information after the transaction would achieve this.

Conveyancing

The NLIS conveyancing application will transform the way property is transacted in England and Wales. Information will be available electronically and this will speed up the property transaction process. Initial services will be developed to computerise the property searches necessary as part of the conveyancing process. This will reduce the length of time between offer, acceptance and exchange of contracts and help to reduce the risk of gazumping (where the seller of a property raises the price of a property after having accepted an offer by an intending buyer) or gazundering (where the purchaser of a property lowers the amount of an offer made to the seller just before exchange of contracts).

NLIS Internet services will be offered via a number of licensed commercial service and data providers known as 'channels' that will provide an on-line link to the various sources of information that are held at disparate locations, including local authorities and the Land Registry. These services will give a quicker service by providing easier access to a wide variety of information on a one-stop basis and in a reliable and secure environment.

Submitting property searches, and receiving results electronically will offer significant time-savings and greater convenience as it will dispense with the need to fill in and post separate forms, along with a cheque for the required fee, for each property search. Instead, searches will be possible via the Internet, with a single statement and invoice being provided for those services used.

The Bristol pilot of the NLIS conveyancing application was launched in April 1998. It allows electronic searches, via the Internet, of the following data:

- large-scale digital maps and addresses with grid references from the OS;
- ownership details, registered mortgages, rights of way, restrictive covenants and registered extents from the Land Registry;
- residential dwelling codes and rateable values for commercial property from the VO and
- Local Land Charges, planning applications and environmental health data from Bristol City Council.

This information is supplemented by data from other providers that are usually required during a conveyance including Bristol Water, British Geological Survey (BGS), the Coal Authority, Companies House, the Environment Agency, the Highways Agency, the Lord Chancellor's Department and Wessex Water. Users can locate land parcels and properties using address or map-based queries, as shown in Figure 4.18.

Figure 4.18 NLIS conveyancing pilot application.

Search requests based on the site or property boundary can be submitted and most are returned within 48 hours. The Water Authority returns a map showing pipes, BGS provides a report on radon or a 'Homebuyer's Report' on geological conditions, the Coal Authority provides a report and the Land Registry displays current entries on its three registers, confirmed by office copies posted the next day.

Users of the Bristol pilot included solicitors, mortgage lenders and surveyors and the application provided conveyancers with an electronic search system for property information that reduced costs and time taken to complete searches.

The NLIS conveyancing application is important because conveyancing is a high profile, critical process, represents a partnership between the public and private sectors and is technically demanding. In 1998 the NLIS Bristol pilot won the British Computer Society award for IT excellence. More detail can be found at www.nlis.org.uk. Clearly there are many challenges ahead. Expansion from a city-based pilot to a national web-based service will present organisational as well as technical difficulties. The NLIS conveyancing application is an initial step towards full electronic conveyancing. Although the revised BS7666 defines the standard for recording land and property boundaries, there is no plan for the NLIS to become a national cadastre. This is perhaps a missed opportunity given the introduction of Mastermap by the OS with polygon-based feature representation.

In the future, online conveyancing services will include electronic searches, a public web site for property that is for sale together with related product and service sales, a databank of demographic and other information such as house prices and school league tables. This will be similar to the system in Canada and Realtor.com in the United States. One of the NLIS channel service providers for conveyancing services is called Searchflow (www.search-flow.co.uk). Searchflow provides conveyancers with information about properties online using a digital map base to integrate various data at the property level. The service works in two ways. First, Searchflow Referencing gives users the means to validate the search description and geographical extent and identify relevant potential sources of information to support the processing of conveyancing searches. Second, Searchflow Handling will provide an automated, online conveyancing search service accessible via NLIS.

Summary

Map production is where GIS began. With the wide variety of map data now available, digital mapping is straightforward and becoming relatively inexpensive. The agreement between OS and central government departments and local government for access to OS digital mapping can be related to the considerable progress being made on the creation of national data sets through the effective use of geographic information. The biggest hurdle that many organisations have had to overcome is cleaning their own data before they can be mapped. This has proved to be expensive and time-consuming. For many public sector organisations it has been difficult to justify the investment and the absence of a national standard for referencing has been a significant barrier to private sector involvement in the mapping of land and property data.

So with the introduction of Mastermap, NLPG and NLIS we should witness a significant expansion of GIS applications in the field of land and property management, similar to that which occurred when the OS completed the digitisation of the national large-scale map-base back in the early 1990s. These initiatives, coupled with competitive pricing that is attractive to small organisations as well as large ones will be particularly important to the property industry which is characterised by small to medium-sized firms. With regard to NLIS services in particular it will be inevitable that many will focus on London given the geographical concentration of market opportunities for data and data-based services in the capital. The challenge will be to ensure that NLIS applications are truly national.

References

Capital and Counties, 1992, *Domesday 2000: A National Land Information System*, University of Cambridge, Cambridge.

Dale, P., 1989, Land information systems and the land parcel, AGI Yearbook, Association of Geographic Information, London, pp. 59–63.

Dale, P., 1991, Domesday 2000: The professional as a politician, AGI Conference Proceedings, Association of Geographic Information, Birmingham.

Department of the Environment, 1987, *Handling Geographic Information*, HMSO, London.

Figueroa, R., 1998, GIS supports property reassessment in Regina, Geo Information Systems, Hune, pp. 32–36.

Fovargue, A., Larner, A., Lopez, X., Ralphs, M., Sabel, C., and Wyatt, P., 1992, Towards a national land information system for Britain: a co-ordinated research strategy, Proceedings of the 1992 AGI Conference, AGI, Birmingham, 24–26 November 1992, 1.31.1–1.31.6.

Heywood, I., Cornelius, S. and Carver, S., 1998, *An Introduction to Geographical Information Systems*, Addison Wesley Longman, Harlow, UK.

Larsson, G., 1991, *Land Registration and Cadastral Systems*, Longman, UK.

Local Government Management Board, 1997, National Land Information Service Feasibility Study – Executive Summary, www.lgmb.gov.uk/im/lpiwg/nlis/ index.html

Manthorpe, J., 2001, *Geomatics: Yesterday, Today and Tomorrow*, University College London, September 21.

Ralphs, M. and Wyatt, P., 1998, Can international lessons be learnt from the development of national land information systems? Proceedings of the FIG Conference, Brighton.

Smith, B., 1994, National land information system pilot project, *Mapping Awareness*, December, pp. 34–38.

Smith, B., 1998, NLIS 1998: faith in the future, *Mapping Awareness*, March, pp. 22–24.

Sundquist, A., 1995, Real property general assessment in Sweden receives GIS support, International Real Estate Society Conference, 28 June–1 July.

5   Property management

Introduction

This chapter examines the use of GIS for property management. This usually involves the linkage of a property asset management database to some form of large-scale digital mapping. GIS-based property management is typically used when many of the decisions that the property management system is designed to support are geographical. Utilities and local authorities are therefore prime candidates but large landowners or landowners who own or manage a significant amount of contiguous land and property can also justify the investment in GIS because the quantity of map data required is less. Property management decisions need to be made on the basis of accurate information and so GIS applications often entail a relatively low level of analysis but a high degree of accuracy. GIS provides a geographical view of data – perhaps revealing links and trends that would not otherwise be apparent in an alphanumeric database. GIS also provides a geographical interface or 'front-end' to the data and, in a large property management system, this can speed up access and search time.

This chapter begins by considering GIS-based property management applications in local government, including land terriers and property management information systems. It then describes how large landowners, many of which are either government agencies or quasi-public organisations, are using GIS for land and property management functions. GIS is increasingly being used for facilities management and two case studies illustrate how the tools and techniques that are familiar to users of land parcel and property-based GIS can be applied at room level too. Finally the chapter considers the use of GIS in agriculture and rural land management.

Local authority property management

This section focuses on the application of GIS to property management functions in local government. It will begin by outlining reasons for recent moves towards the application of IT in local authority property management and argue that information used in property management functions is largely geographically referenced. Such data are handled most effectively by IT solutions that include mapping and geographical analysis capabilities such as GIS.

Local government has been at the vanguard of the implementation of GIS for land and property management. Local authorities have many statutory responsibilities that require the collection and management of a large quantity of property data within discrete geographical areas. Therefore investment in GIS technology and data collection can be justified. The utilities have also found this to be the case with regard to the maintenance and management of their plant, property and infrastructure networks. But unlike utilities, which have an emphasis on infrastructure management, local authorities provide a range of services, and the two tiers of local government mean that they vary in size and functionality considerably. Consequently local authorities collect and store a substantial amount of data, the majority of which is geographically referenced. In order to use these data efficiently and to allow them to be shared beyond the source department, thus avoiding any potential duplication of effort, GIS is regarded by many as a solution. Furthermore, GIS are appropriate for local authorities because they need to consider proximity issues (typically planning and development control) and they are responsible for discrete geographical areas. Local government is one of the largest users of GIS and in 1993, 29 per cent of local authorities had a GIS compared to 16.5 per cent in 1991 (Masser and Campbell, 1994). Later surveys by the Royal Town Planning Institute (RTPI) were carried out in 1995 (Allinson and Weston, 1999) and RTPI (2000). The 2000 survey showed that some 94 per cent of authorities had either implemented, or were implementing, a GIS. This compares with 64 per cent of authorities in 1995 and indicates that GIS has consolidated its position as a mainstream technology within local government. Table 5.1 shows that around 57 per cent of councils considered that they had a fully operational GIS (compared to 30 per cent in 1995). In 1995 8.3 per cent authorities had no plans to introduce GIS; this figure is now 1 per cent. This reduction may in part be attributed to the increased accessibility and affordability of PC based GIS. The 1995 survey showed that 68 per cent of those with GIS had a system shared with other departments. In 2000 the proportion is very similar (69 per cent), although there were wide variations between different types of Council, with National Parks the most common home for corporate systems whereas metropolitan districts and London boroughs favoured departmental systems.

Table 5.1 GIS procurement in local government in 2000 – (in %)

Fully operational GIS – development complete	27
Fully operational GIS with extensions/modifications	30
GIS currently in development	38
No GIS but introduction planned	5
No GIS and no introduction planned	1

Examples of applications include:

- administration of council tax and non-domestic rates;
- administration and searches of local land charge registers;
- recording of planning histories and administration of planning applications;
- land and property terriers;
- property and highways maintenance;
- demographic analysis for the location of public services such as schools, libraries and day centres;
- identification of underutilised property and its proximity to over-utilised property or premises that are in need of refurbishment;
- highway adoptions;
- impact assessment of development proposals;
- space and facilities management, for example, the calculation of grass-cutting area prior to tendering for work;
- mapping vacant and contaminated land;
- modelling the environmental impact of proposals for housing development.

Data sets used in these applications might include land ownership, right-to-buy sales plans, contaminated land, land charges and vacant land registers, street furniture surveys and highway adoptions registers. GIS is often used to superimpose these data above OS digital mapping and in-house large-scale land surveys.

In its memorandum of evidence to the Audit Commission the RICS (1987) recommended that:

- local authorities should adopt a more corporate approach to property management;
- a central property department be established in each authority;
- a central database of property owned and leased by the authority is essential to allow performance indicators to be constructed and to permit more informed decision-making;
- a central property database could be used for estate management, development, and by maintenance and service departments.

Clearly the RICS felt that the property portfolio is a key area of management policy within a local authority and therefore suggested a move towards a 'property' department with a dedicated property information system, rather than having the property functions dispersed among legal, treasurers', environmental services and other departments. This would allow performance indicators to be developed and more informed property decisions to be made. The Chartered Institute for Public Finance and Accountancy, in its report on local authority asset registers, supported the recommendations of the RICS regarding the need for a central database of property owned and occupied by local authorities.

When the Audit Commission published the findings of its review of local authority property management it highlighted a lack of information about property in many authorities (Audit Commission, 1988a). For example, it was commented that many authorities did not know what they owned. It was therefore suggested that a Property Committee should be formed within each authority and one of its main responsibilities would be to bring together information on the authority's property holdings. The Audit Commission also recommended that all authorities bring their land and property terriers up to date and maintain a reliable index of property deeds (Audit Commission, 1988b). The Commission suggested that UPRN should be used to link terrier information to financial and other information systems. The Audit Commission suggested core land and property information that should be recorded in a property database, with links to financial and other information systems using unique identifiers that contain map co-ordinates (a national grid reference, for example) (Audit Commission, 1988a). This offers an opportunity for local authorities to implement a GIS – an opportunity that many authorities have taken up. The Commission stated that 'the importance of an up-to-date and comprehensive terrier cannot be overstated since it is the starting point of much property management work'. Information regarded as essential is illustrated in Figure 5.1.

This type of geographical information is well suited to input and analysis using GIS and many authorities now use this technology to maintain their land terriers. The advantages of a GIS-based land terrier over paper-based systems are described by Lilburne and Rix (1991) and can be summarised as:

- an ability to handle changes in features over time;
- an opportunity to tailor the database to suit different types of user;
- provision of authority-wide access to a corporate resource;
- geographical analysis of property information;
- high-quality map production.

The transition from a paper-based to a GIS-based land and property terrier represents a shift from a static inventory/index system to a dynamic, integrated property management system.

Since the Audit Commission's report on local authority property management in 1988 (Audit Commission, 1988a) there has been substantial investment in GIS by local government. This has been assisted by the Service Level Agreement with the OS, which makes the cost of OS mapping more affordable to local authorities. Many implementations of GIS in local government have been aimed at improving planning-related functions such as planning application procedures, the recording planning histories for each land parcel and the administration of local land charges. But the application of GIS to local authority property management is also regarded as an effective use of

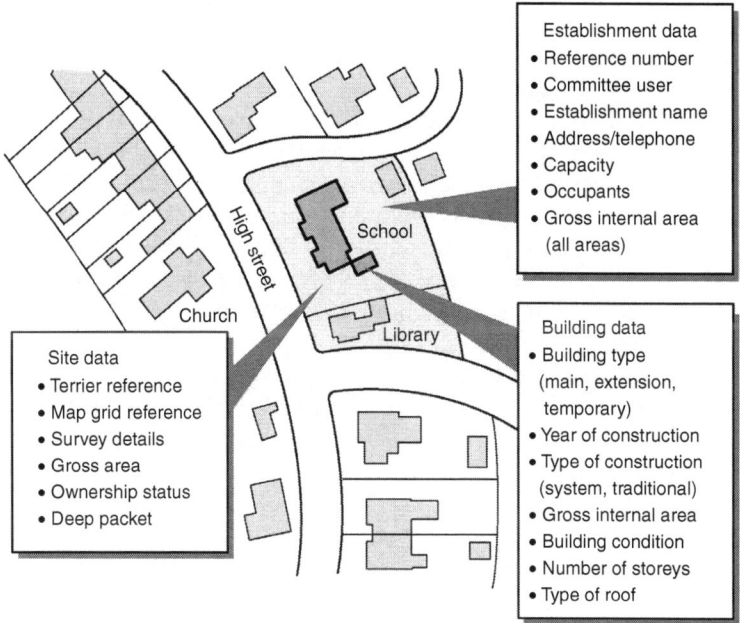

Establishment data
• Reference number
• Committee user
• Establishment name
• Address/telephone
• Capacity
• Occupants
• Gross internal area
  (all areas)

Site data
• Terrier reference
• Map grid reference
• Survey details
• Gross area
• Ownership status
• Deep packet

Building data
• Building type
  (main, extension,
  temporary)
• Year of construction
• Type of construction
  (system, traditional)
• Gross internal area
• Building condition
• Number of storeys
• Type of roof

Figure 5.1 Property information required for local authority property management.
Source: Reproduced from Audit Commission (1988a) with permission.

data collected as part of the statutory functions undertaken by local authorities. Property plays an essential role in the delivery of core local authority
services. It must therefore add value directly (minimum costs and maximum
capital receipts) and indirectly (fit for purpose and right location) (Jenkins
and Kearns, 1999).

The conventional means of recording and monitoring data on local
authority property is the land terrier. A land terrier is a record of council-
owned properties normally managed by the Estates Department or the legal
records section of the Chief Executive's Department (Rix and Lilborne,
1994). The terrier comprises maps and text of property owned and leased
by the local authority. Cross-referenced card files often contain summary
information on the source of the conveyance, land title documents, leases,
any sale and acquisition details, easement and wayleaves, Land Registry
certificate number, financial details of purchase, vendor details, boundary,
covenants, controlling committee, assignments, alienations, subletting, tenant
mortgages, alterations, etc. At its simplest a land terrier comprises

• a street index entry to land terrier maps;
• land terrier map tiles based on large-scale OS mapping;

- deeds registers (basic details from deeds);
- deed packets.

Traditionally paper maps are used to indicate the site boundaries together with ownership, parcels sold, leased or owned and the controlling committee. This leads to updating problems. Digital mapping presents obvious benefits in terms of archiving of historical data, customisable symbols, wider access to data, geographical analysis and high-quality printouts. In support of the evidence submitted to the Audit Commission by the RICS (1987) a survey of local authorities in the Eastern Thames Corridor revealed that the land terrier was regarded as one of the key themes in the plans for a corporate information strategy. Kirkwood (1998) comments that 'the traditional property terrier has all the essential elements of a GIS – maps, data and explicit geographic references linking the two – such paper-based creations provide very limited search facilities and little or no opportunity for data analysis'. A GIS can help monitor the cost of support and maintenance to each department measured by the amount of space they occupy. It can also match people, products and services with property, monitor the location of vacant property and link with property, asset and financial systems.

The Chartered Institute of Public Finance and Accountancy and the Audit Commission recommended the elevation of the land terrier to full asset management system (Rix and Lilborne, 1994). Use of GIS in a land terrier will evolve it from a rather static inventory and index to a more dynamic property management system capable of helping to optimise returns on assets, budgeting, facilities management, valuations, inspections and maintenance. Decision support might include queries like:

- Where are the sites that were subject to unplanned maintenance last year?
- Which properties have been vacant for more than six months?
- Have rents in the Central Business District and business parks changed in the last year?
- Where are the properties that have rent reviews in the next 12 months?

A more sophisticated geographical query might be to list properties that have occupancy under 50 per cent and are within 10 minutes distance of a property with occupancy over 75 per cent. Other benefits of using a GIS include:

- Ease of editing data
- Overlay of minor land and property interests such as easements, leases, committee boundaries
- Council-wide access to the data
- Statistical and geographical review of the property portfolio
- Accurate and up-to-date large-scale map base
- Custom printing and plotting.

A GIS-based land terrier can be used to visualise and spatially analyse patterns in data and integrate disparate data sets such as the terrier and travel patterns (Rix and Lilborne, 1994).

The land parcel is the basic spatial unit of a land terrier but since ownership boundaries are not legally defined in absolute geographical terms in England and Wales (refer to the 'General Boundaries Rule' described in Chapter 4), the capture of legal extents of land parcels may reveal overlaps and gaps in the mosaic of land ownership (Lilburne and Rix, 1991). This problem becomes more apparent as legal extents of land parcels are digitised and recorded in a GIS. To try and reconcile boundary overlaps and gaps some local authorities have commissioned the Land Registry to help. In this way the introduction of GIS has prompted many local authorities to audit and verify the geographical extent of ownership of their land and property assets. In the sections that follow we describe some specific examples of GIS implementation for property management.

Aylesbury Vale District Council

At Aylesbury Vale District Council GIS is the basis for address accuracy and consistency throughout the Council's services. It is based upon a BS7666 compliant LLPG, which contains 78,000 text and graphics records, together with a centrally maintained and updated corporate OS map base that is available to all users. Creating a LLPG that records the geographical extents of land ownership as well as text-based information offers a number of benefits. The LLPG currently supports planning and development control, highway management, local land charges, property management and housing, and will support environmental health and electoral registration in the future.

The GIS-based LLPG is used to support development control through the capture of planning application site extents and automating planning history and planning constraint checks. GIS is also used to help capture building regulation application site extents, input data records to the local land charges process and automate the processing of all local land charge searches. It is the intention of the council to link the land charges system with NLIS.

A property management system was established to capture current and historic land parcels held within the Council's land portfolio and is used as the basis for asset management and property management processing. The system is also used as a reference and as a resource by many other users within the authority, for example, for grass-cutting contract plans. The ability to produce a map of property attributes recorded in the housing system (e.g. three bedrooms) will form the basis of a customer search facility to assist tenants in their choice of a Council accommodation.

At Aylesbury Vale GIS implementation has enabled a large quantity of data to be made available electronically to a wide user base within the

Council, and the LLPG is used as an interface to any query about a property. Furthermore, GIS is used for the capture and analysis of other spatial data associated with leisure and community safety functions, such as the selection of appropriate sites for new facilities and the matching of different spatial data sets from different agencies in the fight against anti-social behaviour. The key factors that have made GIS a success at Aylesbury Vale are:

- the corporate commitment made by the Council and a willingness to stand by this decision, no matter what financial pressures were facing the Council;
- the commitment of a small group of very committed and dedicated personnel;
- a readiness by colleagues, not involved in the GIS Team itself, to embrace the potential of GIS, despite the difficulty of converting systems, data, etc.;
- the development of a funding model, which took the issues of finance away from those working at the coalface, and allowed them to concentrate on getting the job done.

The London Borough of Barking and Dagenham

In 1991/92 the London Borough of Barking and Dagenham (LBBD) devised a strategy for property management incorporating GIS technology. This is ongoing, and involves the implementation of the following components.

Core referencing system   The central feature of the strategy was the development of a framework of common referencing standards. The authority decided that the problems and costs of taking forward a corporate information management strategy were outweighed by the benefits arising from the resulting opportunities for data sharing and reduction of data duplication (London Borough of Barking and Dagenham, 1996a).

The initial phase of GIS development concentrated on the land terrier and corporate master address file. The council used to operate a range of addressing systems, which was seen as a source of potential problems when an integrated system was adopted. The master address file resolves this issue by providing a standard system of address referencing. The addresses are compliant with BS7666.

Computerised land terrier   The LBBD used a GIS to create a digital land terrier and property management system. Prior to this the council had in excess of thirty manual property information registers, indexes and other systems. The paper-based information system was becoming increasingly complex for a number of reasons, not least the inherited property from the Greater London Council and the sales of council-owned residential property under the Right-to-Buy legislation. After several local government reorganisations the authority was operating many manual property information systems. Eighty-five per cent of the authority's property interests were unregistered with the Land Registry. At the end of the 1980s the creation of a corporate property database was recommended, centring on an address file 'hub' facilitated through a BS7666 compliant LLPG. Four developments were recommended:

- land terrier and commercial property portfolio;
- integration of existing and development of new property regulation systems for planning, housing, health, highways and local land charges;
- planned maintenance and building management system;
- integration of existing infrastructure related systems with new systems (MacLellan, 1998).

GIS was regarded as a logical information system for recording long leaseholds in flats, flying freeholds and relatively minor interests such as pram sheds, all of which are difficult to record on a paper-based or alphanumeric database. The land terrier computerisation and Council House Sales data capture are now complete and significant financial benefits have already arisen from the land terrier update through the cleaning of manual records. The GIS-based land terrier means that paper maps do not have to be renewed and the information about each parcel can be queried quickly by many users. For example, vacant land in a certain locality or committee ownership can be identified. The GIS also offers the ability to visualise 'lost' sites and development opportunities and correct ownership boundaries (MacLellan, 1998). The council is aware that if it encounters problems reconciling deeds with reality this could delay a conveyance and cause developers to look elsewhere for opportunities.

The land terrier has already made a significant contribution to the operations of the authority. The effective administration of council house sales information has been a primary source of benefits. The LBBD has 21,000 council house sales operations to administer, each with its own set of restrictive covenants. These properties were sold off variously by the LBBD, the Greater Lender Council (GLC) and the London Borough of Redbridge. By using the master address file and GIS mapping to identify these properties, the LBBD has saved money and generated new income by recognising errors in the title registration process, speeding up the process of inventory by colour coding and mapping properties digitally and flagging up potential benefits from existing covenants on sold properties.

An interesting application of digital mapping arose in the capture of parcel boundaries for the land terrier. Before the Second World War much of the Borough comprised agricultural land. Field boundaries were used to demarcate the extent of many of the council's older land titles and deeds and these have long since been destroyed by urban development. One of the problems faced by the authority was to reconcile ancient land titles and deeds with the existing urban landscape. The solution devised was to utilise digital OS County Series maps from the 1920s and 1930s which show the field boundaries, and to reconcile these with the existing digital cover of the Borough. This approach was found to be a highly effective way of identifying the exact location of older land titles, and has been praised by the Land Registry (LBBD, 1996).

Commercial property portfolio management system   The commercial property portfolio management system was developed for the Estate Management Team in the Property Services Section of the authority. The system is the first application to operate in conjunction with the computerised land terrier. The council has fifty-seven sets of property data and 70 per cent of their information relates to council-owned property. The property management system handles lettings administration of the entire commercial property portfolio owned by the authority.

The authority is engaged in six types of property activity:

(i)   management of council property ownership;
(ii) control and regulation of property related activities;
(iii) council building management;
(iv) infrastructure development and maintenance;
(v)   service planning and marketing; and
(vi) local property taxation and electoral registration.

Further development of the property management system now includes a planned maintenance system and related building management systems and an infrastructure management system.

Land charges administration system   The land charges administration system encompasses information from the Planning, Health and Legal Services departments relating to planning and building controls, highways management and grants available for properties. By bringing together information resources from these different departments, the administration process is made more efficient. The authority is also considering integrating Census of Population information with the terrier database for planning and development purposes.

To summarise, GIS implementation at the LBBD began with the computerisation and merger of the land terrier and local land charges system. GIS now comprises corporate property database, a land and property gazetteer (master address file), planning and building control, local land charges and property management systems (MacLellan and Musgrave, 1999).

Other local authorities

Bristol City Council has adopted a federal approach to GIS implementation where each directorate sets up its own GIS and uses core data from a LLPG. So far, in addition to the core address and land parcel boundary data, data sets captured include electoral ward boundaries, footpaths and other public rights of way, areas of flood risk, Sites of Special Scientific Interest (SSSIs) and map data from the Bristol Local Plan.

Morgan *et al.* (1994) describe how GIS is used to manage approximately 15,000 property interests at Strathclyde Regional Council. GIS implementation grew from grounds maintenance in Strathclyde Country Park and followed an application-led development path. The Buildings and Works department commissioned detailed plans and inventories of maintainable features at 2,500 council-owned properties in the region. The land terrier is also being digitised for use in the GIS. The aim is to develop a property management information system which will include OS digital mapping, political and administrative boundaries, grounds maintenance sites and legal titles.

At the Vale of White Horse District Council a GIS was introduced to manage the implementation of the regulations under the new Home Energy Conservation Act. This act requires councils to produce plans to reduce domestic energy consumption by 30 per cent over 15 years. The council captured building outlines and energy data for approximately 5,000 residential properties owned by the Vale Housing Association. The GIS allows individual properties and estates to be identified where energy consumption is highest.

At Birmingham City Council GIS is used for property management and economic development, including management of regeneration, contaminated land, local land charges, planning and development control and grounds maintenance. With regard to the property functions to which GIS has been applied, internal data sets used in relation to these GIS applications include:

- Terrier records (see Figure 5.2)
- Land availability
- Building condition
- 'Deed' plans
- Employer database.

These are combined with OS digital mapping, address data and rectified aerial photography. Land terrier records support property management, site assembly policy and decisions. They are also used as a key to locate additional records such as title deeds and individual property plans. Planning data in the GIS is used to identify areas of economic need. This helps in the preparation of bids for funding and the measurement of compliance with performance indicators. The GIS is also used to assist the planning of regeneration activity and targeting of support for businesses

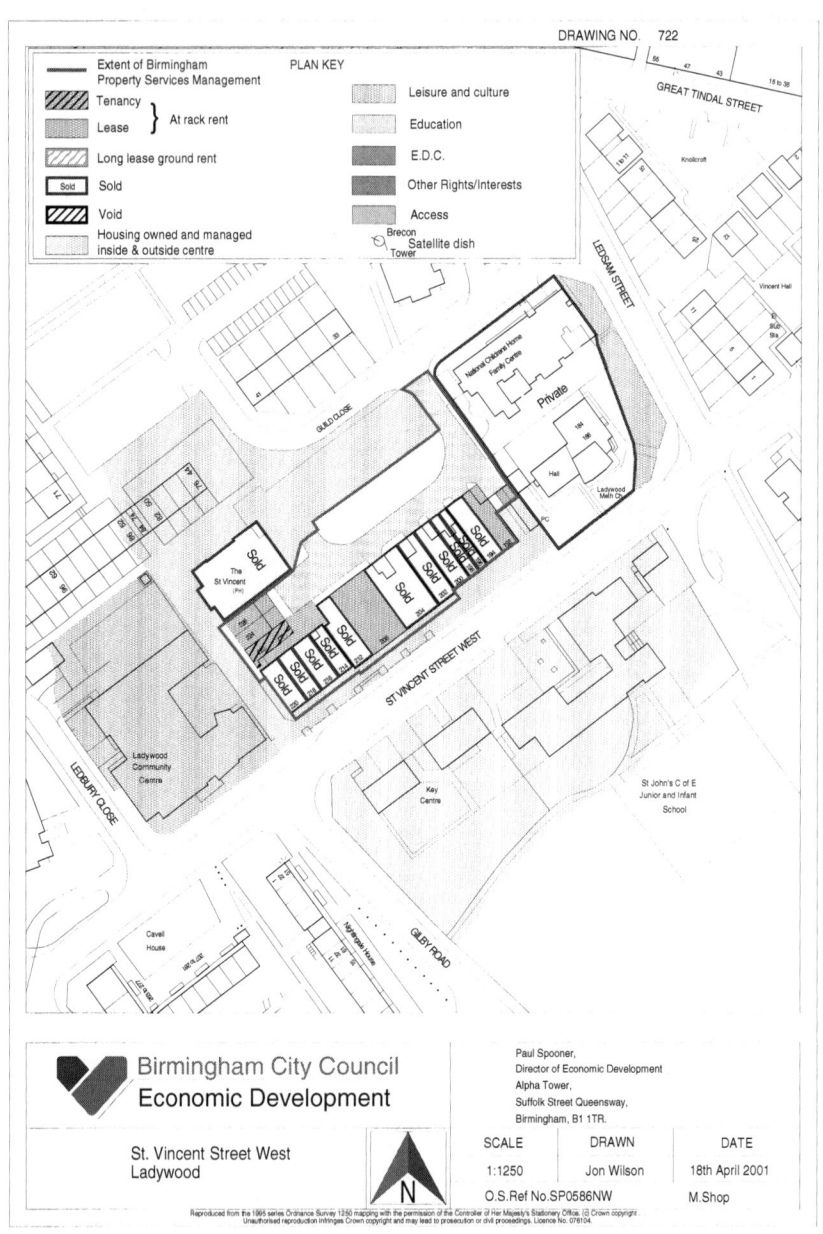

Figure 5.2 Estate Management plan denoting Birmingham City Council departmental responsibilities within a shopping parade and its environs,
produced from a GIS. Reproduced Courtesy of Birmingham Property
Services Divison, Birmingham City Council and © Crown Copyright.

and community initiatives. At a more strategic level geographical areas of economic activity such as tourism can be identified and analysed. The benefits resulting from the application of GIS to these local authority functions include easier targeting of activities within a defined boundary rather than by address (the quality of the latter can be doubtful) and the production of consistent and unambiguous property plans for use in conveyancing and other applications. Cost savings have been identified as a result of the application of GIS in these areas and include the avoidance of wrongly allocated grants (where a map boundary defines extent of eligibility) and general efficiency savings in property management and plan preparation. The only problem encountered in the application of GIS at Birmingham City Council is that users can have difficulty interpreting multi-layered data.

England (1996) reports that at Gloucestershire County Council the corporate GIS strategy was implemented by the county surveyor. Benefits were seen in terms of access to digital mapping, demographic analysis for better targeting of services, 'what if' modelling and a map interface to existing systems. Studies by Sutton London Borough Council and Cardiff City Council estimated the cost of working with maps to be in the order of £400,000 per annum and these studies show that a GIS can reduce time spent on the above activities by more than half. Each department established its information requirements and these were shared to see where duplication occurred. Data requirements are illustrated in Table 5.2.

Table 5.2 Local authority information requirements

Information Requirements:
1. OS mapping
2. Addresses
3. Socio-economic indicators
4. Census data
5. Property database
6. Pupil database
7. Gazetteers (street names, postcodes, parishes, addresses, towns and villages, schools, libraries, pubs, churches, garages, stations and hospitals)
8. Highways and planning data

Applications include:
1. Road maintenance (more than 1.5 million road inventor items are recorded)
2. Accident analysis (accident information from last ten years)
3. Traffic (counts over the last 10 years)
4. Street works and road condition survey
5. Development control and public rights of way
6. Planning (applications, census analysis, sites and monuments, register of industrial premises, waste sites and waste land use, key environmental sites, minerals, strategic policy constraints, vegetation)
7. Police (use the accident system, mapping and address location, link to crime statistics, property information and crime patterns)
8. Fire (hydrant and property location)
9. Social services (postcodes and address information)

One authority has digitised all of the boundaries of the parks and open spaces under its control so that the GIS can calculate the area of grass that needs cutting. This helps the authority when advertising the contract for tender. Some authorities use GIS to assist in their emergency and disaster planning procedures, others use GIS to help manage traffic flow and provide visitors with route information. A key benefit cited by local authority property managers with regard to GIS-based property management systems is access to data at remote sites. Problems typically centre on maintenance of data, hardware and software incompatibility and lack of staff.

Overseas case study

The Economic Development Department of the City of Vallejo in California (www.ci.vallejo.ca.us) has developed a GIS application that lets prospective businesses search for information on commercial sites. The web-based site selection application allows users to locate existing buildings or development land, together with demographic, economic and traffic count data for the surrounding area. These queries can help determine whether the workforce education levels and/or consumer expenditure levels are compatible with the needs of a business. Business listing information can identify possible competitors or find companies that offer agglomeration potential. The Economic Development Department works with local real estate agents and landowners to keep property listings up to date. This saves time for the local authority and puts prospective business in direct contact with agents and owners. Prior to the GIS a prospective company would talk to staff at the Economic Development Department who would manually search for properties that matched the company's requirements in databases that were only updated a couple of times a year. Subsequent to the launch of the GIS a company contacted the local authority to arrange a tour of five properties that it had already shortlisted on the web.

Large landowners

Housing associations

Riverside Housing Association and South Staffordshire Housing Association use GIS to show trends in rent arrears, vacancies and affordability and to map areas of high property turnover. The entire property stock of each housing association can be displayed on a map and geographical trends within the data, such as clusters of rent arrears and areas of high repair frequency, can be highlighted. Any data relating to a property, such as rent information, stock condition information and tenant status can also be symbolised and queried directly through the GIS.

Trends in the data, such as cost of repairs by parish, can be analysed and compared to other statistics about those areas. This helps assess whether underlying diversity in the regional geography causes clusters in the housing association's data. An additional benefit is the ability to produce detailed property location maps for tenancy officers and other third parties such as local authorities.

GIS implementation has enabled the housing associations to access information that was previously either unavailable or difficult to extract. Moreover, by examining various types of data geographically, they are able to make more informed decisions with regard to housing management and development strategy. There are a growing number of applications in housing associations for GIS. For example 'letting packs' can be produced for new tenants, which include detailed maps and photographs of the property and neighbourhood. Profiles of local areas or neighbourhoods can also be constructed by analysing both in-house and external data against boundary information.

Property Advisors to the Civil Estate

The Property Advisers to the Civil Estate (PACE), an executive agency of the Office of Government Commerce in the Treasury, was established in 1996 following a recommendation that Property Holdings be reconstituted as an executive agency. Its aims are to co-ordinate departmental activity on the Civil Estate and to provide departments with advice and support on property issues ranging from facilities management to hiring consultants and finding new accommodation.

Property Advisers to the Civil Estate is developing a Property Information Mapping System for core Civil Estate data. The new database will have the capacity to display the exact location and outline of Civil Estate properties on OS digital maps together with digital photographs. The system is designed to provide departments with shared access to their core data together with remote update facilities via the Internet.

The PACE web site (www.property.gov.uk) provides a detailed description of the plans to implement a GIS-based property information system; the reasons for and method of this implementation are summarised next.

The UK Government collectively owns, leases or has on licence many thousands of land and property interests. Those which are used as offices, for storage, etc. are often recorded on a range of databases within the various 'owning' departments and agencies. However, there is no central framework that enables estate managers or property consultants to identify easily where these land and property interests are and where responsibility or ownership lies. The implications of this are that if an asset is 'lost from sight', it cannot be effectively managed and opportunities to realise capital proceeds may be missed and redevelopment opportunities lost. Furthermore Central Government has a key role in promoting urban regeneration with efficient use of assets in the public sector. Knowledge of linkages and potential marriage value is a key to such innovation. The Government is also committed to improving the management of public assets. This links with the aim to optimise use of the property interests held by the Government. The development of a co-ordinated land and property database could ensure that assets are managed properly and that opportunities are taken to realise the value of unused or under-used facilities. In the past property was acquired for public use and then it was left to local staff to manage. The Government's commitment to the more effective management of public assets, coupled with developments in technology, means that it is now possible to establish an easily accessible and comprehensive national record of government-wide land and property assets.

Property Advisers to the Civil Estate is working independently and with the Highways Agency to take forward pilot studies in relation to current data sets, to identify how these can be brought up to BS7666 standard and thereby integrate with other national initiatives. Once the preliminary viability of these pilots is known, PACE will be inviting Departments to participate in the creation of a proposed national Register of Government Land and Occupational Property. The Register will facilitate tracking of Government's land and property interests and will identify where primary responsibility for these interests lies. The Register will use a GIS and OS digital mapping to ensure a consistent and common referencing framework to underpin land and property information. In the longer term it is proposed that the Register should provide a definitive record of all Government interests in land and property and thereby offer a connection with the many databanks in which more detailed property-specific information is held. There are a number of features of the proposed Register which make it particularly appropriate for Government departments and agencies:

- Every physical property boundary and seed point will be referenced to the national grid and will be shown on large-scale OS maps.
- Information about the status of each property can be represented geographically.
- It is possible to establish that parcels of land are, in fact, physically adjacent and could combine to form a development site whose value is greater than the sum of its parts.

- Each land parcel can be related to other geographical and demographic information.
- Users can establish what property interests are owned by whom in any given area.
- The data can be linked to other databases providing common references are adopted.
- All branches of the Government will be able to access the key locator information stored on the Register via the Government Secure Internet.
- Selected information stored on the Register could, if appropriate, be made available to businesses and the general public via the Internet provided protocols and gateways are established.

Each land or property parcel will also be allocated a unique identifier which can be used as a cross-reference to other land and property information. Ultimately the Register could contain the following core data:

- site area
- local authority area
- land certificate number if the interest is registered
- owner
- responsible organisation
- tenure
- deed reference and location
- postal address and code
- entry date/amendment date
- whether the property is affected by any easements or other registered encumbrances.

Much of this information can then be viewed as overlays, for example, with leaseholds and freeholds shown in different colours.

Using unique identifiers, the Register can link with the estate databases of local authorities to provide, for example, information about planning or local strategies. The Register can also link with the databases of the Land Registry and the Registers of Scotland. This will provide access to title information, indicating, for example, whether a freehold is subject to a restrictive covenant or other encumbrance. The Register can provide access to valuable contextual data relating to property interests by drawing on a wide variety of external information sources such as:

- information about the socio-economic characteristics of a specified area
- contaminated land
- archaeological sites
- underground water courses

- siting of retail and other facilities within a given radius or drive-time,
- identities and contact details of relevant utilities and statutory undertakings.

Each land parcel polygon can be overlaid with geographical information corresponding to the selected search area or a specified buffer around it. This enables the user to search for information relevant to, say, a five-mile buffer around the site boundary. Typical applications of this might include relating a site to zoning information within the local plan, which can be visualised as an overlay.

The principal benefit to Government departments and agencies of the Register is that it will enable them to identify units that are no longer needed for operational use – and therefore make informed decisions as to which are available for disposal. It will also facilitate consideration of options as to the best way to dispose of these properties. By identifying surplus interests against an OS map base the system offers an effective way of establishing that several interests are adjacent and that small pieces of low value land can be put together to form something more valuable or marketable. When properties are viewed against a map base, adjacent Government interests can be identified more easily which allows for smaller interests to be sold to adjoining owners.

The Register can use the GIS software to filter and prioritise. For example, all interests in built-up areas (i.e. those shown on 1:1,250 scale mapping) of an acre or less can be targeted first, or all rural interests (i.e. those shown on 1:2,500 and 1:10,000 scale mapping) of over an acre. Where departments or agencies have a large number of property interests, the majority of which will be impossible to sell or of a nominal value only, the ability of the Register to filter and prioritise will help to design a disposal strategy which can maximise the value of these interests.

The GIS-based Register can relate parcels of land to other geographical data. For example, land can be shown overlaid on the local plan whose zoning/use designations can provide significant clues on how to maximise the potential value of the property. The ability to overlay current OS mapping onto old title plans can provide a clear indication of whether a parcel of acquired land was fully used in connection with a development scheme or whether part or all of that land is surplus and now available for disposal.

Many departments and agencies are likely to find that the bulk of their surplus parcels of land are of little value. These interests are a potential liability and will frequently be worth disposing of irrespective of value. The Register offers the opportunity to design a disposal strategy which can achieve significant proceeds from even small parcels of land, for example, by building a programme of garden land extension sales.

The aim is that the Register database and GIS applications and the digital mapping will be easily accessible to users, via the Government Secure Internet. The Register will provide property owners/managers and occupiers across the public sector with a building block on which to construct a standard-setting land and property management information system.

London Transport Property and Network Rail

London Transport Property has implemented a GIS-based property asset register. Property assets covering an area of London out to the M25 motorway can be searched by the nearest tube, street name or reference number. At present the referencing of property assets is not BS7666 compliant but is based on the Royal Mail's Postcode Address File. The search result displays a map centred on the identified property, which is shaded according to tenure type, with OS large-scale mapping as a backdrop. Attribute data can then be queried from the database and search results can be output to a property asset report which includes a map. The GIS also includes editing tools for the update and maintenance of the register. The benefits of the GIS-based register are the visual perspective on property assets and the removal of reliance on paper-based OS map sheets that were maintained manually and suffered from the usual land terrier problems described earlier in this chapter. Problems encountered during implementation of the GIS application have largely been of a technical nature. For example, legacy data was not easily transferable to the new system and the GIS database adopted was not relational.

Aerial photography has been used as a background for property identification relating to specific development projects. A particular problem that London Transport Property faces is that some of the property assets are underground and therefore OS mapping does not provide a meaningful backdrop to the location of these assets. Consequently underground assets are not geo-referenced.

A further GIS development at London Transport Property is the compulsory acquisition system that automatically issues Notices to Treat to property owners who are affected by new development programmes. The notices include A4 plots of the area affected by the development at the land parcel level. Explanatory letters can also be automatically sent to relevant parties within a specified proximity to the development proposal.

Network Rail has invested in GIS for the compilation of their asset register and to assist with planning, designing and information sharing on major infrastructure projects. Bespoke software was developed to link legacy asset management data to GIS software. Features can be located geographically and attribute data queried, and vice versa. The GIS therefore assists Network Rail in the management of their property portfolio of approximately 7,000 properties, together with incident management.

Ministry of Defence

The Defence Estate comprises 225,000 hectares of freehold property, 15,000 hectares of leasehold property and rights over 124,000 hectares of danger areas and training. This includes 48,000 structures in 2,600 sites in 1,800 locations in the UK (including airfields, docks, masts, offices, warehouses and retail) which is valued at £14 billion. According to Wooden *et al.* (1999) GIS is used to manage the OS digital mapping, training area maps, farm plans and lease information for this estate. Plans are produced for the acquisition, disposal and development of land. GIS can help identify cost-effective methods of managing and rehabilitating infrastructure and facilities, environmental compliance monitoring, cleanup and remediation. GIS is also used to manage environmental and conservation information and rights of way. The system is now available to approximately 900 users over an intranet.

Valuation data are collected on a five-year rolling programme and, together with other land and property data, this forms the basis of the Defence Property Register. All land and property interests have been geocoded and a map interface can be used to access the database. Simple geographical query tools are available to show, for example, sites with a value of more than £10 million and within a radius of 10 kilometres of a selected site. There are plans to transfer the land terrier to the GIS with links to scanned title deeds and other documents.

Facilities management

GIS tools can be applied to room-size assets as well as land parcels and properties. Information about facilities is used to support a number of business processes such as maintenance and operations, property management (acquisitions, disposals, refurbishment and redevelopment), human resources, capital planning, inventory services and information services. GIS offers advantages over traditional CAD and Computer Aided Facilities Management (CAFM) systems due to the ability of GIS to perform complex geographical analysis and graphical navigation. GIS also provides the opportunity to integrate infrastructure management with room-level detail. This is not possible with traditional CAFM systems. The map interface and topological data structure in GIS provide a powerful and flexible means of organising, analysing and presenting facilities information at every level of a company's infrastructure. For example, using GIS for space management, a planner can identify the impact of infrastructure projects on building occupants, pinpointing, say, the building support, public bathrooms or other areas that will be taken out of service when underground utilities that service that facility require maintenance (Gondeck-Becker, 1999). By implementing a spatial model for facilities information, relationships can be developed among these business processes that could not be achieved previously. For example, a common spatial reference of 'office number' in both human resource data (personnel name, position and office number) and in maintenance management data (equipment name, service call, service date and office number) allows correlation of personnel to equipment and service calls. This spatial reference provides a consistent vocabulary for locating facility assets and enables business analysis to cross departmental boundaries (Gondeck-Becker, 1999).

The following case studies illustrate the application of GIS to facilities management. They show how the structure and data management schema of a GIS can be used to extract, query and analyse facilities information and present the results of an analysis in map form. They provide an appreciation of the potential advantages and disadvantages of a GIS in a facilities management context.

University of Bristol Healthcare Trust

The internal room layout of one of the buildings occupied by the University of Bristol Healthcare Trust (UBHT) was mapped and attribute data collected in 1995 in order to assist with space planning at the hospital. The five-storey building comprises a basement, ground floor and three upper floors. The floor plans and room-specific attribute data can be viewed and analysed using a GIS. Figure 5.3 shows that the rooms for each floor of the building were mapped in separate layers within the GIS and floor 3 is displayed, together with the attribute data for room 176. This allows adjacency and the proximity of compatible or incompatible room uses to be examined visually. Maintenance and repair programmes can be constructed using the attribute data in such a way that teams can work on logical groups of rooms before moving on.

Similarly, thematic or choropleth maps can be created to illustrate room use (Figure 5.4), occupancy level (Figure 5.5) and any of the attribute data stored in the database for each floor. Attribute data can be used to extract subsets of the data. For example, Figure 5.6 shows rooms on floor 2 which

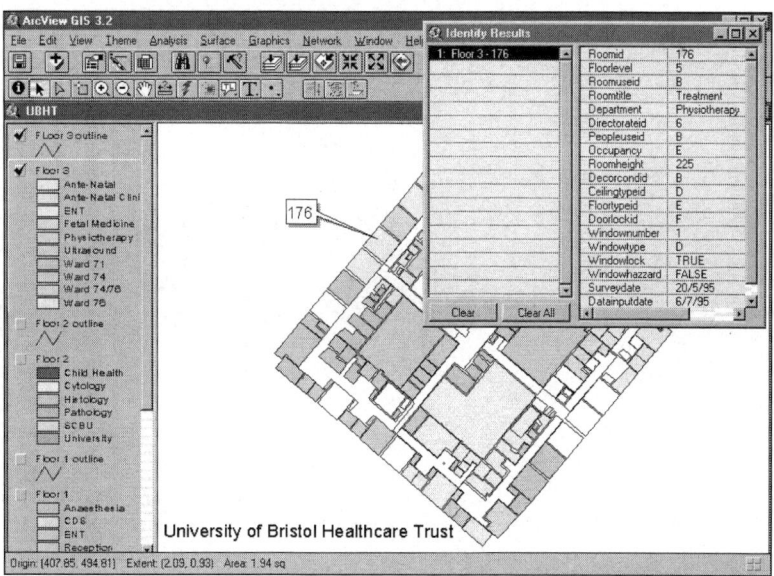

Figure 5.3 Floor 3 rooms classified by departmental responsibility.

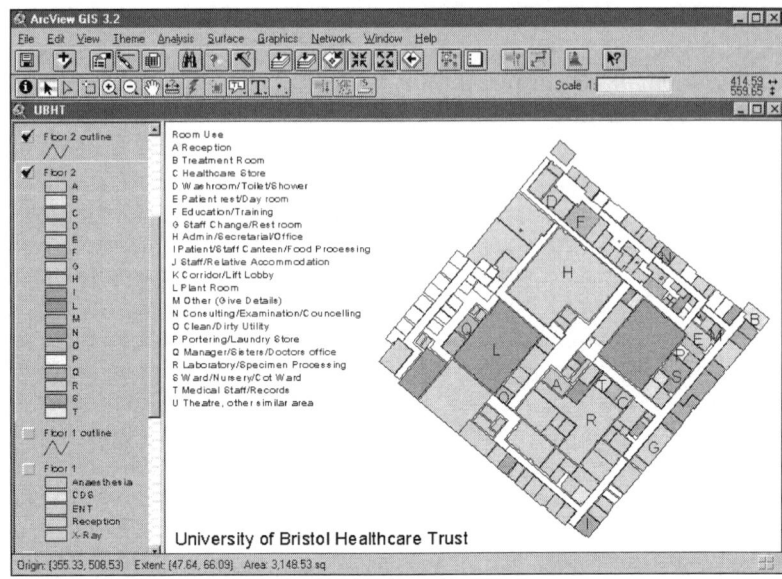

Figure 5.4 Rooms on floor 2 classified by use.

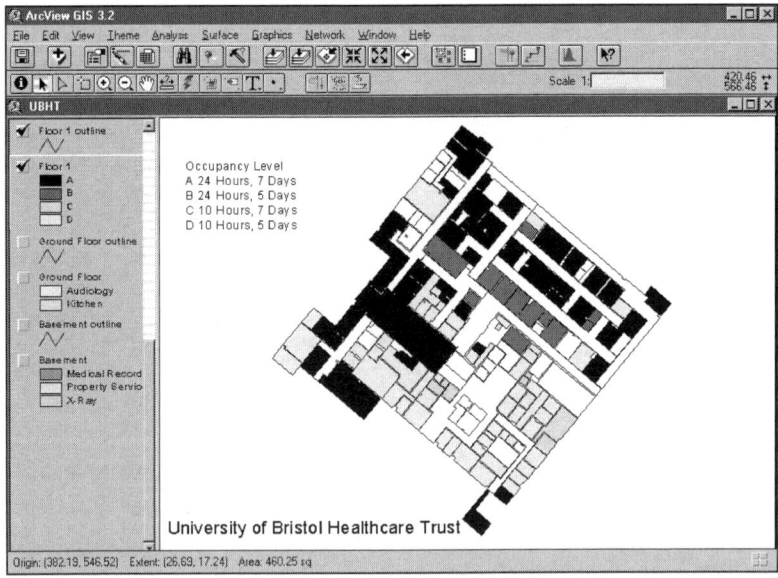

Figure 5.5 Rooms on floor 1 classified by occupancy level.

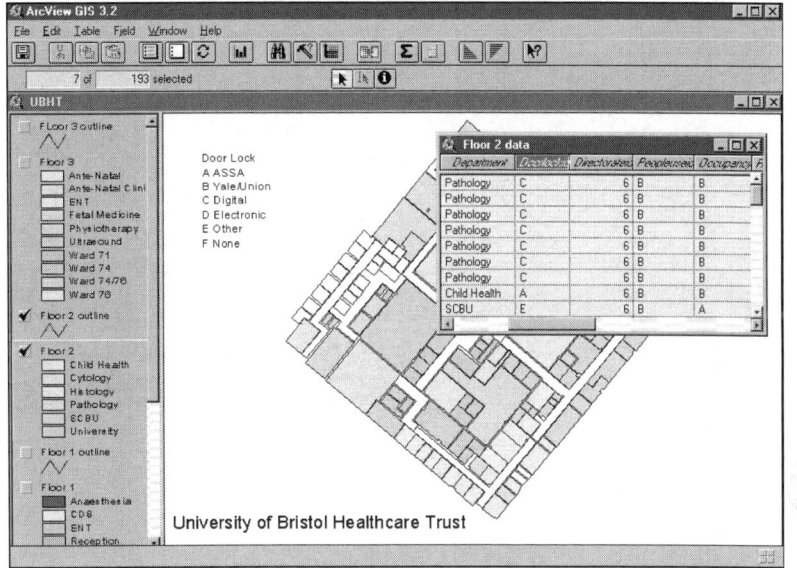

Figure 5.6 Rooms on floor 2 used by the Pathology Department and have a digital door-lock.

are used by the Pathology Department *and* have a digital door-lock. Attribute data relating to facilities that are not necessarily room-dependent can be stored in relational databases and linked to the room data using unique identifiers. This is how the facilities data for the UBHT building were managed and Table 5.3 illustrates the attribute data that were recorded in databases for each floor of the building and the database structure for associated facilities data. The room_id field is the link between the attribute table for each floor and the related 'facilities' databases. A GIS is able to use these relational links between databases when analysing data. For example, Figure 5.7 highlights those rooms on floor 3 that are used by the Ante-Natal Department *and* Ante-Natal Clinic *and* have tap temperatures that are 60 degrees or greater. To perform this query the water fittings database must be joined to the floor 3 database.

What sets GIS apart from other facilities management software is the ability to perform spatial analysis. Using a simple overlay technique, for example, it is possible to identify rooms where the X-ray Department on floor 1 is directly above the kitchen area on the ground floor. Similarly, proximity analysis can be used to determine which rooms on floor 1 are within a distance of say one metre of rooms used by the Ear, Nose and Throat Department.

Table 5.3 Building facilities databases

Attributes of floor rooms	Associated facilities	
room_id (code)	*Water fittings*	*Light fittings*
room title (text)	**room_id**	**room_id**
department (text)	watertitting	lighting type
directorate_id code	taptype_id	number
peopleuse_id (code)	taptemp	comments
occupancy_id (code)	tapref	
room height (value)	*Radiators*	*Ventilation*
decorcond_id (code)	**room_id**	**room_id**
ceilingtype_id (code)	radiator type	ventilation type
floortype_id (code)	number	number
windownumber (value)	comments	comments
windowtype (code)	*Pipework*	
windowlock (T or F)	**room_id**	
windowhazard (T or F)	pipe type	
surveydate (date)	diameter	
datainputdata (data)	length	

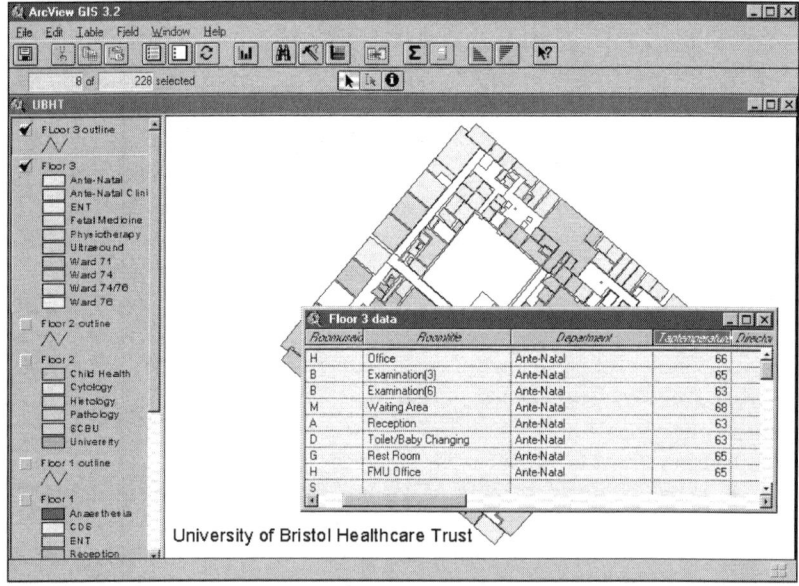

Figure 5.7 Rooms on floor 3 used by the Ante-Natal Department and Ante-Natal Clinic and have tap temperatures 60 degrees or greater.

Overseas case study

In the USA a GIS is used as a space management system at the University of Minnesota. The university has 80,000 students and staff, 24 million square feet of floor space and 1,000 buildings valued at more than three billion dollars. The GIS provides information on occupancy, size, use, programs, accessibility, etc. and can be linked with other departmental systems. It was designed to help human resources, inventory services, telecommunications and other departments that use spatial identifiers (building or room numbers) as a primary method of organising data. A graphical navigation and query interface links 1,300 floor plans and campus maps with over 60,000 database records. The campus maps provide information on buildings and the floor plans provide information on rooms. Colour-coded plans can be created using database data and infrastructure management data can be integrated with room level detail so a planner can identify the impact on building occupants when underground utilities require maintenance. The finance department uses the GIS to analyse building and operating costs and data can be analysed to identify buildings and departments with high operating costs (Jordani, 1998).

Gondeck-Becker (1999) reported that implementation of the Space Management System (SPACE) at the University of Minnesota began with an analysis of requirements, management procedures and related business processes to identify space management needs and requirements. Key factors uncovered during the analysis phase included:

- Effective business decisions require access to and synthesis of information from a variety of information systems across the University including facilities management.
- Location is an organising concept for a variety of departmental data including human resources, security, class scheduling, inventory and research. These databases include information on attributes ranging from staff names and position to lists of controlled substances, equipment and hazardous waste. Associating a building and or room number to organise departmental data is a common practice.
- Departments need information about the facilities that accommodate people, equipment and activities tracked in departmental databases. The information they need includes data about the location, size, condition, accessibility, assignment of space and other factors.
- Lack of an enterprise-wide spatial model (e.g. site, building and room numbering scheme) causes other users to develop their own. This leads to multiple database maintenance and inconsistent data.

- Departments need access to accurate, up-to-date drawings of facilities.
- Departments need robust and flexible data access and reporting tools that incorporate graphical navigation, query and reporting.
- Accurate space data has a significant economic impact on Indirect Cost Recovery (as related to federally funded research), occupancy and deferred renewal.
- Legacy space management systems that make data difficult to access frustrate efforts by academic units to make effective decisions regarding assignment of space. With the existing system integrity was questionable, data was not current and the system was not used by some groups. Difficulty in use and maintenance of data led to a backlog of incomplete revisions.

User requirements dictated the selection of an enterprise-wide facilities management system rather than a departmental solution. The Project Team determined this approach would add business value to many departments in the University, not just facilities management. The system needed to provide an intuitive interface for finding, analysing, consolidating and communicating facilities and related data and present data graphically via maps, floor plans and other data. It also needed to allow users to integrate facilities data with information from other university systems such as classroom scheduling, inventory services, human resources and departmental databases, spreadsheets and other tools used to manage departmental data. The system needed to be easy to maintain and provide accurate, timely and auditable data.

The SPACE System allows users to connect institutional data, such as occupancy and use, with data that departments maintain for their own requirements such as instructional activities, research projects, controlled substances and hazardous waste. User-defined data can be analysed together with institutional data, floor plan drawings and maps. Responsibility for collecting and updating institutional data resides with the facilities management department while user-defined data maintenance is the responsibility of the individual department. In order to connect user-defined to institutional data for analysis, the data sources must have a common identifier (e.g. building or unique room number). The data are organised in a spatial hierarchy by: (1) Campus, (2) Zone and/or District (if one exists), (3) Building, (4) Floor, (5) Room and (6) Room Detail. Most of the information about space is attached to the room and room detail records, but it is possible to capture information about an element at any level of the hierarchy. The room record stores all of the information that is valid for the entire room, such as size, capacity and accessibility. User-defined data can be connected at any level. Floor plan drawings can be used as a tool for navigating and reporting to a variety of data. Users can conduct 'what if' scenarios for planning and design.

Campus maps illustrate major geographic and facility features; roads, sidewalks and building outlines have been incorporated. Plans for future releases include the incorporation of utility and other infrastructure data. Floor plans illustrate exterior and interior walls, window locations and door openings. Major structural, circulation elements (stairs, elevators) and mechanical shafts are also noted.

As an example of application, Residential Housing merged its database of room assignment characteristics (gender preference, smoking preference, visitation, etc.) with institutional data. The combined data was used to produce drawings used for student registration. In another example, the group responsible for facilities maintenance is integrating their maintenance management system with the SPACE System. By sharing data that describe the location hierarchy, maintenance planners have access to institutional space data to schedule preventative and unplanned maintenance. The Finance Department is using SPACE to help analyse the building and operational costs for University facilities. Once the cost data are linked to SPACE they can be queried to identify buildings and then departments and activities in those buildings that have high operating costs. The Facilities Management department is using the system to create campus maps that are colour coded to reflect the costs of different utilities.

The SPACE System uses GIS technology to provide accurate, up-to-date drawings and facilities data to support the activities of a broad spectrum of users and business processes. A university-wide solution requires more collaborative effort during implementation but the result can add value to institutional data, facilitate the flow of information between business functions and support integration across departmental boundaries. It is a powerful management and analysis system that is having a much broader impact on the efficiency of the organisation.

Also in the US, the Davis Campus of the University of California uses a GIS for physical, environmental and capital planning. The campus comprises 3,500 acres, over 1,400 buildings and 24,000 students and the GIS is used for long-term development of the campus (Boyd and Rainbolt, 1998).

Finally, in the District of Columbia, a GIS has been developed to manage the facilities of 164 school and administrative land and buildings, totalling some 31 million square feet of space. The GIS was implemented to manage information on ward boundaries, attendance zones, rivers and streets. School information such as names, educational programmes, average class size, student enrolment, capacity, community accessibility, comfort rating, handicap accessibility, room closures and physical details of buildings and facilities are also input. The GIS is used as a decision tool for maintenance programmes (Kilical and Kilical, 1996).

Rural land management

Rural estate management

GIS can be used in rural estate management (Donald, 1999) for the following purposes:

- to produce land use maps;
- area management for subsidy claims;
- link database of cropping records, nutrient status, soil type, drainage works, yields to maps;
- accurately map field boundary changes;
- produce conservation maps to accompany bids for grants;
- help valuations, forecasts and budgets for forestry suppliers and the Forestry Commission.

As an example the Boughton Estate comprises 4,500 hectares in Northamptonshire plus additional land and minerals in Cumbria. The estate includes 1,000 hectares of woodland, 11 tenanted farms and one in-hand farm, five conservation villages with over 200 properties, a variety of commercial and other buildings (many of which are listed), wayleaves, easements and mineral extraction rights. As well as requiring access to accurate and up-to-date estate records, managers need to be able to capture and communicate estate information geographically. They also need to be able to produce customised plans and perform land use modelling. A GIS, together with GPS data capture facilities, was regarded as the solution (GI News, 2001). Estate data, such as property agreements, sales and access rights, are held as overlays and tables in the GIS, which is used for:

- Estate modelling – to create maps and schedules for the forest design plan, showing woodland composition, felling and re-stocking.
- Dispute resolution – one cottage on the estate was sold in 1980s and the conveyed boundaries were incorrectly fenced, resulting in a dispute between the owner and the estate when the property was resold in 2000. The boundary was therefore resurveyed and superimposed over the

conveyed boundary to identify discrepancies and create plans for rectification of documents to satisfy solicitors and the Land Registry.

- Development land sales – the estate was granted planning permission for housing on a large area of land and a GPS was used to subdivide the land into plots and to monitor land registration.
- Area-based grant applications – throughout the European Union there is a requirement to continuously update agricultural maps for the Integrated Administration and Control System (IACS) that monitors the disbursement of funds under the Common Agricultural Policy. Only a 2 per cent error in areas recorded is permitted. GPS and GIS are used to ensure correct payments for agriculture and woodland practices by accurately measuring field and wood areas. 'Some disused quarry land on the estate was identified for landfill, resulting in a boundary change. This adjusted the area of an agricultural field previously registered for IACS. GPS was used to survey the new fence line. The data were transferred into the GIS where they were used to create a new field plan, centre point and area for application to MAFF, without having to wait for the OS to resurvey' (GI News, 2001).

Future applications are to include the digitisation of utility infrastructure on the estate to allow wayleave payments to be reviewed, and the identification and survey of potential adverse possession of estate land.

Agriculture

With regard to agriculture, GIS is used to assist precision farming, balancing the need between the economic return from a crop with the environmental impact. An increasing number of farmers are investing in GPS receivers which can pinpoint precise locations by locking onto a network of satellites. Combining this information with OS digital mapping using a GIS allows the farmer to store, analyse and display a wide range of data – from crop yields and fertiliser requirements to tracking machinery and locating staff. Farmers can map yield, weed or insect infestations. This information can then be used to apply herbicides and pesticides more effectively. GPS can also be used to record agronomy variables (nutrient and moisture levels) at grid intervals (alternatively satellite imagery is used for data collection over wide regions). These location data are input into a GIS and integrated with other data geographically. This helps determine where chemicals and pesticides should be applied.

Shuttleworth Farms in Bedfordshire make extensive use of both GPS equipment and digital mapping. GPS is used to determine the amount of fertiliser, chemical or spray to use. Identifying a combine harvester's exact position using GPS enables farmers to plot yield maps by logging the precise yield from crops passing through the machine at any particular moment. The farmer can then produce a map showing the yield across a field. This information can be used to work out how much fertiliser to use and precisely where it needs to be spread. Farmers can also see which fields produce the best returns, helping them to plan their planting for maximum results and use their resources more efficiently. The progress of vehicles and machinery can also be followed on-screen using computerised mapping that shows the layout of the farm. Messages can then be passed to staff on the ground, such as how close a grain trailer is and how long it will take to reach them.

In the US, GIS is used in agriculture by:

- retailers to assist in the marketing and transportation of fertilisers and chemicals;
- agronomists for monitoring production trends;
- agricultural lenders and crop insurers to rate and market crop insurance policies;
- agricultural manufacturers for marketing, operations and distribution;
- agribusiness for wholesale distribution.

Overseas case study

Sunkist growers (a citrus marketing co-operative in California) use GIS to estimate the quantity of fruit available each season. This is done by tagging and measuring a sample of oranges in orchards and noting the position of the source tree in the orchard and the region. This information is combined with satellite imagery and aerial photographs. The system also allocates growers to specific counting houses, thus avoiding potential miscounts. The system will include data from the National Weather Service in the future to pinpoint areas of potential crop damage due to hail or freeze. Demographic data will also be added for marketing purposes.

The Gordon Family Ranch is a vineyard in California and it uses GIS for planting, fertilising, harvesting, preparing reports to government and maintaining data on experimental rootstock plantations. GIS is also used to create a soil nutrient gradient for planting and for the addition of lime in acidic areas. With regard to harvesting, different parts of the vineyard ripen at different times and therefore sugar samples are taken to create harvest-scheduling maps. GIS is also used to define areas needing special soil treatments rather than applying them uniformly to the whole vineyard.

Land management

English Nature promotes the conservation of England's wildlife and natural features and is the Government's advisor on nature conservation. Amongst its other responsibilities English Nature designates the most

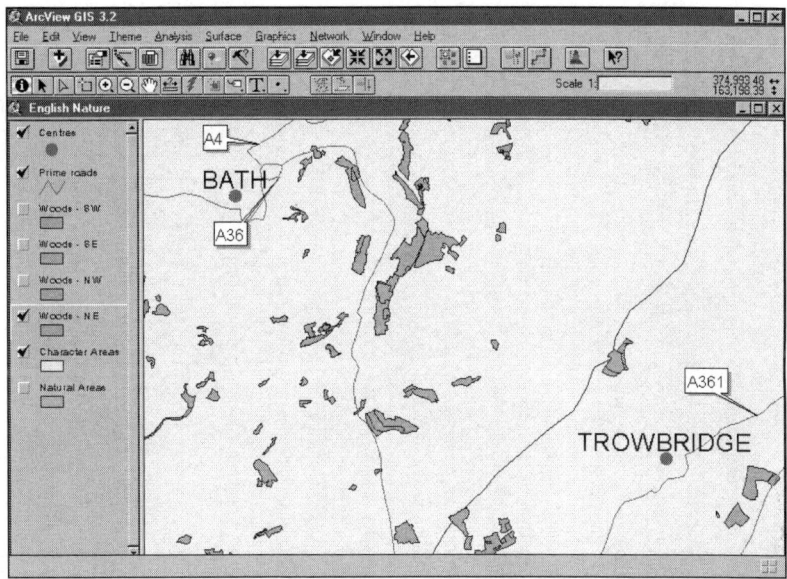

Figure 5.8 Areas of Ancient Woodland.

Source: English Nature, used with permission.

important areas for wildlife and natural features as Sites of Special Scientific Interest (SSSI) and secures the sustainable management of these sites. It also establishes and manages National Nature Reserves (NNRs) and Marine Nature Reserves. In order to assist the undertaking of these tasks English Nature has created digital boundary data for SSSIs, NNRs, Special Areas of Conservation and other areas of specific natural phenomena.

Figure 5.8 illustrates the ancient woodland areas around Bath and Trowbridge in the south west of England. The data recorded about each wood include its grid reference, area in hectares, how much is semi-natural or replanted, whether any of the wood has been cleared (since 1920 approximately), public ownership details where known, and any conservation status. Prior to digitisation of the boundaries, only paper maps depicting each ancient wood at 1:50,000 scale were available.

Figure 5.9 illustrates Character Areas for Devon and Cornwall. There are 159 of these areas in England, each of which is a distinctive division of landscape and cohesive countryside character, on which strategies for both ecological and landscape issues can be based. The Character Area framework is used extensively by the Countryside Agency to describe and shape objectives for the countryside, its planning and management. These areas are not derived from administrative boundaries, but follow variations in the character of the landscape. As they are based upon the distribution of

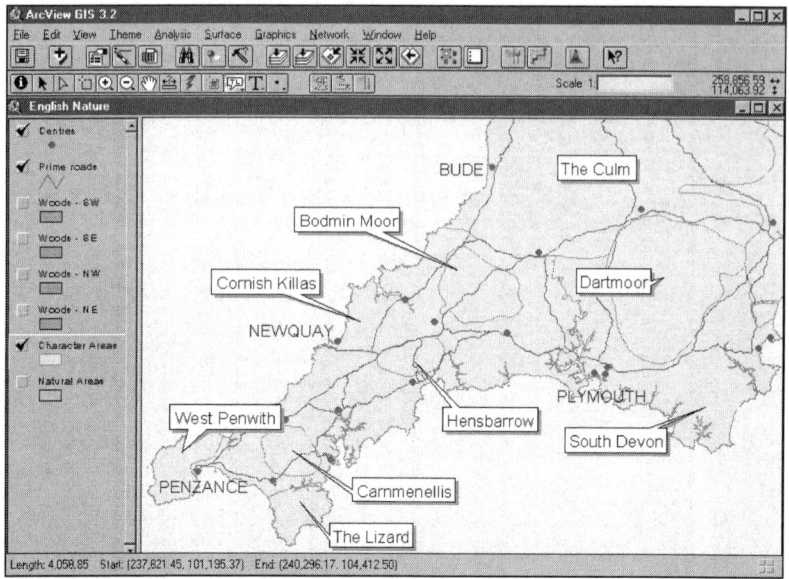

Figure 5.9 Character Areas.

Source: English Nature, used with permission.

wildlife and natural features, the land use pattern and human history of each area, the boundaries are difficult to define precisely because of the gradation from one area to another. Consequently, the boundaries should be regarded as the approximate limits of the areas involved (www.englishnature.org.uk).

Natural Areas are bio-geographic zones that reflect the geological foundation, natural systems and processes and the wildlife in different parts of England. There are 120 Natural Areas, many of which are coincident with Character Areas; the remainder comprise one or more Character Areas. Natural Areas are a sensible scale at which to view the wildlife resource, from both a national and local perspective, and they are used by English Nature as an ecologically coherent framework for setting objectives for nature conservation. Figure 5.10 shows the Natural Areas for part of southwest England (www.englishnature.org.uk).

At Nottinghamshire County Council GIS is used to aid landscape protection, conservation and management and the development of a departmental environmental records centre (Shalaby and Ford, 1995). GIS is used to identify ancient woods, permanent grasslands, heathlands, historic parks and mature river courses. Some 46,000 polygons have been digitised and other environmental data are overlaid. The GIS is used to help determine landscape character by analysing components such as distribution of woodland, ecological habitat and features of designated conservation interest. GIS is also used to

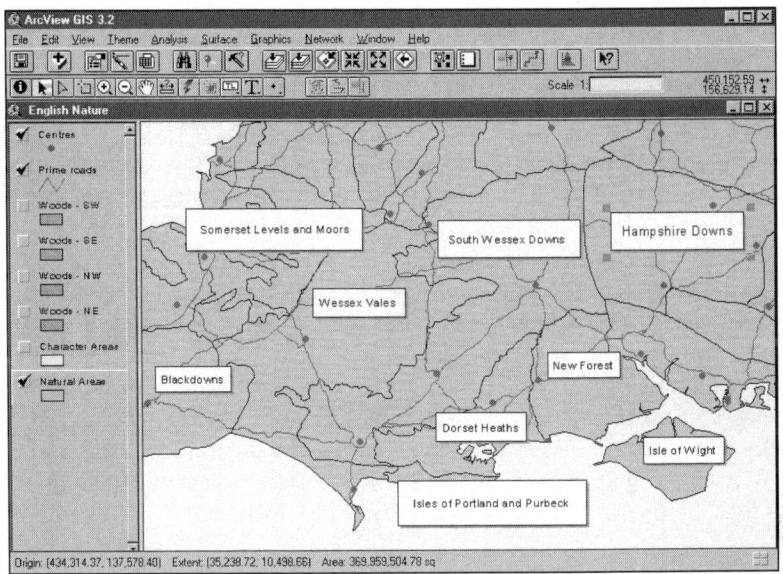

Figure 5.10 Natural Areas.
Source: English Nature, used with permission.

integrate land use and land cover data for each of the landscape types in the Environment Agency's river-catchment management plans. Local authority officers have calculated existing levels of woodland cover, grass, arable land, urban land and mineral extraction and use the data to determine planting and land use policies for each landscape type. GIS is also used to identify areas where, say, woodland planting would not be appropriate. 'GIS has thus helped to ensure that an initiative that will have a major effect on the future landscape takes account of sensitivities and builds upon the established character of the area' (Shalaby and Ford, 1995). The GIS records:

- rural land use;
- derelict land and mineral extraction sites;
- sites of ecological importance, ancient woods and mature landscape areas;
- SSSIs and sites of biological and geological importance;
- wildlife trust, local nature reserves and other designated wildlife sites;
- regional character areas, common land and village greens;
- scheduled ancient monuments, listed buildings and conservation areas.

Applications include:

- landscape protection, conservation management and major landscape initiatives;
- nature conservation management and planning, re-surveying of sites;

- information source for the 'State of the Environment' report;
- land use change and monitoring, land reclamation, minerals planning;
- highway and transportation schemes;
- informing planning policy, processing applications and briefs;
- local planning enquiries and general research purposes.

In Wales the Environment Agency needed to improve the quality of maps and reports for internal and external users such as the recently formed National Assembly for Wales. There was also a need to increase the credibility and utility of environmental information for other partner organisations. The Environment Agency in Wales has therefore implemented a GIS to help meet this objective, the key benefits of which include the ability to quickly find a location and view accurate and precise information and the ability to analyse complex data to assist decision support and policy formulation. Much of the Agency's work is catchment based, with some of the data extending beyond Wales (e.g. the Dee, Seven and Wye river catchments) and is held and managed by external agencies such as the Environment Agency (Midlands). Data sets used by the Environment Agency in Wales include:

- designated site boundaries, habitat surveys, etc.
- scheduled ancient monuments
- agro-environment schemes
- woodland areas
- development plans, planning proposals
- flood plain data
- OS data
- landfill site locations
- flood warning areas
- local environment action plans.

All functions within the agency utilise a standard set of tools, enabling users to view, plot, query and share spatial data. In addition applications are being developed to meet the specialised needs of some functions such as:

- Flood Defence – indicative flood plain mapping, flood warning analysis, maintenance of Agency assets and infrastructure.
- Water Resources and Ground Water Protection – river catchment modelling, water abstraction licence monitoring.
- Customer Services and Planning Liaison – screening of planning applications, staff/customer queries, constraint checking.

In another application the current landscape of the UK has a historic dimension which needs to be recognised and presented in a readily understood format. This facilitates appropriate landscape management regarding evaluation, conservation and preservation of the historic environment. 'Historic Landscape Assessment or Characterisation' is a GIS application designed to assess the 'time-depth' of a landscape, that is, its historic origin (www.agi.org.uk). On completion the project will cover Bedfordshire, Cambridgeshire, Essex, Hertfordshire, Norfolk and Suffolk. The brief, set by English Heritage and the relevant Regional Authorities, was to map the landscape according to its various historic attributes including:

- Field systems: various types of pre-eighteenth-, nineteenth- and twentieth-century origin,
- Parklands and Policies: their date of origin and subsequent changes of use, for example, to golf courses/hospitals, educational institutions, etc.
- Woodlands: 'ancient', nineteenth and twentieth century.

The information allows the historic dimension of the landscape rather than site-specific assessments to be taken into account for strategic planning and in individual planning applications. The GIS application has been used in Hertfordshire for landscape character assessment, planning strategy and policy including minerals, development control issues and the management of archaeological/historical resources.

The application uses a rural, broad-brush approach, where historic maps and current paper and digital data are assessed at 1:25,000 and digitised at 1:10,000 scale. The assessment is applied on a field-by-field basis and is a desk-based approach with minimal field verification. A seamless current landscape layer is created with additional but discontinuous layers within the database of historic events relating to individual land parcels. In some areas the current landscape may date back to the sixteenth century or earlier. Other areas have undergone many changes which may have great 'time-depth' of events, but a more recent history. As an example a 'prairie field' created after 1950 may have been enclosed in 1880, but still retains relic elements of its pre-eighteenth century origins, whereas an adjacent field may date back to 1588.

Data sets used are classified as either current or historic. Current mapping includes OS Landline and 1:10,000 raster map data. Historic maps include Ordnance Survey (1950 sheets), First Edition (nineteenth century), County base maps (eighteenth/nineteenth century), Tithe maps (nineteenth century), Enclosure Maps (nineteenth century) and Estate maps (dating from the sixteenth century). Maps can be constructed, recording the diversity within a landscape, enabling traditional spot site data to be nested within its historic landscape (which may or may not be contemporary). Associated databases allow searches, enabling the enquirer to reconstitute those past landscapes that have survived, or to carry out complex analysis within and with other appropriate GIS data sets.

Another project under way in the UK seeks to create a Land Cover Map of the country. This will be a census of the countryside of Britain in the form of digital maps and databases plus derived products for use in GIS and statistical packages. Land cover will be classified as one of twenty-eight different types in a vector map, thus segmenting the landscape into land parcels and recording dominant land cover and attribute data for each

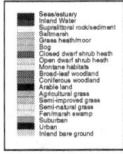

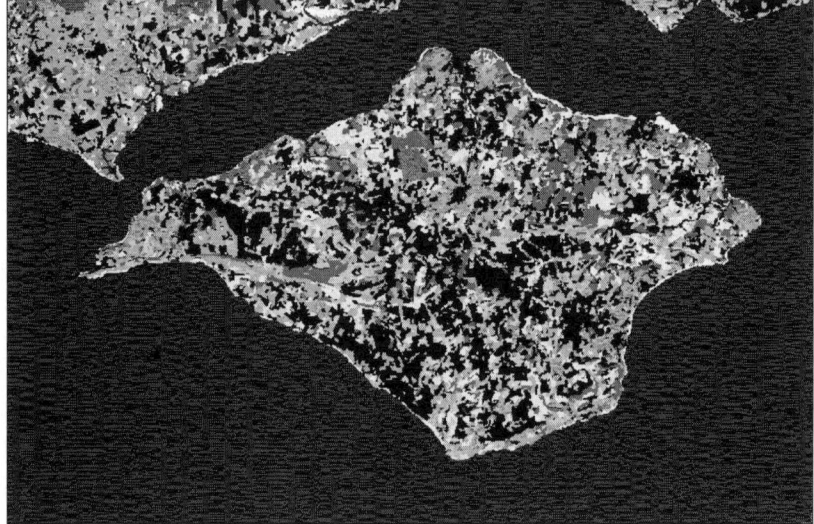

Figure 5.11 Extract from the Land Cover of Great Britain (1990) showing the Isle of Wight, © NERC.

parcel. There will also be a raster coverage on a 25-metre grid that will record land cover and summary data will be produced that will state dominant cover together with its percentage cover for each square kilometre (www.ceh.ac.uk). Figure 5.11 shows a sample of the Land Cover Map for the Isle of Wight, off the south coast of England.

Finally, in Essex the Historic Towns Survey Project funded by English Heritage used GIS to link to Essex Sites and Monuments Records (ESMR). ESMR shows the location of sites and archaelogical finds on a backdrop of OS map data.

Summary

This chapter has demonstrated that the use of GIS for land and property management began in organisations whose functions are geographically discrete, such as local authorities. This was due to data cost and the 'project' nature of GIS applications in the early days of the development of the technology. This latter point also helps explain the application-led development of GIS in local authorities, whereas a corporate approach might have been more effective in the long term. The creation of an authority-wide land and property gazetteer makes sense but departmental/single application implementation is easier to justify financially.

Cost and availability of data have decreased and increased respectively. Consequently the use of GIS for geographically dispersed applications is now cost effective and they have increased in number. The use of GIS by large landowners and organisations responsible for a more expansive geography of property are now finding that GIS offers real benefits. GIS have been used in property management ranging from crop and land parcel level to natural areas of several thousand square kilometres in organisations as diverse as Defence Estates to English Nature. The utilities have not been considered here because their use of GIS focuses on the management of infrastructure networks rather than property management. The experience of GIS implementation in many of these organisations has shown that it is a case of balancing the need for data accuracy and completeness against fitness for purpose. Sometimes the balance is wrong – what might be an acceptable level of accuracy for the initial application may not be sufficient for subsequent applications.

The initial phase of GIS implementation for many organisations was a data audit, which often revealed many problems. Data cleaning was the next stage and this can be very expensive and requires prioritisation depending on the nature of the application(s) envisaged. Often the first application was map production – the ability to produce maps on demand for many users without a drop in output quality was justification enough for some. The power of GIS is the value it can add to data collected for one purpose but, through integration and geo-referencing, is useful for another purpose. Those organisations that have adopted an application-led approach may find that data collection policies need to be reviewed in the light of wider GIS use. Those organisations yet to tread this path would be wise to undertake an application audit at the same time as a data audit.

Perhaps the strongest tangible benefit noted by many users of GIS for land and property management was the ability to visualise relationships between property and the surrounding geography by mapping land and property assets. This geographical view of assets aids property decision-making. Perhaps the strongest intangible benefit was the data audit and cleaning process referred to above. At a more technical level, the move from point data to ownership extents delineated by polygons on a map helps to identify ownership, proximity and other neighbourhood issues at the land parcel level. The use of land parcel extents rather than seed points will undoubtedly increase following the introduction of OS Mastermap discussed in Chapter 4.

References

Allinson, J. and Weston, J., 1999, Information technology literacy survey, Royal Town Planning Institute, London.

Audit Commission, 1988a, *Local Authority Property: A Management Handbook*, HMSO, London.

Audit Commission, 1988b, *Local Authority Property: A Management Overview*. HMSO, London.

Boyd, B. and Rainbolt, E., 1998, GIS for university and campus planning and FM, ESRI User Conference Proceedings, www.esri.com

Donald, J., 1999, Technology holds the future for rural surveyors, *Chartered Surveyor Monthly*, April, 52.

England, J., 1996, The road to a corporate GIS in Gloucestershire, Mapping Awareness, May, pp. 20–24 and June, pp. 34–37.

Gondeck-Becker, D., 1999, Implementing an enterprise-wide space management system – a case study at the University of Minnesota, ESRI User Conference Proceedings, www.esri.com

GI News, 2001, GPS and GIS in rural estate management, GI News, July/August, pp. 38–41.

Jenkins, W. and Kearns, C., 1999, Everybody likes a smart asset, AGI Conference, 4.5.1–4.5.4.

Jordani, D., 1998, GIS is integral to the University of Minnesota's new space management system, *ESRI Arc News*, Fall, 18.

Kilical, H. and Kilical, A., 1996, District of Columbia public school system (DCPS) facilities master plan using GIS, ESRI User Conference Proceedings, www.esri.com

Kirkwood, J., 1998, GIS insight on site, *Estates Gazette*, 9847, November 21, pp. 130–131.

Lilburne, L. and Rix, D., 1991, The use of GIS in local government property records. Mapping Awareness and GIS Europe, 5(10), pp. 16–21.

London Borough of Barking and Dagenham, 1996, The Property Database Project, Legal and Property Division, Chief Executive's Department, LBBD, London.

MacLellan, J., 1998, A London borough property database project. In Wyatt, P. and Fisher, P. (eds) *Property Information Today and Tomorrow*, Royal Institution of Chartered Surveyors, London.

MacLellan, J. and Musgrave, T., 1999, GIS and best value – a corporate approach, AGI Conference, 4.4.1–4.4.5.

Masser, I. and Campbell, H., 1994, The take-up of GIS in local government, AGI Conference, 14.2.1–14.2.6.

Morgan, R., McKay, I., Kelly, A., Clark, J., Culpan, S., Steward, D. and Donaldson, A., 1994, Estate management through GIS – a Strathclyde experience. In: Geographical Information, Taylor & Francis, pp. 150–154.

RICS, 1987, Property management in local authorities: evidence to the Audit Commission by the RICS, Royal Institution of Chartered Surveyors, London.

Rix, D. and Lilborne, L., 1994, Towards a spatially based property management system for UK local government. In: *Geographical Information*, Taylor & Francis, pp. 155–163.

RTPI, 2000, IT in local planning authorities, Royal Town Planning Institute, London.

Shalaby, M.T. and Ford., P. 1995, Integration: the way for corporate GIS in Nottingham city council, Joint European Conference and Exhibition on Geographical Information Proceedings, Basel: JEC-GI, 2: 36–41.

Wooden, S., Greathead, N. and Meggs, M., 1999, Delivering estate solutions to defence needs, AGI Conference, 4.3.1.–4.3.3.

6 Planning and development

Introduction

Architects, developers, surveyors and planners make decisions on how geographic resources should be utilised. The results of these decisions are perhaps most clearly observed when they relate to the development and redevelopment of the built environment. Since 1947 all land and property development activity has been subject to regulation in the form of planning and building control. The volume, breadth and sophistication of this regulation have risen markedly over the last fifty years and consequently the requirement for information systems to assist in the interpretation of planning and development policy has similarly increased. The number of parties involved in planning and development decisions has risen too, making for a more complex dissemination and cross-checking process, for which GIS is a solution.

This chapter focuses on GIS applications for planning and development functions. Some of the earliest GIS applications in the UK were developed in local authority planning departments and this chapter considers the role of GIS in government planning first. The use of GIS for planning within local government has tended to focus on the automation and improvement of statutory operations such as searches of local land charges and the processing of planning applications. The following section then turns to GIS applications within central government planning. Here the planning functions are more strategic and relate to the formulation of planning policy and issuance of guidance for regional and local government.

In terms of development, the decision-making process begins with site identification and appraisal and GIS has been used in a number of innovative ways to assist these functions. Of particular relevance to UK property development activity at the moment is the reuse of previously developed land or 'brownfield' sites as they are more commonly known. The potential for environmental risks and liabilities is greater on such sites and therefore development decisions tend to be more carefully considered than for undeveloped or 'greenfield' sites. GIS has been used to great effect in the synthesis and presentation of environmental and other land use information for development purposes.

Finally, and perhaps most imaginatively, GIS has become a favoured technology on which to base urban design and visualisation applications. Some of the advantages that GIS offers over CAD and CAFM software were described in Chapter 5. Similarly, for urban design, the ability of GIS to store and link attribute data to geographical or spatial features such as a new building or road means that much more can be asked about how these features fit in to the landscape and relate to neighbouring features. In short, the impact of a design can be analysed rather than merely visualised.

Planning

Planning policy

'Planning is the art and science of ordering the use of land and the character and siting of buildings and communication routes so as to secure the maximum practicable degree of economy, convenience and beauty' (Keeble, 1969). On its web site the Office of the Deputy Prime Minister (ODPM) states that 'our overall objective is to create a fair and efficient land-use planning system that represents regional differences and promotes development which is of a high quality and sustainable' (www.dtlr.gov.uk). For the purposes of this chapter and in keeping with Birkin *et al.* (1996), a more generalised definition of planning is adopted, which includes activities such as social services, labour market and transport planning, as well as the more conventional town and country planning responsibilities of a local authority planning department.

At the national level planning involves the publication of strategic policy and guidance for implementation at the regional and local levels. The formulation of this policy and guidance relies on the collection and dissemination of accurate and up-to-date data and statistics. Because planning is inherently geographical GIS has been identified as a suitable tool for managing, analysing and presenting data and statistics in support of central government planning functions. A good example of the way in which GIS is being used to help central government in this way is the definition of town centre boundaries for statistical data collection and analysis. It illustrates very clearly how GIS can be used to help collect data from disparate sources, integrate and analyse those data and present aggregated statistics for planning policy and guidance purposes. The second example of the way in which central government is using GIS to manage data for planning policy formulation is the creation of the National Land Use Database (NLUD). A key planning policy role for the NLUD is to provide regularly updated data on the reuse of previously developed land and, in particular, a statistical measure of the amount of housing that is being built on 'brownfield' sites. These two examples are described next.

*Defining town centre boundaries for statistical data collection
and analysis*

'Systematic data collection on the UK's town centres has not been undertaken for almost 30 years, yet they have witnessed dramatic change and the
potential impacts of further change need to be assessed' (Hall and
Thurstain-Goodwin, 2000). For example, the lack of town centre statistics
has meant that is has been impossible to quantify the effects of guidance
contained in Planning Policy Guidance Note 6 (PPG6) that precludes retail
development in off-centre locations. For such analysis to occur it is necessary to develop a consistent set of town centre boundaries so that relevant
statistics may be collected and aggregated for comparison purposes
(Thurstain-Goodwin and Unwin, 2000). ODPM has for several years been
undertaking a programme of work to improve the availability and quality
of retail statistics. A Retail Statistics Working Group supported the
approach of making the fullest use of existing sources of data and limiting
new data collection to those areas not already covered. This was reiterated
in the Government Response to the Parliamentary Select Committee Report
on Shopping Centres and their Future in 1995, after the report had recommended the development of 'a nationally consistent system of retail data
collection to be published at regular intervals', which 'should reduce significantly the costs being incurred in Public Inquiries and impact studies'. GIS
was used to help meet this objective.

The key factors that characterise a town centre were identified as follows:

- Economy: land use and employment traditionally associated with town
  centres.
- Property: building densities.
- Diversity of use: land use mix and intensity.
- Visitor attractions: land use and employment data.

However, there are no consistent definitions or boundaries for town centres
and shopping centres for which statistics from these key factors can be
produced. Therefore the Centre for Advanced Spatial Analysis (CASA) at
University College, London, and the Urban and Economic Development
Group (URBED) were commissioned by the ODPM to develop a GIS-based
approach to modelling areas of town centre activity, using fine-scale data,
to produce a continuous spatial index of 'town-centredness'. The way in
which the key factors helped define an index of 'town-centredness' is illustrated in Figure 6.1. This index is then used to define a set of Central
Statistical Areas for all town centres in the UK within which consistent
statistics such as employment, retail sales turnover and floor-space can
be derived.

A computer-based model was developed incorporating as many of the
town centre characteristics as possible. The principal statistics required on

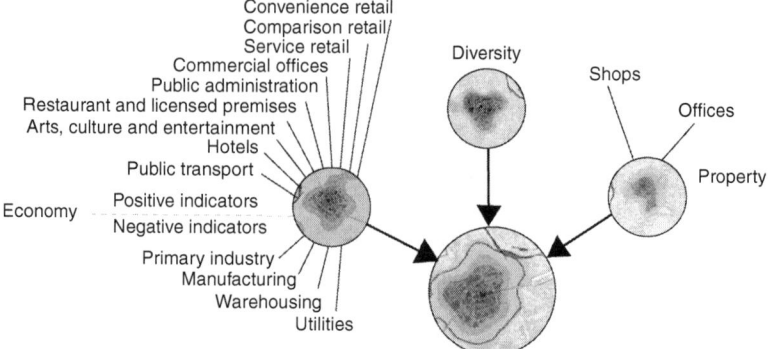

Figure 6.1 The construction of an index of town centre activity.

Source: The Office of the Deputy Prime Minister, used with permission.

a town centre basis are employment, turnover and floor-space data. Despite the evolving nature of postcode geography and some problems with this method of geo-referencing property data, unit postcodes proved to be a viable basis for defining areas of town centre activity and statistics for them. The Inter Departmental Business Register (produced by the Office for National Statistics) and property descriptions and values from the Valuation Office were therefore referenced by postcode. The unit postcode can be represented by a centroid reflecting the spatial average of the delivery points as opposed to a defined area with a set boundary. A GIS was used to model each of the key factors for a particular town's urban area, based on a fine (20 metre square) grid and each grid square was assigned a value. These values were then used to generate a surface or series of contours that represented the graduation of the factor throughout the study area, illustrated in Figure 6.2. The surfaces were then overlaid to produce an 'index of town-centredness' for the study area, a 3D representation of which is shown in Figure 6.3. Analysis of the peaks on this composite surface allows the selection of key contours for delineation of town centre boundaries (Thurstain-Goodwin and Unwin, 2000).

The methodology was tested on ten urban areas in England and Wales that represented a broad range of town centres. In each case the computer model was able to locate the town centre and produce a graduated surface of 'town centredness' from the data. A key 'contour' was selected from the composite surface, which represented the town centre boundary. Each area's local authority and an advisory group of experts reviewed and endorsed the results. Once the key contour is established, the set of unit postcode centroids that fall within the Central Statistical Area (area defined by the key contour) are identified. It is from these that the aggregate town centre statistics are generated. Figure 6.4 shows the contour that represents the boundary of the Central Statistical Area for Bristol, taken from an

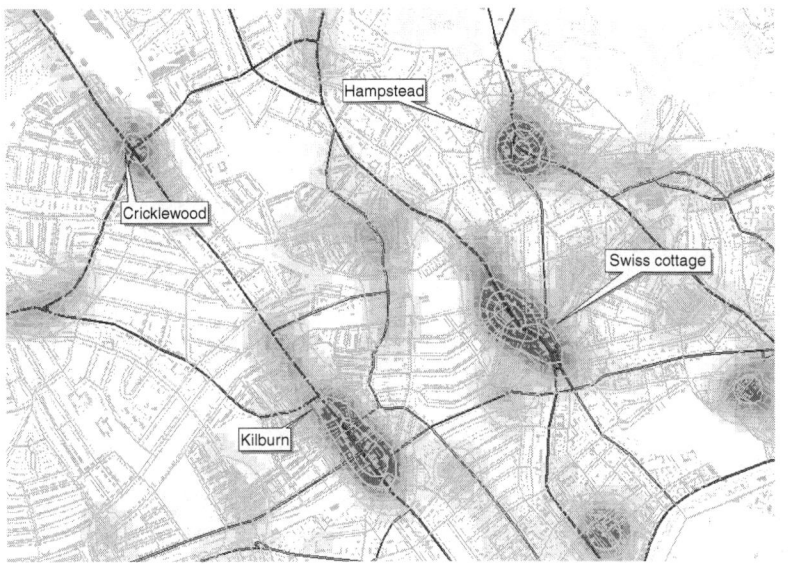

Figure 6.2 Contours representing levels of town centre activity.
Source: The Office of the Deputy Prime Minister, used with permission.

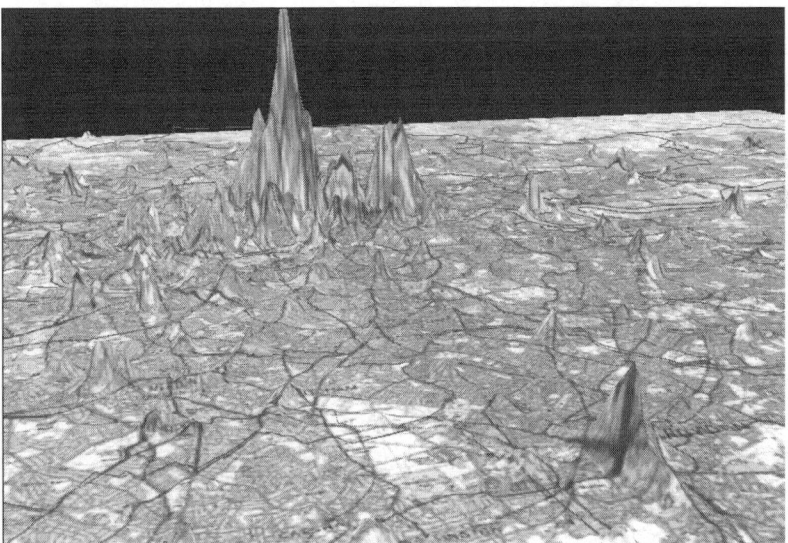

Figure 6.3 3D visualisation of the index of town centre activity.
Source: The Office of the Deputy Prime Minister, used with permission.

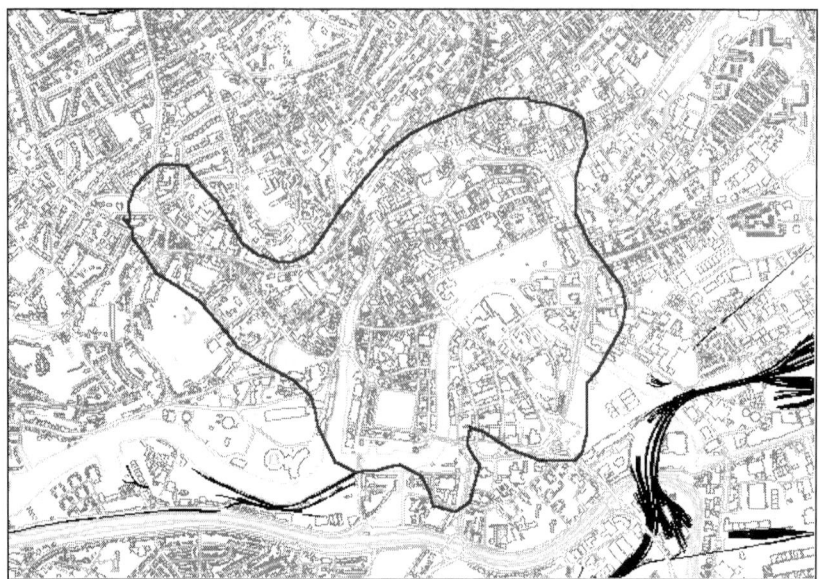

Figure 6.4 Area of town centre activity.
Source: The Office of the Deputy Prime Minister, used with permission.

earlier, 1997, study. It was found that no one surface was sufficient to define an acceptable Central Statistical Area so it was necessary to layer and combine all of the surfaces.

In summary 'GIS provides a particularly powerful medium through which to communicate the complex issues associated with the compilation of statistics ... Town centres are spatial objects and the data used to define them are spatial. Yet it is not until the various aspects of the model are presented through a GIS that people are able to engage with the concepts that underpin the statistics' (Hall and Thurstain-Goodwin, 2000).

National Land Use Database (NLUD)

Under the Environment Act 1995 local authorities are required to identify, assess and ensure remediation (where the contamination is causing unacceptable risk to human health or to the wider environment) of areas of contaminated land within their boundary. The regime places specific duties on local authorities to inspect their areas to identify land falling within this definition and, where they do, to require its remediation in line with the 'suitable for use' approach. The regime also provides detailed rules for assigning liabilities for contaminated land, based on the 'polluter pays' principle. The new regime complements local authorities' long-standing duties to identify particular environmental problems, including those resulting from land contamination, and to require their abatement.

Data sources that are useful in identifying contaminated land include current and historical land use maps. These land use maps are also useful for identifying 'receptors' of contamination such as schools, housing, etc. In 1985 the then DoE commissioned a feasibility study for a National Land Use Stock Survey (Roger Tym and Partners, 1985). In 1991 the DoE commissioned another study published by Dunn and Harrison (1992). The main recommendation was a census of land use collected and maintained in collaboration with the OS, using large-scale digital map data as a base. In 1994 the Department funded further research into 'Preparatory Work for a Land Use Stock System' (Dunn and Harrison, 1994 and 1995). This consisted of three linked research projects that examined:

- the methodology for development, undertaken by Dunn and Harrison;
- the technical aspects of data capture, storage and manipulation methods, undertaken by the OS;
- the business case for a Land Use Stock System, undertaken by Coopers and Lybrand.

The ODPM has a major policy interest in recycling derelict land and facilitating greater use of 'brownfield' sites. Consequently, the ODPM, working with English Partnerships, the local government Improvement and Development Agency (IDeA) and the OS, has commenced work on a consistent assessment of vacant and derelict sites and other previously developed land throughout England that may be available for housing or other development. This register of previously developed land forms the first phase of the NLUD that will eventually cover all land uses in England down to site level and will be regularly updated. When this data set is available and used in conjunction with other land and property data, which detail contaminated land and the underlying geology, the value of NLUD will be clear. The completed NLUD will potentially enable the accurate monitoring of shifts in land use and help in the debates on land availability, urban expansion and protection of the countryside. NLUD represents a major step forward in establishing a comprehensive record of land use. The results provide a basis for strategic planning at national and regional levels and a rigorous and consistent basis for planning at the local level.

Phase 1 of the NLUD, which commenced in 1998, involved 344 local authorities providing information on over 30,000 previously developed sites. Provisional statistics based on Phase 1 were published by the former Department of the Environment Transport and the Regions (DETR) in 1999 at national, regional and local authority levels. The statistics revealed the amount of previously developed vacant and derelict land and other land that might be available for redevelopment. In 2000 DETR published the final statistics for the 1998 snapshot. In June 2000 the 'Explore NLUD

Table 6.1 NLUD sites within the administrative boundary of Cotswold District Council

AREA	EASTING	NORTHING	PAO_DESCRIPTION	NLUD_STREET_DESCRIPTION	NLUD_LOCALITY_DESCRIPTION	NLUD_TOWN_DESCRIPTION	NLUD_ADMIN_AREA_DESCRIPTION	POSTCODE
0.07	420152	235703	HALL	CHURCH FARM LANE	ASTON MAGNA	MORETON IN MARSH	GLOUCESTERSHIRE	GL56 9QN
20.28	413140	235719	POLISH CAMP	A44	CHIPPING CAMPDEN	CHIPPING CAMPDEN	GLOUCESTERSHIRE	
6.45	416323	244618	SEYFRIED SITE	STRATFORD ROAD	MICKLETON	MICKLETON	GLOS	
0.17	420244	232137	FORMER COUNCIL DEPOT	PARKERS LANE	MORETON IN MARSH	MORETON IN MARSH	GLOUCESTERSHIRE	
0.04	420612	232594	LAND ADJOINING BRITISH LEGION CLUB	NEW ROAD	MORETON IN MARSH	MORETON IN MARSH	GLOUCESTERSHIRE	
0.07	420836	232490	THE OLD GAS WORKS SITE	LONDON ROAD	MORETON IN MARSH	MORETON IN MARSH	GLOUCESTERSHIRE	
0.2	420935	232487	OLD LAUNDRY	LONDON ROAD	MORETON IN MARSH	MORETON IN MARSH	GLOUCESTERSHIRE	
0.7	420714	232624	RAILWAY SIDINGS MORETON IN MARSH RAILWAY STATION	STATION ROAD	MORETON IN MARSH	MORETON IN MARSH	GLOUCESTERSHIRE	
0.16	400601	219477	SEVERN TRENT DEPOT	UPPER DOWDESWELL ROAD	DOWDESWELL	ANDOVERSFORD	GLOUCESTERSHIRE	
1.21	418916	212581	WINDRUSH CAFE	A40	WINDRUSH	BURFORD	GLOUCESTERSHIRE	
0.53	386654	199179	LONGFORD MILLS	AVENING ROAD	AVENING	STROUD	GLOUCESTERSHIRE	GL6 9AN
1.16	394007	198001	THE FORGE	RODMARTON	RODMARTON	CIRENCESTER	GLOUCESTERSHIRE	
0.71	389346	193268	THE OLD RAILWAY DEPOT	GUMSTOOL HILL	TETBURY	TETBURY	GLOUCESTERSHIRE	
0.19	389284	193272	OLD CATTLE MARKET NORTH	GUMSTOOL HILL	TETBURY	TETBURY	GLOUCESTERSHIRE	
0.06	389259	193306	OLD CATTLE MARKET NORTH	GUMSTOOL HILL	TETBURY	TETBURY	GLOUCESTERSHIRE	

5.15	404841	196367	LAND AT EVERGREEN INDUSTRIAL PARK	BROADWAY LANE	SOUTH CERNEY	CIRENCESTER	GLOUCESTERSHIRE	
1.43	421862	200495	COAL DEPOT	STATION ROAD	LECHLADE	LECHLADE	GLOUCESTERSHIRE	
3.28	417420	200515	FORMER ARC CONCRETE WORKS	WHELFORD ROAD	FAIRFORD	FAIRFORD	GLOUCESTERSHIRE	
1.2	416382	200864	OLD RAILWAY LINE	A417	FAIRFORD	FAIRFORD	GLOUCESTERSHIRE	
0.38	402111	201839	MEMORIAL HOSPITAL AND CAR PARK	SHEEP STREET	CIRENCESTER	CIRENCESTER	GLOUCESTERSHIRE	GL7 1QW
0.85	401862	201695	CATTLE MARKET	TETBURY ROAD	CIRENCESTER	CIRENCESTER	GLOUCESTERSHIRE	
0.56	402222	201939	LAND AT BREWERY CAR PARK	FARRELL CLOSE	CIRENCESTER	CIRENCESTER	GLOUCESTERSHIRE	
1.19	402625	202049	THE WATERLOO CAR PARK AND ARGOS STORE	DYER STREET	CIRENCESTER	CIRENCESTER	GLOUCESTERSHIRE	GL7 2PP
0.15	402219	202049	SWAN YARD AND POST OFFICE COMPOUND	BLACK JACK STREET	CIRENCESTER	CIRENCESTER	GLOUCESTERSHIRE	GL7 2NH
0.32	402369	201988	CORN HALL	MARKET PLACE	CIRENCESTER	CIRENCESTER	GLOUCESTERSHIRE	GL7 2NW
17.4	416971	236838	NORTHWICK BUSINESS CENTRE	BROAD CAMPDEN ROAD	BROAD CAMPDEN	CHIPPING CAMPDEN	GLOUCESTERSHIRE	GL56 9RF
5.51	418219	236880	BLOCKLEY BRICKWORKS	B4479	PAXFORD	MORETON IN	GLOUCESTERSHIRE MARSH	
1.3	417003	221144	WEST MIDLAND FARMERS	STATION ROAD	BOURTON ON	CHELTENHAM THE WATER	GLOUCESTERSHIRE	GL54 2EP
0.5	402313	219751	CATTLE MARKET	STATION ROAD	ANDOVERSFORD	CHELTENHAM	GLOUCESTERSHIRE	GL54 4HP
1.09	404977	196954	ECC WORKS	STATION ROAD	SOUTH CERNEY	CIRENCESTER	GLOUCESTERSHIRE	
4.11	405266	196076	ECC WORKS	BROADWAY LANE	SOUTH CERNEY	CIRENCESTER	GLOUCESTERSHIRE	

Phase 1 Sites' facility was launched on the Internet at www.nlud.org.uk. Sites can be searched via local authority name and the service to date includes over 9,600 sites from 141 authorities. For example, the search on Cotswold District Council found the thirty-one sites listed in Table 6.1. Additional data are collected for each site and include:

- Site reference
- Land type and use
- Dereliction
- Owner details, tenure
- Market availability, agent, price
- Proposed use, planning status, constraints
- Services available
- Various land and property identifier codes
- Grid reference
- Housing capacity, density and suitability.

Figure 6.5 shows the location of some of these sites.

An update of the 1998 snapshot began in late 2000 and site polygons are due to be captured wherever possible. Annual updates of NLUD Phase 1 are planned in order to provide information for monitoring regeneration policies and the reuse of previously developed land for housing. Also, local

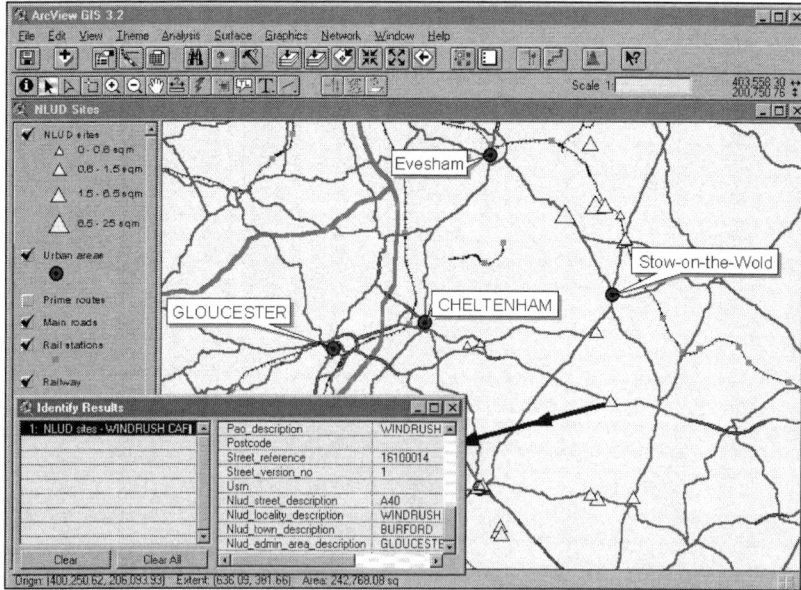

Figure 6.5 NLUD Phase 1 sites in the Cotswolds (Cotswold District Council).

authorities are required to undertake Urban Capacity Studies as part of the equation for calculating housing demand relative to land availability and NLUD will help here.

Regional planning

Strategic planning at the regional level is becoming increasingly important because of the significance the regions have in terms of European Union policy and grant aid. Lothian Regional Council is using a GIS to filter applications for school places and relate these to schools and school catchments. The council is also using GIS to calculate home-to-school travel routes and distances, analyse population, transport and other long-term trends and generate other statistics and forecasts. In another authority, a GIS is used by the housing department for the selection of target areas for urban renewal investment by filtering small areas (enumeration districts) to check for areas of multiple deprivation, certain census characteristics (including tenure), house prices, benefit levels and crime (Ferrari, 1999). These are very similar operations to the Scottish Homes GIS which is used to analyse waiting lists for social housing. The Housing Department calculates notional waiting times for given areas by assessing current letting rates. These are based on 140 Housing Allocation Districts across the city. These districts are not natural units for strategic analysis, management information or political interest – information is usually requested for local housing office areas or electoral wards. Prior to the introduction of GIS, Housing Allocation Districts were wholly attributed to analysis areas manually. Using the GIS, areal interpolation is undertaken and stock information is used to weight intersecting regions. Also, areas irrelevant to the population under scrutiny are excluded (such as industrial estates or parks) so the polygons are tailored to best describe features under scrutiny (Ferrari, 1999). GIS is also used to perform geographical calculations on lettings information, for example 80 per cent tenants moving to extra-care sheltered housing schemes originate from within a 2-kilometre radius. Here, as with other strategic planning applications, GIS is primarily used as a research tool rather than an operational management system (Ferrari, 1999).

Planning procedure

At the local level town and country planning usually involves two processes; the preparation of plans for the implementation of planning policy and development control to ensure that planning policy has been implemented according to the plan. But in a wider sense local authority planning includes a wide range of responsibilities, listed in Table 6.2. Another illustration of the range of functions that GIS is used to support in local government can be obtained by looking at the software developed by Innogistic GIS, a supplier of GIS solutions to local authorities, listed in Table 6.3.

Table 6.2 Local authority planning responsibilities

Service provision
Housing
Waste management
Education
Social services
Environmental protection (against pollution and contamination of air, land, water)
Risk/disaster management and emergency planning

Infrastructure
Transport planning
Development control (planning applications and building regulations)
Conservation

Economic development
Labour skills
Inward investment
Urban regeneration

Much of local government planning involves the collection, processing, analysis, presentation and use of land and property information and therefore local government planning departments have benefited particularly from GIS implementation. But other users have benefited too, including those functions that control and record how we use our environments: planners, certainly, who are concerned with land use and its governance; but also environmental health officers, housing officials, land charges clerks and highways engineers. All of these functions require that records of the use of land or buildings are collected, referenced, maintained, analysed and controlled and all can benefit from a system that allows this in an easy and organised fashion. GIS, with its versatility of applications and use, can provide such a system, and is increasingly doing so. How, then, has GIS been introduced to town planning, what has it been used for, and what have been the problems and pitfalls in its implementation? This section outlines, in a town-planning context, the nature of GIS technology, the policy framework and applications in practice.

The important uses of geographic information in planning and local government, as identified by the Department of the Environment (DoE) (1987), include:

- monitoring (of land, buildings, economic, social, demographic and environmental matters);
- forecasting changes (housing, schooling, travel, economic, community services);
- service planning (scale of location of need and provision);
- resource management (building maintenance, refuse, libraries, social services);

Table 6.3 Local government applications of GIS

Building control
The Building Control Management System includes applications processing (including full plan checking and consultation processing and building notices), dangerous structures, demolitions (including spatial generation of notices), unauthorised works (management and processing of unsatisfactory building works), initial notices and cavity fills.

Street Gazetteer and Land and Property Gazetteer (BS7666)
Street data can be imported from OSCAR and other street network data sets. Address data can be imported from ADDRESS-POINT or BS7666 CSV from Intelligent Addressing and mapping allows BLPUs and LPIs to be created, displayed and managed. Text and map-based search facilities are available.

Development control
The Development Control Management System includes development control (calculation and tracking of statutory deadlines and committee dates, spatial-based neighbour notification and constraint checking), appeals (processing from receipt to decision, questionnaire generation), enforcement and complaint handling, agenda creation for planning application records.

Grounds maintenance
The Grounds Maintenance Management System includes task scheduling, financial management, bill of quantities, contract control and penalty monitoring and valuation orders for the maintenance of trees, grassed areas, gardens, hedges, sports facilities, cemeteries, rights of way, council-owned open spaces, etc.

Land charges
The Land Charges Management System includes searches (processing of personal, LLC1 and CON29 searches and configuration of LLC1 and CON29 reports), departmental answers to CON29 Parts I, II and III, and register (GIS polygon selection of properties when adding charges to multiple addresses, spatial and textual land charges search). Full integration with BS7666 LPG ensures a corporate approach to property information management.

- transport network management (highway provision and maintenance, public transport, school transport, cleaning);
- public protection and security services (police operations, fire/incident logs);
- property development and investment (preparation of development plans, assessing land potential, property registers, industrial and rural resource management);
- education (data for teaching purposes);

Add to this the ability to produce high-quality hard copy reports and maps, and the ease of information retrieval, it is not surprising that GIS is considered a useful tool by local planning authorities. Ideally, most planning authorities would like to use GIS to perform, or at least ease the performance of, most of their functions. These may be divided into the holding of data; the output of data; the processing of data; and the analysis of data. Each of these functions is considered in turn together with the possible ways in which a GIS could assist with their discharge.

Holding   Local planning authorities hold vast amounts of data. The planning system as we currently understand it has operated since 1947, and over that time, vast numbers of planning applications have been submitted and decided; structure plans and local plans have been produced; and reports have been written. A great many of these are, of course, out of date, but it is a bold planning authority that throws away any of its data! Legal cases relating to decisions made in the dim and distant past are rare, but they do occur. The history of the use of land is very germane to the carrying out of the planning function. For example, although past planning decisions on a plot of land should not determine the decision of a current planning application on that same plot of land, they, and the reasons for them, would be of interest to the officer making the current recommendation.

So there is a well-recognised need in town planning, as in many other local authority functions, to maintain records; usually over long periods of time. However, there is obviously a price to pay. Keeping records requires space. It also requires some standards of custodianship as records tend to deteriorate over time. Planning authorities have tended to solve the first problem by using technologies such as microfiche, which involves taking photographs of documents and storing these photographs on tiny strips of film which may be viewed in black and white and (usually) as negatives on a special viewer. This certainly saves space, but it is not ideal. The most immediate problem is that it is rather awkward; the strips of film, or fiche, are difficult to handle, easy to lose and can only be viewed with special equipment. However, there is a more serious problem which microfiches present to the planning function; it is that the 'fiched' images are difficult to interpret. Scale diagrams, for example, are obviously no longer at an easily identifiable scale when photographed and reduced; and scale diagrams are vitally important to town planning, for example, to ascertain whether a development that has been constructed is in line with what was approved. Fiche images are in black and white, which can present some difficulties where, as any planner will tell you, a planning application map must indicate the site of the proposed development outlined in red and any other land owned by the applicant in blue!

A further problem, of deterioration of data sources over time, can also be solved by techniques such as microfiche, which do not deteriorate like paper maps or letters. Another common solution in planning departments is to hold maps on acetates, which are then hung vertically in steel cabinets; this alleviates the necessity to extract, unfold, refold and file a paper map.

But even polyester film can deteriorate; particularly if they are used as 'plotting sheets' to show all the planning applications in a district over a long period of time.

It is not surprising, perhaps, that planning authorities have been eager to utilise the information holding capabilities of GIS. A GIS can hold enormous amounts of data digitally in tiny spaces; furthermore, these data should be available at the touch of a button, and of course, it does not deteriorate. It can also provide rather better representations of maps and drawings than microfiche, and the incorporation of scale and colour do not present problems. Clearly, there are implications as to whether a GIS should hold planning data 'from this point forward', or an authority should input its historic data into the system, and a decision based on resources and the utility of increasingly historic data would need to be made. But even if the former course of action is taken, an authority can be confident that its stores of manual data are not going to be (as significantly) added to.

Output   GIS have undoubtedly improved the quality of map-based data in planning departments. Both updating and access can be greatly simplified by GIS. Appropriately detailed, up-to-date maps can be produced quickly where conventional paper map sheets would have been considered dispensable or too cumbersome to provide in the past, and by far the most common application of GIS in planning is the production of good quality hard copy maps. It may seem rather strange that a computerised mapping system is used to produce paper maps but it must be remembered that a lot of planning business is still paper based. Hard copy maps can be used in a number of contexts: from the very simple purpose of finding one's way to a development site to the plotting of a layout on an up-to-date OS map. Policy work demands the production of thematic maps and reference maps showing, for example, the extent of conservation areas, agricultural land quality or population density. These outputs can find their way into reports, policy documents and statutory documents such as the local plan.

It may also seem a little odd to use the power of a GIS for such an apparently rudimentary application, but consider for a moment the process by which one of these examples of hard copy would be provided under the manual systems that the GIS has replaced. A paper map would need to be obtained and possibly photocopied if the original was not to be used for annotation. A technician would need to transcribe site details onto the map using pens, letter and shading transfers and, possibly, colour wash. The finished product would need to be sealed to prevent the fading of inks or prints and large format photocopiers would need to be used to provide multiple copies. Compare this with the accessing of a map file on screen, the on-screen digitising of a polygon and the printing of multiple copies on an inkjet or laser printer, and it is clearly apparent that the GIS has introduced a much easier, more efficient and less error-prone process.

Processing   Many planning authorities will, however, be looking to GIS to do more than produce good quality maps. Much of the day-to-day job of town planning involves the processing of information: applications are received, registered and acknowledged, consultations carried out, responses received and processed, advertisements placed, decisions issued. A lot of this work is routine and is able to be carried out quite easily by a computer system. Planning application processing includes assisting development control staff in identifying all of the factors affecting a particular building or site. On receipt of a planning application, site reports are generated from one council's system, giving details of local plan policies, constraints, hazards and previous planning applications. This is passed to the case officer to assist the decision-making process. Planning application processing systems can hold relevant details and issue standard documents (such as acknowledgement letters). Planning authorities have implemented such systems for many years, but the advent of GIS gives new potential to the organisation of such information geographically and the interface of planning application processing with map-based interrogation and retrieval.

Geographical Information System can, and indeed probably should, be central to the handling of large and complex data sets, which may grow by many thousands of planning applications in a normal year. Potentially, a GIS can provide a seamless link between the plotting of a development site on a map to the retrieval, processing and output of all the information and correct formats for the carrying out of the planning function. However, to do this, it needs data; comprehensive and reliable data that may extend beyond the pure planning function. The complexity and variety of data sets typically used is illustrated by the itemisation of data volumes used for Birmingham City Council's operations in Table 6.4.

Clearly, management of these large and growing data sets is essential. It may be that some of them are contained in other sources which can be

Table 6.4 Volume of data held and used by Birmingham City Council

Volume	Element
1,500	1 : 1,250 OS map sheets
2,000	Census enumeration districts
1,25,000	Council tenants
2,000	Property transactions p.a.
500	Schools and colleges
2,15,000	Pupils and students
75,000	Social services referrals p.a.
5,000	Planning applications per year
11,000	Building regulations applications p.a.
2,100	Kilometres of adopted roads
150 m	Capital schemes p.a.
2,000+	Sites available at any one time
4,65,000+	Rateable hereditaments

integrated or interfaced with a GIS; it may also be that some of them are held manually and need to be captured, or alternatively run in parallel as a manual supporting system. In many circumstances, it is necessary to capture the data needed for a GIS implementation from manual sources or other computer databases. GIS implementation should therefore be accompanied by strategies for information gathering and management. Other readily available data sets are generated at national level by central government and its agencies. They are often important locally as base data or comparative data.

Ultimately, a GIS should be a support for good decision-making and, if emphasis is placed on the decision support role of GIS, there are implications for the approach to information management not just within a planning department, but also throughout the local authority. The ideal corporate information strategy would make sure an authority's data are:

- held once only;
- frequently updated;
- captured efficiently and at the best source;
- immediately available to any user in an appropriate format.

Under such a corporate information strategy, transaction processing systems, such as planning applications, would automatically contribute data towards the information system. The corporate use of a single map reference, based on OS digital maps, helps to reduce errors, redundancy and duplication. The adoption of a common set of digital maps linked by a unique land and property referencing strategy will help reduce the number of disparate maps and overlays in town planning, highways and technical services departments. It should also be noted that other departments, such as Education and Social Services, may be less dependent on the use of maps but are also important users and providers of geographically referenced information such as demographic data. GIS has the greatest potential under a corporate information strategy where such disparate data sets are linked together.

Analysis   We have referred to GIS as a holder, producer and processor of information. It is also a research tool, in that a great deal of geographical analysis can be carried out once data are geo-referenced. A GIS can allow the statistical analysis of maps to provide new policy information. For example, a map showing contour data can be overlaid with one showing rainfall data to produce a new map showing areas liable to flooding and thus unsuitable for housing development; an important input to the land allocations made in the local plan. The resultant map could be overlaid on another map showing existing land use to project the possible effects of flooding on the built environment. In this example, a new set of information has evolved from a number of disparate sets through a process that would have taken many hours using manual methods.

As in all GIS applications, the success of such a piece of analysis depends on the system having the correct data to begin with. However, there is evidence to suggest (Allinson and Weston, 1999) that these sorts of policy applications are rather more common in planning departments than the management of large sets of planning application data referred to above. The reason for this is almost certainly because it is reasonably straightforward to assemble the necessary data sets. Planners can carry out all sorts of analysis and relatively sophisticated modelling, using land use information, site area and environmental data. They can produce sieve maps overlaying such information as constraints (slope, drainage, land quality) to identify land with development potential; the least-cost or least-distance route between two points (either for travel planning or installing infrastructure); and site development histories, with attendant house type and price details.

The implementation of GIS in local government planning departments

In 2000, the Royal Town Planning Institute (RTPI) undertook a GIS survey to establish the level of GIS implementation in British planning authorities and the range of applications GIS is used to support. The report of the survey was published in 2001 (RTPI, 2000). Essentially, this was an update of a similar survey carried out five years previously (Allinson and Weston, 1999), and it allowed comparison of GIS implementation and practice over time. The 2000 survey reported that the key data sets that have been captured are:

- Electoral boundaries
- Listed buildings
- Planning policies
- Scheduled ancient monuments
- Tree preservation orders.

Some 56 per cent of authorities have either completed, or are engaged in, the capture of Land and Property directories, although more than half is not compliant with the data standard BS7666 described in Chapter 4. This is perhaps surprising, as the standard has been in place for more than five years, showing the time lag between introduction and widespread adoption of the standard. Notably 19 per cent of authorities have no plans to create a local land and property gazetteer, as defined in Chapter 4.

The 1995 survey showed that the principal use being made of GIS was the production of good quality maps (reported by 34 per cent of respondents) with social analysis (23 per cent), planning constraints and thematic mapping (22 per cent) slightly behind. Map production is also the most significant application in the 2000 survey, but the proportion had risen significantly. Either linked to other applications or stand-alone, 83 per cent of authorities have this facility. Only 3 per cent of authorities have no plans to introduce this application. The next most frequently cited applications are planning application systems, where 68 per cent have a system and another 26 per cent are in development or planned. Only 4 per cent of authorities have no plans to introduce an application processing system. This is in marked contrast to the situation in 1995, when only 7 per cent of authorities named this as a current application and 9 per cent had plans to introduce it. Development plan policies and proposals maps are similarly well developed, with 40 per cent of authorities having a fully operational system and a further 54 per cent having an application in development or planned. The corresponding figures for 1995 were 7 and 10 per cent, respectively. These findings would suggest that, although there is still some way to go, GIS is increasing and broadening its application base amongst local authority planning functions.

In 1995, GIS use in local planning authorities was rather patchy, but by 2000 GIS had become fully embedded as an IT application. The nature of GIS is evolving with the growth of networked computing and an increase in the number of seats. However, the evolution is far from complete with 75 per cent of authorities having plans for further development. This is hardly surprising given the rather open-ended nature of the systems. Although data capture for most planning related entities has been undertaken by many authorities, there is still a significant number where substantial data capture needs to be undertaken. In terms of applications map production is the most universal, followed by planning applications. Development Plan and Local Land and Property Gazetteers are still relatively moderate in numbers but all have increased significantly over the past five years.

Local government GIS in practice

In 1998, the RTPI carried out a series of case studies, which highlighted some of the aspects of GIS implementation and GIS in practice. The studies considered what lessons can be learned from the experience of a number of planning authorities which were among the first to acquire, implement and develop GIS. It focused on GIS implementation in the planning departments of four different local authorities: Brent, Swansea, Wakefield and Gordon. Worthy of particular interest were the following enquiries:

(a) the catalyst for the initial decision to adopt GIS and in what areas of the different authorities the systems were first applied;
(b) the data that are used by the various GIS and how the authorities input or accessed these data;
(c) the applications that were developed in the authorities, in the planning departments and elsewhere, and the benefits that were realised;
(d) the authorities' plans for future development of the systems;
(e) the problems that were encountered during the implementation, how the authorities coped with these, and what recommendations they would make to others treading the same path.

SOFTWARE

In the main, the councils seemed to favour 'mainstream' packages. The days of proprietary systems developed specifically for a council by consultants or its in-house staff seem to be over as councils seek standardisation, support and the prospect of future (compatible) upgrades from industry-standard providers. This could be seen to confirm one of the findings of the RTPI's GIS surveys reported earlier, of a slow but steady process of standardisation. Fortunately, these standard packages also appeared to offer the ability to link with a range of other databases that were in existence at the authorities.

DATA

The experience of data preparation, processing and incorporation seemed to vary between those councils that claimed to be 'ahead of the game', having taken early decisions to develop property databases or digital map coverage which were then 'grafted onto' a GIS: and those that were grappling with large-scale data capture exercises to achieve full-functional GIS implementations. A modular approach seemed to be the most successful, where the authority established a suite of databases covering its major information sources (e.g. listed buildings, tree preservation orders) and 'bolted' these on to the GIS in sequence as data was captured. This could be very rapid, with one council reporting that new simple systems could be added to its GIS within two person-weeks from initial analysis to pilot for testing. Both in-house and external data capture approaches were used, and challenges included the need for a strong contractual relationship with data capture companies to ensure data was captured to proper standards, and the poor quality of some source documentation; for example, deteriorating paper maps.

APPLICATIONS

Applications fit neatly into the above typology of holding, output, processing and analysis. On the holding front, the addition of zonal information to land and properties can be quite straightforward once databases are established: it is simply a matter of digitising a polygon; the relevant information will then be added to the records of all sites and properties within the polygon. One council used its GIS as essentially a set of 'digital plotting sheets', which have completely replaced paper sheets. On the output front, GIS allows the provision of: good quality up-to-date maps and site plans; provision of planning histories on-screen for interrogation by officers or by members of the public at the reception desk; and the display of local land charges through their attachment to address polygons, which has had the effect of dramatically reducing search turnaround times.

In addition to the RTPI case studies, further examples of how GIS is being used within local authorities can be found. These 'success stories' of

GIS implementation in local government tend to originate from those authorities that have had significant experience of GIS and now host some mature GIS applications. It is certainly not the case that every GIS implementation has been a success in the local government sector – political, budgetary, staff and bureaucratic constraints can and often do serve to frustrate and sometimes prevent successful GIS implementation. Therefore, after considering some examples of successful GIS implementations in local government in the remainder of this section, the following section offers some lessons that have been learnt.

Powys County Council's GIS provides a range of services from strategic planning and regeneration to the management of street furniture and holes in the roads (www.ordnancesurvey.gov.uk). The GIS contains information from a variety of sources, such as the Post Office, the Environment Agency in Wales and utility companies, as well as the council's own data sets. Underpinning these layers of information is OS mapping. Each of the council's departments can attach their own information to the mapping. In the planning department the GIS is used to record the outlines of planning applications and to find out whether there are any restrictions on the development of a particular site such as rights of way, sites of special scientific interest or listed buildings. It can also be used to assess the impact of proposed tall buildings, wind generators or incinerators on visibility and amenity. In addition, statistics such as unemployment and population figures can be mapped and analysed for use in local development plans. Other GIS applications within the council include keeping records of the condition of council property. It is also being used in emergency planning by mapping resources, hazardous installations such as chemical and explosive stores and main pipelines, and even to help monitor areas affected by earth tremors.

West Oxfordshire District Council justified GIS implementation on the basis of digital mapping alone and GIS applications were initially developed in areas where the contribution was immediate, such as the land terrier and housing databases. The council has compiled a central property register that integrates with GIS and various planning systems giving, for example, access to seven years of planning applications through GIS maps. It is possible to undertake searches of constraints as well as conditions relating to previous applications in the vicinity. This reduces the processing time for applications and raises the consistency in decisions (Peel, 1995). The holding of geographically referenced data has facilitated all sorts of analysis, such as the distances travelled by pupils to their local schools, and the numbers of households remote from access to basic facilities. The areas of residential land availability and thematic mapping from the Census have been explored and have been useful aids to such processes as review of local plans and bids for European assistance.

At Horsham District Council a corporate GIS strategy has been implemented but initial application development concentrated on planning application processing. Planning constraint data totalling some 28,000 polygons in sixty layers were digitised, representing sewer lines, gas lines, tree preservation orders, planning zones, footpaths and 140,000 planning history records. These data overlaid 700 OS large-scale maps and the process took three years (Wagner, 2000). The time taken to process approximately 2,600 planning applications each year has been reduced from three days to fifteen minutes for each application. The procedure is as follows: a planning officer digitises the subject property, the system then allocates a reference number and runs a constraint search and identifies consultations that will be required. It then produces a plot on an OS map indicating the constraints. Nearby owners are also plotted and notification letters automatically generated. After the implementation of the planning application service, the council developed a corporate property database and now other GIS applications that have been developed include an automated land terrier, grounds maintenance and land charges.

At the London Borough of Richmond, a GIS contains large-scale map data, planning histories and related information. It is possible to determine whether a site is within a conservation area, what listed buildings it may contain and which ward it is in. Details of every planning policy area affecting the site can be attached. Specific addresses can be sourced for public consultation and there is a link to aerial photographs, scanned documents and details of tree preservation orders and street lighting.

The Kingston London Borough Council has created a web site that provides public access to planning history dating back to 1947 and building control history dating back to the 1880s for each land parcel in the borough. The information can be searched and displayed using OS large-scale mapping as a backdrop over which the extent of selected land parcels are displayed. Information about nearest council facilities such as schools, leisure centres and recycling centres is also available together with maps showing the location of parking restrictions, listed buildings and aerial photographs.

Bristol City Council has introduced a corporate land policy to ensure a consistent approach to land management. The fundamental review of property management began with the identification of property interests. GIS-based computerisation of the local land charges service was the first step, the second was the computerisation of legal interests in property (Musgrave and Flack, 2000). There were several problems to overcome: separate land terriers for Bristol City and Avon County Council, tens of thousands of deed packets (in one case 450 for one site), four miles of shelving in the archives of documents relating to land acquisition and disposal. Consequently the council commissioned the Land Registry to register council-owned property so that clear title could be established and obsolete encumbrances removed. The registration process began with priority areas such as the main shopping area of Broadmead which has an historical legacy of acquisition, disposal and re-acquisition over many decades. Registration of this area was regarded as a prerequisite for regeneration and redevelopment. Five hundred and twenty deed packets comprising five to six thousand documents will be reduced to one land certificate (Musgrave and Flack, 2000). Other benefits of registration included the control and limitation of encroachment, adverse possession and ransom strips, increased speed of conveyancing (which assists grant aid and regeneration and brings forward capital receipts), identification of surplus assets and the widening of access to records. The GIS-based local land charges system at Bristol City Council has halved the processing time of the annual 13,000 searches to less than five days without increasing costs. Thirty data sets have been captured including the land charges index, planning applications, listed buildings, conservation areas and highway schemes. The land charges system maintains an up-to-date register, automates searches (including the creation of registration schedules and search certificates), satisfies CON29 queries and includes plots that show the search extent. Access to the BS7666 compliant LLPG and other data sets is possible via the intranet to other departments. For example, the leisure services department has installed a grounds maintenance system.

A GIS was installed at Swansea City Council in 1992 and is used for development control and forward planning (Weston, 1995). In the development control department GIS is used to produce reports on the specific development sites in the early stages of the planning application. The development control officer receives the report covering local plan policies, constraints and current applications from a GIS search of the area around the site. The Estates Department has transferred the land terrier onto the GIS to include an asset register for all council-owned property in Swansea. The Leisure Services Department has also mapped grounds maintenance data. The Engineers Department holds sewerage network data digitally and is extending its use of GIS to plot the locations of lampposts, bollards and other street furniture. Census data at the Enumeration District level has been transferred to the GIS and a digitised version of the local plan was also added. Including policy proposals at an early stage speeds up the review of the local plan. For example, proposed housing allocations were added in the review of the local plan. This involved taking existing housing allocations, identifying those which had been developed and adding new allocations by automatically selecting current applications for housing on another GIS layer. These selections were combined to produce a final housing-allocation plan for the review.

At Rugby Borough Council GIS is used to support highways and traffic land charge searches. Proposed highway improvements are mapped so that a layer holds the proposed improvements and these can be viewed to enable answering of highway related questions on land charge searches, for example, if a property falls within an improvement scheme. The benefits resulting from the application of GIS to highways improvement are speed and accuracy and information is available to other sections without the need to arrange an office visit. It is estimated that the use of GIS to assist this one function leads to a saving of more that £2,000 annually in the highways department alone.

The East Lindsey District Council covers more than 1,800 square kilometres in Lincolnshire. Geographically, it is England's third largest council. The local authority has a GIS containing records on around 65,000 properties in 196 parishes. It is based on OS large-scale mapping on which a database of planning histories and details on building control and ownership are recorded. Online access is available to around 100 users throughout the council departments, from planning and estate management to grounds maintenance.

GIS use at Norfolk County Council started in the Department of Planning and Transportation as a graphics facility, mapping and highways data. A data audit in the council unearthed a substantial number of data sets and several versions of some of them – there were twenty versions of the parish register! Standardisation of spatial references and data sharing across departments were thus seen as primary objectives. As an example, the streetlights database had double the number of lights because existing ones and their replacements had been included. It was only when they were mapped that this was realised and thus reduced the energy bill which is calculated on the number of lights!

Lessons learnt

All councils point to the provision of better quality, more up-to-date, more consistent and more accessible information as the main benefits of their GIS but no GIS implementation is without its problems and, in general, they can be categorised as follows:

- Organisational problems; or the management of change in the way officers and members use, access, hold, and deal with data, which has threatened some entrenched positions. One council recommends the formation of a user group, which should meet regularly, to assist this process.
- Hardware problems; such as the lack of sufficient memory on GIS network servers (GIS data files are often extremely large), and the lack of large plotters to obtain hard copies of maps and map-based analyses.
- Software problems, such as the limited number of user licences for the GIS.
- Perhaps most importantly, data problems. One council described its GIS implementation as 'traumatic', primarily due to the large data capture exercise that was undertaken over a relatively short time period. Inaccuracies in manual sources were computerised and then needed to be corrected; another council warned that the amount of time allocated to data capture was far in excess of expectations, and that this should never be underestimated.

- There is also an implication for continuing maintenance of the data, which can mean diverting staff from other tasks, and bring along a requirement for staff development and training.

It was recommended by more than one council that expectations of benefits from a GIS are contained; systems have high start-up costs, and these are only paid back gradually, and sometimes in a non-quantifiable way. Convincing users of the value of GIS may involve a 'culture change', which is seldom a 'road to Damascus' conversion but is instead based on sound procedures and programmes which gain widespread agreement. These issues will be discussed in more detail in Part III of this book.

Grimshaw (2000) argues that, in urban planning, important decisions have to be made on what are the best locations for facilities given variations in local demand. It is the support to such decision-making that GIS can be of value. Several local authorities are using GIS to help identify target groups for their services and this represents a move from the application of GIS to administrative tasks to more strategic use of GIS. Grimshaw (2000) also points to evidence that as experience with GIS grows then so does the level of corporate interdependence. Often the departmental implementation of GIS hinders corporate implementation because it is fragmented, is often technology-led and is purchased for the job in hand rather than a wider corporate information strategy.

This overview of the use of GIS in local government planning has shown the potential of such technology to improve many aspects of geographical data handling. If there is a general conclusion from this review of local planning authorities' experience of GIS implementation, it is, first, that it pays to be realistic in terms of what is involved in successful implementation and what a GIS can reasonably achieve, when and at what cost. Even for a small sub-area of a council's activities, implementation of a GIS involves a vast exercise of data capture, checking, correction and updating. This is not only an 'up-front' commitment, it is also an ongoing investment by the authority. The second point is that the key to successful implementation of a GIS appears to be agreement to methods and applications by both senior and user staff. This must be painstakingly achieved through the use of information strategies, work programmes and good project management. There is a further need to attend to human relationships and organisational culture, and to have data and systems well linked to business priorities. In short, a GIS is just a tool to do a job and councils must take care to define the job properly.

Property development

Some of the earliest examples of GIS applications were developed to assist the identification of underground oil reserves. Preliminary desktop studies typically involved a 'sieve' analysis of paper maps of geology, land use and ownership, transport infrastructure, protected areas and populated centres.

Areas of potential interest were traced onto each map and these were over-laid to see where they overlapped. The method is time-consuming and only a handful of sites could be considered at any one time. Using a GIS, geo-logical data were stored on a computer in a co-ordinate system. Other geo-graphical data that were associated with the existence of oil would be referenced to the same co-ordinate system. By 'overlaying' these data it was possible to identify potential areas where oil would have a measured prob-ability of being found. This type of sieve analysis is familiar to those involved with the early stages of the development process. This section describes examples of GIS applications that focus on pre-construction investigations that often need to be made, rather than the construction process itself. The following application areas are considered:

- Site appraisal
- Flood risk assessment
- Contaminated land assessment
- Geology analysis.

Site appraisal

Highly detailed, accurate mapping plays a crucial role in the planning process by helping to identify suitable sites for development. Planners, architects and environmental scientists from the planning consultancy, Terence O'Rourke, use OS mapping to show the layout of the local coun-tryside and towns using colour to mark different features of the landscape, such as woodland and areas of water (www.ordnancesurvey.gov.uk). A GIS processes and simplifies the large-scale mapping and integrates it with other OS data such as the national road network and height contours. One way that this detailed mapping assists the consultants is to help them find a spe-cific location, but it also has a wide range of more specialised uses. It is used in the first stage of project work before moving on to more detailed types of OS mapping. The detail means that specific features, such as the outlines of individual buildings or the layout of agricultural fields, can be easily spotted. This indicates to planners what the land surrounding a potential development site is being used for. Consultants carrying out site visits also make notes and observations on to their printouts. Features, such as vege-tation, height information, water features, landforms and existing build-ings, are also shown using colour. Charles Planning Associates use similar mapping to identify potential development sites in the southeast of England. The design of the maps allows the town planners to compare and contrast different types of land to identify potential areas for residential development.

There are many stages to registering land title for new development and they involve numerous organisations and manual procedures which are slow and costly. Wakefield Council undertook an eighteen-month trial of a

Pre-Build Information Service (Ford, 1995). The service allows a developer to deposit original and updated plans with the pre-build service in paper or digital form. The service then digitises or edits the plans to a standard specification and format and these digital data are incorporated into revisions of large-scale 1 : 1,250 or 1 : 2,500 OS maps. The title boundary from the Land Registry is digitally overlaid and anomalies between the site boundary, OS map detail and the legal title boundary are highlighted and resolved, by site inspection if necessary. The service then distributes copies of the boundary and layout details to appropriate organisations in digital or paper form. Currently the OS use the development plans from several large housing developers and infrastructure projects as change intelligence for the purposes of revision of large-scale mapping. The OS has an agreement with the large housing developers for digital provision of their development plans and these are held as a suppressed layer in the National Topographic Database (described in Chapter 4) until the data has been verified.

English Partnerships have produced a plan of the mixed use development envisaged for the Greenwich peninsula and GIS is used to integrate the vast quantity of geographical data including:

- Ground investigation records
- New and existing services
- Hard and soft landscape materials
- Infrastructure
- Land ownership
- Surface levels for 3D modelling
- Remaining underground structures
- Ecology features and development features such as the Dome.

Also, the Canary Wharf Group has purchased satellite imagery to record construction and development of the area until completion.

But GIS can also be used for more sophisticated development appraisal. It is possible to generate 'what if' scenarios in a geographical context for predictive analysis. For example, GIS can be used for a site suitability study – to identify locations suitable for a proposed development by analysing the interactions and spatial coincidence of development objectives as well as environmental, demographic, economic, and political factors. In this way, GIS development applications typically involve sieve analysis and map overlay. Tomlin and Johnston (1990) undertook a land use allocation project (sieve analysis) using GIS for the siting of a research and development facility in the US. The process outlined in Table 6.5 is typical of development sieve analysis. For example, also in the US, GIS has been used to determine the optimum location for rural suburban residential

Table 6.5 Development sieve analysis
(Tomlin and Johnston, 1990)

Process
Define problem
Inventory of site
Establish land use siting criteria
Refine criteria
Develop database
Create suitability maps
Allocate land uses
Analyse land use/land use relationships
Adjust suitability weightings
Create schematic designs
Refine designs

communities (Barnett and Okoruwa, 1993). Sieve analysis was used to sift following constraints:

- Area with low to moderate traffic (extract 2-lane highways and buffered 1,200 metres and buffer entire road network to 250 metres)
- No zoning restrictions (extract suitable land use zones)
- Gently sloping terrain (extract suitable soil conditions)
- Near small stream (buffer 1,000 metres small streams)
- No railroad noise (buffer 1,000 metres)
- Access to major employment, retail, leisure and schools (distance analysis).

Geographical Information System can also be used at other stages of the development process. For example, developers of commercial property can analyse lease information, expiry dates and rental values of properties that they own within a particular geographical area and then project potential client needs for expansion. A GIS could display the location of properties with upcoming rental reviews. Shopping centre developers are using GIS to identify trends, compare locations, project sales and estimate population growth to help make decisions on size and design of a retail development and the optimum mix of the stores. These sorts of applications will be discussed in more detail in Chapter 7.

Flood risk assessment

Using a GIS it is now possible to predict, at postcode level, which areas around rivers and estuaries will be affected when they flood. This information is invaluable for insurance companies, construction and transport industries. Willis Risk Management Consultants have produced a flood risk assessment system. Risk assessment may be property by property for rating purposes or the total risk for a portfolio of properties. A GIS uses a high resolution Digital Elevation Model (DEM) with orthorectified images able to distinguish land use at a 2.5-metre resolution. Historical flood data for the River Thames were combined with engineering information on structures such as weirs and bridges. The GIS was used to project the DEM onto the National Grid and delineate the area between the river level and the 10-metre contour. Flood heights for fifty-five key points along the river were added for periods ranging from 50–1,000 years. An approximate flood surface was then generated between the points. The flood risk assessment system is now in use and covers tidal and non-tidal return events of up to 1,000 years. It is possible to examine building development proposals and rate individual properties using their postcodes as a locator. The system can also help developers and planners consider the risks of building on flood plains.

Contaminated land

Landmark Information Group (www.landmark-information.co.uk) is a supplier of land use information, digital mapping, property and environmental risk information and has created what is claimed to be the largest GIS in Europe, with information from a wide range of sources, including the Environment Agency, OS, the Scottish Environment Protection Agency, British Geological Survey, English Nature and the Centre for Ecology & Hydrology. This information is used in conjunction with over 600,000 historical OS maps, dating back to 1850 (the year), which show some 400,000 industrial developments that may have left contaminative residues, such as old gas works, tanneries or foundries, and a further 275,000 areas of unknown fill. Examples of infilled land include old quarries, gravel works, canals, ponds and old streams. Data on landfills, pollution incidents, water abstraction points, sites of special scientific interest, etc. are geographically referenced so that contamination reports locate environmental features on both current and historical OS maps. The entire collection of historical OS paper maps at both 1:10,560/1:10,000 scale and County Series 1:2,500 scales as well as 1:1,250 and 1:500 town plans have been scanned and geo-referenced so that there are five complete maps of Britain, each representing a thirty-year epoch. Land use features can be identified, recorded and accurately placed on each successive map. This allows 'time layered mapping' to be undertaken. For example, an old gas works or landfill site that has disappeared from view can be re-projected on a modern map. The planning of a new housing estate will benefit from knowledge of when and where previous land uses occurred both on the site itself and in the surrounding area. Landmark provide a range of services which include:

(a) A site-specific report for environmental and geo-technical consultants, environmental managers and engineers undertaking site assessments and desk studies to determine environmental risks and liabilities. The report comprises a site-centred map covering a radius of one kilometre overlaid with environmental data. Small-scale maps on the environmental setting, ground water vulnerability and river quality are also provided. Site-centred historic maps are provided to illustrate historical features located accurately on current map data.

(b) A site-specific environmental risk report designed to help homebuyers find out if the property they are buying is built on contaminated land or affected by environmental risk. The report covers a radius up to 500 metres and provides information that is not covered in the local authority search. This information helps identify registered landfill sites, waste transfer, treatment or disposal sites, current industrial polluting processes and prosecutions and premises licensed to keep radioactive materials and other hazardous substances. Additional information is provided in the report on overhead transmission lines and pylons, the risk of flooding and subsidence and whether or not the property is in a coal mining area or area affected by radon gas.

(c) An environmental due-diligence report for property professionals such as surveyors, lawyers, property developers and landowners. The site-specific report includes an analysis of a specific location and its vicinity for potential environmental risks and liabilities. It incorporates a site summary and risk categorisation, a site inspection visit and visit to the Local Authority planning office, an environmentally sensitive land use map, current and historical land use information and large-scale historical mapping. The report lists trade directory information, registered landfill sites, waste treatment and disposal sites, premises licensed to keep radioactive materials, statutory authorisations and other current land use information on the site and up to a radius of 250 metres.

(d) The digital mapping system Promap has been combined with environmental data from Landmark, enabling property professionals to access environmental reports via a map interface for individual sites in Britain. Promap, provides access to large and small scale OS mapping and provides the ability to drill through various map scales from an overview of Britain down to a detailed street map. Features include map customisation, full colour display, export to GIS, CAD and other digital mapping software and printing. The map data is updated at six-monthly intervals, with monthly environmental data updates. Site selection can be made by address, postcode, national grid reference, land parcel number, OS map tile or place name.

GIS is also being used to help forecast and manage current and future risk of contamination of land and property. The overseas case study below provides an example.

Overseas case study

There is increasing concern surrounding the manufacture, storage, transportation and disposal of hazardous materials (McMaster, 1988). Therefore, anticipatory hazard management is suggested as a means of anticipating an area's hazard distribution and what might be done to prevent a serious accident. A pilot project was undertaken in the US to identify hazards present in a community and define a hazard zone associated with each. The population distribution and demographics can then be superimposed. A grid-based GIS was used to build a spatial model for risk assessment, focusing on airborne toxic releases. Nathaniel and Nathaniel (1994) undertook a similar study but focused on methane hazard. The study involved the development of spatial risk assessment procedures to be used during site studies. Simulations of the probability of excessive gas concentrations were undertaken and summarised using GIS and combined with soil information to quantify the hazard. Risk to nearby housing was assessed by comparing hazard distribution with the location of houses.

Geology

British Geological Survey has developed a GIS application called 'Address-Linked Geological Inventory' which is used for interrogating digital geological maps. This GIS aids the identification of difficult ground conditions, old mines and shafts, landfill sites, caves quarries, areas prone to flooding, subsidence and land-slip and radon emission. An address-based search can be performed using OS Address-Point for addressable properties or grid references for non-addressable sites. The system interrogates multiple layers of digital map data for any specified address or grid-referenced site. Automatically generated reports then summarise a range of geological information and give an indication of ground conditions. The reports come in two forms: a Ground Conditions Report for prospective purchasers, which includes an inventory of geology around the property / site and describes artificial deposits (man-made), mass movement deposits (land slips), superficial deposits (fairly recent in geological terms!), deposits such as river sediments and solid geology (bedrock). It also provides information on the composition of deposits where known, a radon potential report and other considerations such as known mines, areas at risk from non-mine subsidence, areas at risk from slope instability, flood, etc. The second report format is a Geological Report which includes the above information plus a colour geological map extract.

Heywood *et al.* (1998) describe how the Nuclear Industry Radioactive Waste Executive, which has responsibility for finding radioactive waste disposal sites, use GIS to undertake sieve analysis of potential sites. Maps showing geology, land use, land ownership, protected areas and population densities stored as thematic layers in a GIS. These layers are then overlaid and selection criteria input (suitable geology, specific distance from main roads, outside protected areas and areas of high population density). These criteria can then be altered and the selection process repeated as required.

Urban design, 3D modelling and interaction

'There are now many techniques which can be used to construct two-dimensional (2D) and three-dimensional (3D) models of cities and their characteristics, which are useful for various types of visualisation. Until quite recently the main approach was based on rendering wire frame models of buildings based on Computer Aided Architectural Design. However, with the widespread development of GIS, there has been a move from the 2D map to the 3D block model which is based on extruding building plots, street line data, and basic topography' (Hudson-Smith and Evans, 2001). As we have seen, GIS applications in planning are widespread. Planning is a function that involves public and private sectors, business and home alike. With increased use of the Internet, planning needs to be more accessible and understandable to offer a credible and socially inclusive decision-making environment.

'The emergence of affordable virtual reality and Internet GIS is providing the fundamental infrastructure to begin building virtual cities which can provide an interactive simulation and analysis environment for planning real urban places. The virtual city on the Internet will provide planners with a computer environment to interface with the myriad of complex physical and social data needed to plan and manage cities, along with necessary tools to explore and analyse that data in meaningful and intuitive ways' (Dodge *et al.*, 1998). The authors argue that the planning community is driving the development of GIS as a tool to widen and deepen access to spatial information together with appropriate functionality delivered via the Internet.

Figure 6.6 shows an image of Virtual Berlin, usable among other things as a context for judging the effectiveness of new development proposals in context. Other similar models in the UK include Bath, Glasgow and Edinburgh Old and New Town models. Various Australian city models exist, and in

Figure 6.6 Spittelmarkt, Berlin; rendering of the new office building (ART + COM, Germany).

some cities it is a requirement to submit a 3D model for integration into the model and visualisation in order to get planning approval. In the UK the Bristol 'Harbourside' development has been modelled and Figure 6.7 shows the underlying map data in MapInfo GIS. Figure 6.8 is a plan view of the 3D model while Figure 6.9 is a ground-level view of the 3D model.

Computer Aided Design becomes GIS if spatial analytical tools can be used on it. Hence the shift from purely CAD models to Virtual Reality Modelling Language (VRML) interactive ones is blurring the boundaries between CAD and GIS and making them interactive tools for urban planning and design as well. With the meteoric development of the Internet, map images with limited interactivity (Putz, 1994) have been surpassed by and replaced with GIS as a critical component in the development of virtual cities (Smith, 1998). The Centre for Advanced Spatial Analysis (CASA) at University College London was established to develop computer technologies in several disciplines that deal with geography, space, location and the built environment. The technologies include GIS, CAD, spatial analysis and simulation and methodologies of planning and decision support. CASA is currently undertaking several research projects that apply GIS to urban design and some of these are described below.

The Virtual Environments for Urban Environments (VENUE) project was concerned with developing computer-based technologies for urban designers. The tools were developed around the application of GIS to large scale urban environments which are represented in 2D map form. The

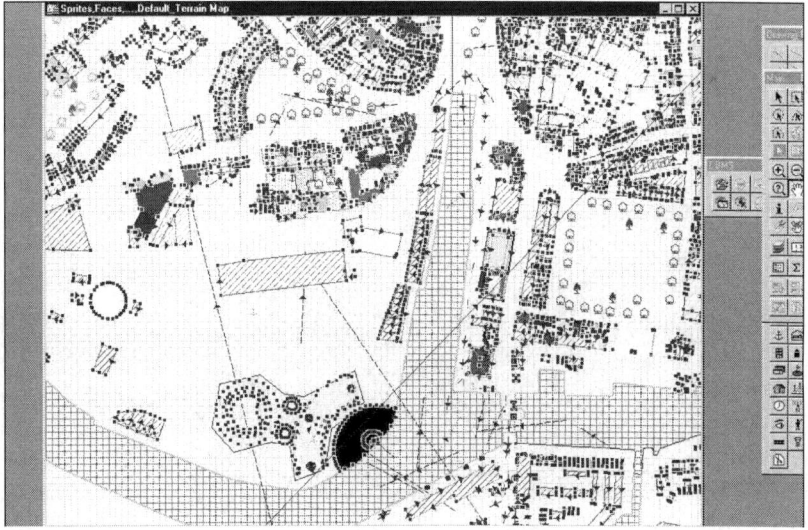

Figure 6.7 Bristol Harbourside, modelled in MapInfo GIS.
Source: Based on OS mapping and data gathered by BCC for the 'Bristol 2000' project.

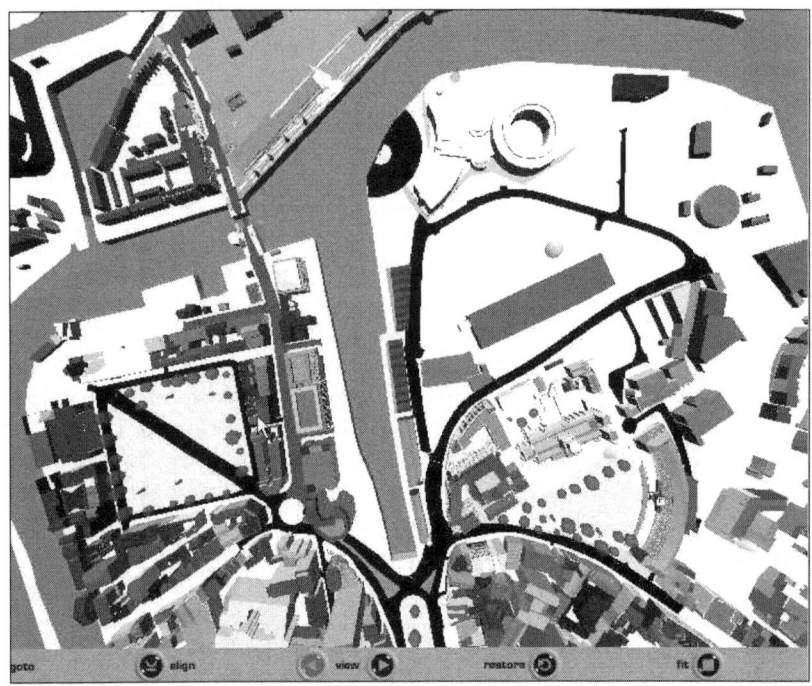

Figure 6.8 Bristol Harbourside, plan view of 3D model.

Source: Based on OS mapping and data gathered by BCC for the 'Bristol 2000' project.

Figure 6.9 Bristol Harbourside, virtual reality image.

Source: Based on OS mapping and data gathered by BCC for the 'Bristol 2000' project.

project sought to adapt proprietary GIS to this domain and to link these tools to the 3D environment, and hence to CAD. Adding height to conventional 2D GIS thematic data allows the creation of 3D urban models in which features such as buildings can have attribute and multimedia data attached. VENUE also demonstrated the application of GIS at the urban design scale of land parcels and individual buildings. This is a level of geographical detail not commonly handled by GIS, tending to be the preserve of CAD software. Batty *et al.* (1998) argue that urban design has seen little development in IT terms, but as more large-scale digital map data become available and the link between the information management capability of GIS and visualisation techniques available in CAD software strengthens, this will change. At the moment, the display and manipulation of 3D building blocks with associated attribute data lends itself more to urban development control than urban design.

The ability to rapidly sketch and visualise design ideas is an important task in urban design. CASA has experimented with links between GIS and 3D visualisation tools. Functionality was added to ArcView GIS to produce block model visualisations in VRML of built features. Macros were written to enable designers to edit the 3D scene by introducing new blocks and to convert 2D plan sketches into 3D format, thus establishing a link between 2D and 3D GIS. Attributes from the 2D GIS database are used to determine the characteristics of the 3D model, for example height and colour. Production of 3D models of the built environment requires building footprints, roof morphology, elevational photographs and height data. Building footprints are widely available, most commonly from OS large-scale data. Roof morphology and height data are more difficult to obtain but it is possible to overlay an OS map with aerial photography at a resolution of up to 12.5 centimetres and thereby visually determine the roof morphology. In Figures 6.10 and 6.11 a set of urban features are visualised in 3D. The building block outlines for the central area of Wolverhampton in the UK were derived from OS large-scale data and are stored and represented as 2D polygons in the GIS. Each polygon also has attributes that determine its colour and height in the 3D VRML model output. The circular polygon around the outside is Wolverhampton's ring road.

In 2000 the Corporation of London commissioned CASA to carry out a review of 3D models of cities. The models ranged from CAD models through various 3D GIS to VRML Web content and related simulations. The primary benefits of such 3D models (when compared to 2D) is their visual impact, persuasiveness and sense of presence when a virtual reality interface is used. There is a significant interest in such models from telecommunications companies who require 3D urban morphologies to determine the ideal allocation for base stations. It was generally found that there is a wide acceptance of the role of 3D city modelling. However there is no one preferred strategy for model development and no one strategy emerged as

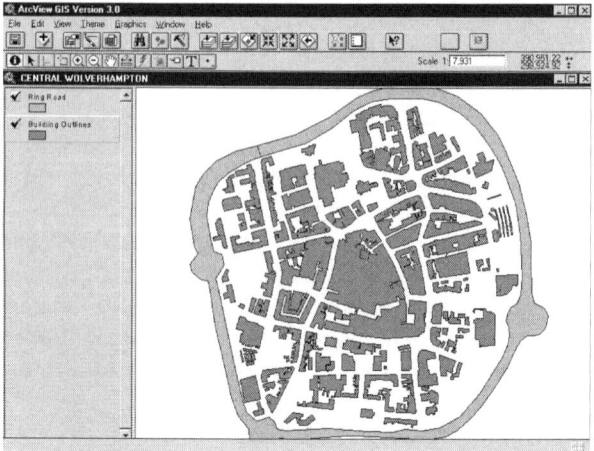

Figure 6.10 Building outlines and ring-road for Wolverhampton, UK.
Source: Centre for Advanced Spatial Analysis, used with permission.

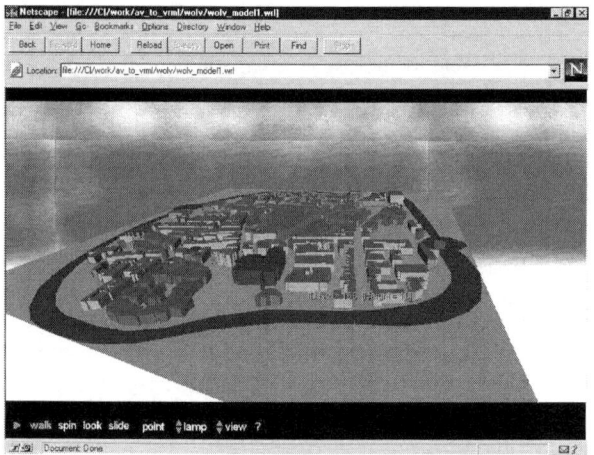

Figure 6.11 3D visualisation of Figure 6.10.
Source: Centre for Advanced Spatial Analysis, used with permission.

being the most appropriate. The team identified three distinct approaches
to 3D city modelling, each from very different view points based on different skills. These are:

* traditional 3D computer aided architectural design;
* engineering approach based on photogrammetric analysis and surveying;
* geographic information systems.

It was largely agreed that those 3D models that were strongly coupled with or derived from a GIS were the models that were the most 'information rich' and were 'built to last'.

At CASA the next stage of this research is to build a 3D model of London – Virtual London – with the support of the Architecture Foundation and Hackney Building Exploratory. The aim is to deliver mapping and 3D visualisation, rendered sufficiently accurately to provide users with a feel that 'virtual' London is 'real' London (Hudson-Smith and Evans, 2001). This model will be a visualisation of all buildings within inner London through which users can navigate at street level as well as fly across in panoramic fashion. It will be possible to query the data within the model by simply pointing at and clicking on buildings and streets to reveal data concerning floor-space, land use, rents, traffic volumes etc. Users will also be able to make proposals and seek answers to 'what if?' questions involving placement and visualisation of new buildings, demolition and changes to transport links. The 3D model will be available over the internet and it will use GIS, CAD and a variety of photorealistic imaging techniques and photogrammetric methods of data capture. CASA is currently developing a prototype virtual urban information system for London. Figure 6.12 shows a section of Oxford Street in London where the ringed building has been moved to a new position as shown in Figure 6.13. The development of such an ambitious virtual model assumes, of course, that an agency can affordably maintain a major 3D city-wide model at a level of detail that can be useful to most stakeholders. One means of reducing the total cost of use of such models is to broaden their appeal and thus spread the cost of initial

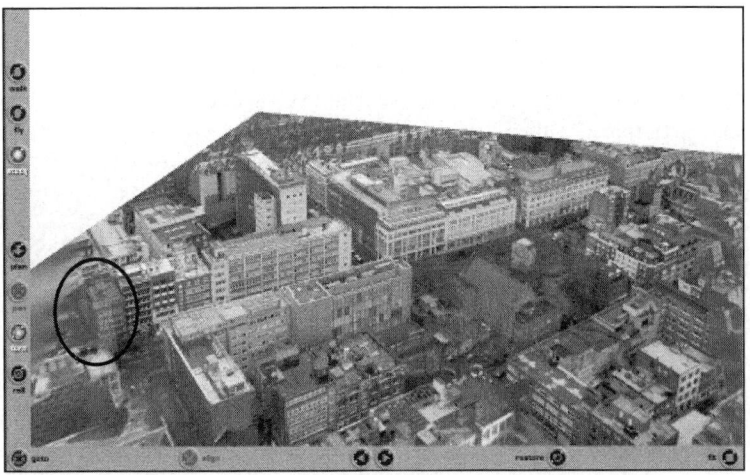

Figure 6.12 GIS-based photographic image of Oxford Street in London.
Source: Centre for Advanced Spatial Analysis, used with permission.

Figure 6.13 Ringed building from Figure 6.12 has been repositioned.
Source: Centre for Advanced Spatial Analysis, used with permission.

creation and subsequent amendment and use across a wider range of applications. Mitchell (1995) defined designing a building or development as fundamentally a collaborative, interdisciplinary, geographically distributed multimedia activity. If this collaborative digital multimedia activity is integrated and disseminated over the Internet, and encompasses a large enough urban area, it could provide the context and constraints for future development and serve to inform and engage the public (Smith, 1998).

The CASA has also developed a range of interactive demonstrations for Hackney Building Exploratory on CD and on the web. The aim of this work is to promote online public participation in urban design and redevelopment of the Woodberry Down residential estate (Hudson-Smith and Evans, 2001). Participants can be posed with the kinds of questions planners and decision-makers face. For example, what happens if we change the traffic flow or knock down this building? Planners and politicians – the decision-makers – can see the results of making changes. A series of panoramas from around Hackney were photographed to create a virtual tour of the borough. The user is able to navigate the scene from a central point. Panoramic imaging technology has been used for the first time by CASA in a public planning inquiry. This allowed the members of the court to visualise before and after scenarios of the proposed development. Each panorama was augmented with a CAD model to provide key view points of the development. By using a range of techniques it is possible to significantly reduce the complexity of 3D and virtual reality models, making them suitable for Internet-based distribution. This is an important step when considering how to convey geographic information to a mass audience.

Overseas case study

A local authority in Denmark wanted to zone an area of sloping terrain for residential development but was keen on preserving the scenic views of an inlet from the higher points. To preserve these views, the land parcels, building heights and elevations were input into a 3D GIS. The GIS was able to provide a picture of the possible layouts for the design of the development as a whole and document views of the inlet from each of the parcels in the final plan. With a visual model of the scenic possibilities of the development and each land parcel, local government officials were able to give better guidance to potential buyers (GEOEurope, 2001).

Summary

Planning was one of the earliest applications of GIS in property in the UK. The mushrooming body of planning legislation, guidance and statutory control over recent years has led to a significant increase in demand for information systems that can assist in the collection and interpretation of data, automation of operational procedures and the monitoring of policy effects. There are many examples of good practice and lessons learnt in local government up and down the country. In this chapter some of the ways in which GIS has been used to assist planning functions at the strategic, regional and, perhaps predominantly to date, operational level have been described. Use of GIS for operational procedures and application-specific implementations of GIS in planning within local government have tended to dominate for technical (data cost, project nature of GIS in early days) and organisational (GIS champions, departmental budget financing and cost justification) reasons. But as the cost of data declines and data availability improves the value of integrating and sharing data becomes more apparent and we should witness the development of more regional and national GIS planning applications. We have also looked at the use of GIS for pre-construction investigations and urban design functions. Some of the lessons learnt from GIS implementation in a planning context have been outlined. They include organisational, technological, data and staffing problems. A recurring problem highlighted by local government staff involved in the application of GIS for planning is the significant skills shortage. Many authorities have great difficulty in retaining skilled staff in this field.

Development decisions are inherently geographical. The adage 'location, location, location' pertains directly to property development decisions. From site identification and evaluation, use/economic/planning/legal/ financial development appraisal through to construction, marketing and disposal, the reliance on geographical data is overarching. Some of these areas of application we have described in this chapter. Others that focus on market research, accessibility modelling and demographic analysis will be considered in Chapters 7 and 8.

Planning and development decisions require accurate and reliable data. Landmark is a good example of an organisation committed to the integration and dissemination of data to assist decision-makers in this area. But of equal importance is the way in which these data are perceived and visualised geographically. This chapter therefore ended with an overview of the way in which leading edge research is challenging the 2D representation of the built environment and is experimenting with 3D visualisation and virtual models of development and design decisions.

References

Allinson, J. and Weston, J., 1999, Information technology literacy survey, Royal Town Planning Institute, London.

Barnett, A. and Okoruwa, A., 1993, Application of geographic information systems in site selection and location analysis, *The Appraisal Journal*, April, 245–253.

Batty, M., Dodge, M., Jiang, B. and Smith, A., 1998, GIS and urban design, Working Paper Series, Paper 3, Centre for Advanced Spatial Analysis, University College London.

Birkin, M., Clarke, G., Clarke, M. and Wilson, A., 1996, Intelligent GIS: location decisions and strategic planning, GeoInformation International, Cambridge, UK.

Dodge, M., Doyle, S., Smith, A. and Fleetwood, S., 1998, Towards the virtual city: VR and internet GIS for urban planning, www.casa.ucl.ac.uk/publications/ birkbeck/vrcity.html

DoE, 1987, Handling geographic information, Department of the Environment, HMSO, London.

Dunn, R. and Harrison, A., 1992, A feasibility study for a national land use stock survey, AGI Conference, Birmingham, 1.13.1–1.13.4.

Dunn, R. and Harrison, A., 1994, Working towards a national land use stock system, AGI Conference, 8.1.1–8.1.5.

Dunn, R. and Harrison, A., 1995, Preparatory Work for a National Land Use Stock System (Stage 1), Final Report to the Department of the Environment.

Ferrari, E., 1999, From points and polygons to housing strategies, AGI Conference, 3.3.1–3.3.5.

Ford, N., 1995, The best laid plans...a blueprint for improving the planning process, *Mapping Awareness*, March, pp. 35–37.

GEOEurope, 2001, editorial, 46, November.

Grimshaw, D., 2000, *Bringing Geographical Information Systems into Business* (2nd edition), John Wiley & Sons, New York.

Hall, S. and Thurstain-Goodwin, M., 2000, Geographic information – policy driving – policy driven – providing statistics for the town centres, *Statistical Journal of the United Nations*, 125–132.

Heywood, I., Cornelius, S. and Carver, S., 1998, *An Introduction to Geographical Information Systems*, Addison Wesley Longman, Harlow, UK.

Hudson-Smith, A. and Evans, S., 2001, A collaborative three dimensional geographic information system for London: Phase 1 Woodberry Down, Proceedings of the GISRUK Conference, Glamorgan, UK.

Keeble, L., 1969, *Principles and Practice of Town and Country Planning* (4th edition), Estates Gazette Ltd, Birmingham, UK.

McMaster, R., 1988, Modeling community variability to hazardous materials using geographic information systems, Proceedings of the third international symposium of spatial data handling, Sydney, pp. 143–156.

Mitchell, W.J., 1995, 'Virtual Design Studio', ed. Wojtowicz, J., Hong Kong University Press, London, UK.

Musgrave, T. and Flack, J., 2000, 1,000 years of ownership: Registering the City of Bristol, AGI Conference, London.

Nathaniel, A. and Nathaniel, J., 1994, Spatial risk assessment for contaminated land, Proceedings of AGI Conference, Birmingham, 11.3.1–11.3.7.

Peel, R., 1995, Fast-track to GIS, *Mapping Awareness*, pp. 30–33.

Putz, S., 1994, Interactive information services using the world-wide web hypertext, Paper presented at the First International World-Wide Web Conference.

Roger Tym and Partners, 1985, National Land Use Stock Survey: A Feasibility Study for the Department of the Environment.

RTPI, 2000, IT in local planning authorities, Royal Town Planning Institute, London.

Smith, A., 1998, Virtual cities – towards the metaverse, Virtual Cities Resource Centre, www.casa.ucl.ac.uk/planning/virtualcities.html

Smith, A., 1998, Metaworlds and virtual space – towards the collaborative virtual design studio, In Proceedings of the International Journal of Design Computing DCNET'98, www.arch.usyd.edu.au/kcdc/journal/vol1/dcnet/

Thurstain-Goodwin, M. and Unwin, D., 2000, Defining and delineating the central areas of towns for statistical monitoring using continuous surface representations, Transactions in GIS, 4(4), 305–317.

Tomlin, C.D. and Johnston, K.M., 1990, An experiment in land use allocation with a geographic information system. In Peuquet, D. and Marble, D. (eds) *Introductory Readings in Geographic In-formation Systems*, Taylor and Francis, London.

Wagner, M., 2000, Connecting to the future at Horsham, *Mapping Awareness*, May, pp. 42–44.

Weston, J., 1995, Planning is paramount: GIS in district councils, *Mapping Awareness*, September, pp. 32–35.

7 Retail and financial market research

Introduction

Businesses compete in many ways; product differentiation, price and non-price competition for example, but market share can vary dramatically from locality to locality. Many businesses, particularly retailers, must establish branch networks at local, regional and national scales in order to penetrate markets and maximise revenue. Business market research in support of commercial property decisions is an established discipline but it is one that has benefited significantly in recent years from GIS tools and techniques. 'Many important business decisions are now made utilising business mapping as one of the key tools' (Wicks, 1995). Consultants GeoBusiness Solutions argue that GIS is part of the armoury through which businesses compete to gain competitive advantage. For example, GIS is used for site screening and selection, catchment area definition (using customer or expenditure estimates and local demographics), store performance evaluation, marketing and customer profiling. In specific market sectors, geographical analysis is used to support decision-making in the following ways:

- Retail: sales analysis and site development, new store sales forecasts, impact analysis, store performance analysis, dealer network planning and after sales forecasting, loyalty card and customer database analysis.
- Leisure: demand elasticity estimation, local marketing, site development and support for licensing.
- Finance: local market demand strategy, optimal networks (branches and automated teller machines), service planning and merger impact analysis.
- Property: feasibility studies, planning research, catchment area definition and analysis.

Examples of decision support include:

- How many outlets should we have in region X?
- What will happen if we close outlet A in centre Y?

- What will be the result of refurbishing our outlet network in X?
- How does actual performance compare with forecast?
- Where should we expand/contract?
- Where will changes in market size and demographic profile affect performance?
- Where are our competitors opening/closing outlets?

Pioneers of GIS for location modelling and marketing had a limited choice of systems. They were usually expensive, required bespoke development and support, suffered from a lack of skilled personnel and produced mixed results. GIS software is now more user-friendly and includes desktop mapping and viewing tools, drive-time analysis tools, area analysis and reporting tools. Importantly for market research – where demographic, property and business data are combined – the software is now capable of much better data exchange. With more proprietory GIS software packages available for market research, competitive advantage increasingly depends on the quality of models and analysis developed with the GIS and the way in which these tools are used to support decisions.

This chapter describes how GIS has enhanced demographic analysis, retail and business location planning, and insurance and financial risk analysis. The introduction of GIS has added not only a geographical dimension to these business market research activities but has also revolutionised the way in which customer, sales and market data are combined, analysed and reported as an aid to decision-making.

Geo-demographic analysis

A fundamental resource for market research is demographic data. Demographic data are analysed to produce statistical information about population groups. These statistics are used for customer profiling, store location planning and branch development. Geo-demographic data and geo-demographic analysis refer to the mapping and geographical analysis of demographic data.

Many businesses hold postcode address data sets of their customers which, through the application of postcode geography, allow the relationship between customers and their location to be established. This information can be mapped to analyse customer distribution or can be combined with other geo-demographic or business data sets for wider analysis. Market research has incorporated demography for a long time but it is the introduction of GIS that has allowed geo-demographic analysis to be undertaken easily and on a desktop PC. Because branch network and store location decisions are expensive it is important to get them right and geo-demographic data are vital to this decision-making process. As with all other statistical analysis however, care is needed in structuring the question, performing the analysis and interpreting the results. These issues will be explored in greater depth in Part III but the level of sophistication evident in the geographical analysis of demographic data is now high and this will be appreciated from the example applications described in this chapter.

As a simple example of the way in which GIS has been used to enhance market research, catchment areas were traditionally defined either by scribing a circle of a given radius around the subject site or by driving around the locality at various times of the day to establish drive-times along particular roads. Many GIS now incorporate drive-time simulation algorithms as an enhanced measure of accessibility. These make use of readily available route networks and allow the input of speeds and impedances for any stretch of road. Drive-times can then be modelled using these networks and the results are regarded as a more accurate representation of travel than Euclidean distance measures. Figures 7.1 and 7.2 illustrate the different sizes of catchment area when the two methods are used to calculate a 500-metre drive-time around an office in Queen Square in the centre of Bristol. The 500-metre radius, measured using Euclidean distance, clearly contains a much greater catchment than the 500-metre drive-time, particularly in this case where there is a river acting as a barrier. These more sophisticated catchment areas can be used to identify how many potential customers, competing businesses or development sites are in a specified area. For example, it is possible to calculate the number of potential customers within, say, 500 metre of a proposed retail site to determine how many are within walking distance. This could strengthen the argument at a planning enquiry for the site as a sustainable location for a new store. Again, the analyst must exercise caution. Different specifications for drive-time models lead to different results. There is no national consistency with regard to road speed or type (it would not be possible to travel along an A-road in Central London at the same average speed as an A-road in rural Scotland). Care must be taken therefore in defining drive-times so that they reflect real world travel speeds.

Retail location planning

In the retail sector GIS is used primarily to identify new sites and to measure the performance of existing ones. The types of analysis undertaken often include:

- demand assessment; catchment area analysis (population characteristics, family incomes, expenditure and buying preferences, future trends);
- supply and competition analysis (relative attractiveness of a proposed store);
- existing store appraisal (customer penetration);

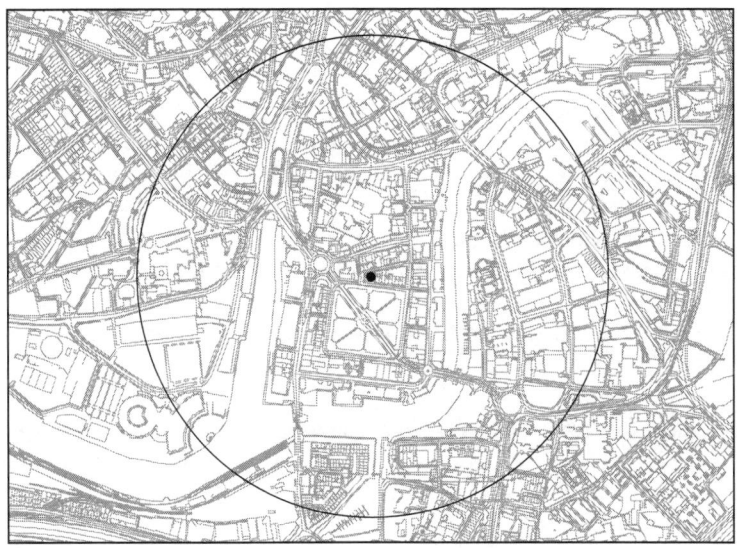

Figure 7.1 A 500-metre drive-time defined by Euclidean distance.

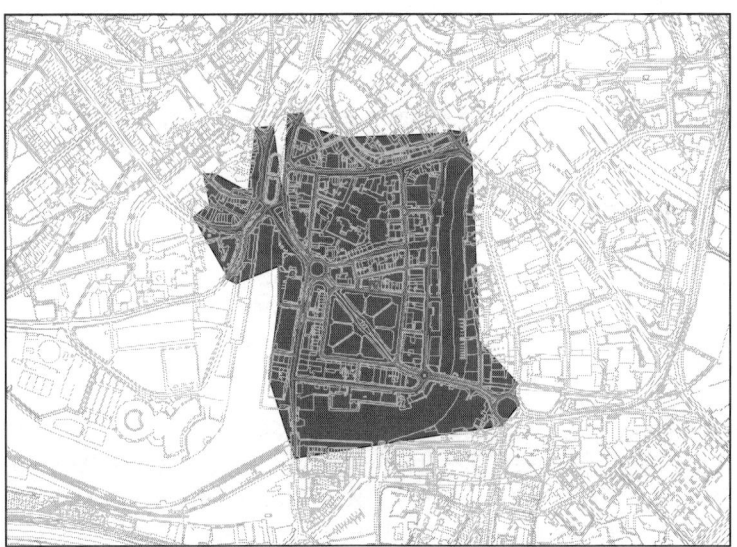

Figure 7.2 A 500-metre drive-time defined by road network distance.

- new site appraisal; identifying and ranking the location of new sites in terms of accessibility, trade potential, size, shape, parking, prominence;
- integration of demographic, marketing and geographic data to produce a turnover estimate, for example, overlay buying power of population (market), location of competitors (competition), road links and drive-times (accessibility).

Retail location analysis is becoming increasingly complex. Components of GIS for store location planning include:

- Mapping facilities.
- Catchment demographics; usually aggregated to postcode sector level.
- Store accessibility; distance, drive-times, travel times by public transport and on foot.
- Competition and agglomeration of specific business types.
- Trading area composition; measures of the attraction of retail centres, for example.
- Outlet characteristics; size, frontage, number of staff, etc. (Grimshaw, 2000).

Geographical Information Systems can be used to select suitable comparison stores in order to estimate turnover for new outlets. It is possible to take the output from GIS queries of geo-demographic data sets and convert them into predictions of key performance indicators. Using statistical techniques such as multiple regression analysis, GIS can interpolate area variables and proximity variables. Gravity modelling is used to predict levels of customer trips between residential zones and outlets or centres and is used by retailers for planning store networks and predicting expected revenue from new sites. It is also increasingly used by retail financial services, leisure and health sectors. Unlike multiple regression analysis, gravity modelling can conduct impact analysis and network optimisation (catchment area analysis).

Mature retail branch networks mean that new site development is an unrealistic means of increasing sales because new outlets will cannibalise existing sales (Walker, 2000). Therefore selective opening, closure and relocation of outlets to optimise the network becomes the principle, property-related means of increasing sales. An optimum store network might be defined in terms of sales, proximity to the maximum number of customers or achieving an even geographical coverage. This is done using gravity models and is regarded as better than assessing store's performance on a store-by-store basis.

According to Goodwin (1997) catchment areas based on postcode sectors and Goad Plans are analysed in combination with in-house data on sales areas, turnover and profit. GIS has integrated these functions and it is now possible to study population growth, average spending and the demographic status of emerging urban and retail areas. By using gravity modelling to model customer flows between possible retail destinations and then using GIS to determine how resources should be allocated to various sites or centres based on their strengths (size, attractiveness, store layout, stock), it is possible to model market potential in an area and estimate how it might be affected by changes in accessibility (road or pedestrian layout).

The large food retailers such as Sainsbury, Tesco, Safeway, Asda and Somerfield, who operate substantial supermarket chains, use GIS for performance forecasting of new stores together with demographic and lifestyle data for postcode sectors surrounding existing ones. Non-food retailers and financial service providers such as Boots, WH Smith, Debenhams, Leeds Building Society, Lloyds TSB and the Halifax use GIS for site selection and the modelling of population change, spending profiles and accessibility. Some of these retail applications of GIS are described in the following sections.

Food retailers

Food retailing is a location-specific business. Safeway use a GIS as part of their store location strategy to define catchment areas of shoppers who fit the profile of the highest spending customers at Safeway stores, overlaid with existing and planned superstore developments (Wilkinson, 1991). Approximately £20 million investment goes into each store. 'Competition in the British grocery industry is evident in the accelerating new store development programmes of the major retailers. The escalating site cost and the heavier levels of capital investment have resulted in the development of a more sophisticated approach to evaluating the sales potential of both new sites and existing stores' (Wilkinson, 1991). Key determinants of a successful site are sales potential and likely profitability. These are dependent on site characteristics (such as size, shape, parking, ancillary facilities), infrastructure (that determines accessibility relative to competitors and population), competition and population (size, demographics, age, etc.). Customer, store, product, census and competition data are analysed as follows:

(a) Outlet strategy: evaluate the profile of high spending customers at existing stores to target population with a propensity to shop at Safeway, map locations of these target shoppers and overlay existing/planned superstore developments. Draw catchment areas, rank and aggregate specific locations to produce a regional and national development programme. Outlet strategy determines resource allocation, acquisition programme and filters opportunities presented by developers.
(b) Trade potential reporting: this is a more detailed examination of a specific site. Catchment area definition is based on drive-times, geographical and psychological barriers, physical capacity, accessibility and visibility of the site, scale and condition of competitors, and the location of target customers. Once the catchment area is defined it is mapped and examined for market penetration, competitor activity and impact analysis (own and competitors') under different scenarios of population growth, infrastructure changes and catchment expansion.
(c) The investment decision: mapping can add value to boardroom decision.
(d) Tailoring the store: behavioural data are examined geographically to highlight, by customer postcode, patronage of in-store departments. This information can then be cross-referenced to geo-demographic

indicators to establish rules on buying habits by population type, which can then be applied to the target catchment of the proposed store.

(e)   Local advertising: geographical targeting.

(f)   Post-opening evaluation: compare actual with projected performance.

'A GIS has an extremely important role to play in store planning. It establishes a single geographical base against which to analyse any number of different data sets. ... It is enormously effective as a communication tool...' (Wilkinson, 1991).

Non-food retailers

A GIS was used to model new shopper behaviour at the Meadowhall out-of-town retail centre in Sheffield, mapping home locations of shoppers together with isochrones (lines that join places of equal travel-time to or from a given destination) to illustrate the extent of the attraction of the centre geographically. Results showed that the attraction was especially strong along the motorways and demographically from young affluent families. Marks and Spencer have used GIS to analyse exit surveys of customers in order to define catchment areas of stores. This allows the development of predictive algorithms for catchment forecasts for new stores. The road network is mapped to create drive-times around stores and the market share in an area is analysed to determine whether new business can be generated. Debenhams also used a GIS to assess potential store locations, monitor the performance of existing stores and assist in in-store development. Geographical analysis includes the following tasks:

- catchment area definition (using credit card penetration levels to define catchment areas for existing stores and drive-time analysis for new stores);
- area profiling (demographic make-up of surrounding population for specific goods); and
- retail potential (estimate annual spending potential within the catchment area for merchandise groups).

Longley and Clarke (1996) describe how a GIS was used by a home electronics rental company to assess the impact of opening a new store. Costs of opening a new store were estimated (such as property, staff, stock, etc.) and a GIS was used to predict revenue over the next five years based on geo-demographic data for surrounding postcodes. The GIS model revealed that much of this revenue would be redirected from the organisation's other stores in the area.

Lloyds Chemists also use GIS for customer profiling, site location and competitor research, analysis of market activities and the planning of marketing strategies (Goodwin, 1997). They are considering showing the location of GPs in any area to locate stores in easy reach of surgeries.

Geographical Information Systems have seen significant application in the building society sector, particularly amongst the large national operators. The main functions are target mailing, assessing market potential, catchment area and branch location analysis, insurance rating and credit assessment. The majority of users regard GIS as a strategic tool (Grimshaw, 2000). This contrasts with evidence from the local government sector where it is regarded as an operational tool in the main. The Nationwide Building Society uses GIS to match products and capacity to local demand and to consider the hierarchy of the branch network (Hornby, 1991). Data include a customer database with postcodes, census and geo-demographic data and a catchment area database, which includes the locations of competitors. Catchment area analysis and competition analysis are combined to forecast customer penetration per head of population. This is then used to determine the size and location of the branch network. For a particular area the objective is to deliver a distribution plan of Nationwide products. The first stage is to analyse the existing market in terms of competition and market share by product within specific customer segments. The results are then fed into a capacity model to calculate the distribution supply requirement from the market demand. Longley and Clarke (1996) describe how a GIS was used by another building society to decide whether to proceed with a refurbishment programme for existing branches or to develop the branch network in regions of the country that were currently under-represented.

Car dealerships use GIS to define catchment areas, create thematic maps of dealer performance and areas of dealer potential. In the automobile industry franchised dealerships are allocated discrete geographical territories. Evidence shows that the more dealerships that a car manufacturer appoints the greater the likely market share, although diminishing returns have a consequent effect on profitability for each additional dealership. A GIS can help achieve a balance between maximising the number of dealerships and maintaining profitability for each one (Longley and Clarke, 1996). Clark (1991) describes how a GIS has been used by Toyota GB to:

- appraise Toyota and competitor performance in the UK at different spatial scales;
- measure the performance of individual dealers for target setting;
- identify new market opportunities (taking into account predicted levels of business and competitor activity);
- assign areas of responsibility to new dealers;
- assess the impact of dealer terminations; and
- identify the optimum dealer network under different scenarios.

Grimshaw (2000) describes how a European car dealer network used GIS to define dealer catchments. These were large areas as the dealer operated in the specialist car market. GIS output included thematic maps of individual dealer performance and areas of dealer potential.

'Catalist' is an independent company specialising in the petrol retailing market. The company provides information about petrol stations in Europe and data analysis techniques provide decision support tools for the planning and operation of a petrol retail network. Computer-based maps and GIS are used to optimise the planning and operations for clients who include BP, Esso, Shell, Tesco, Sainsburys, Granada and Welcome Break. Typical applications are:

- Optimising locations for petrol stations, depots, shops, branches, restaurants, etc.
- Analysing census data, road networks, drive-times and other factors to establish the most efficient sales territory network.
- Generating maps and reports.
- Defining catchment areas based on customer locations, census profiles, drive-times or other factors.
- Mapping the optimum distribution network between multiple supply points, depots and delivery points.

Overseas case study

In the US a mortgage company in South Carolina uses GIS to assess the feasibility of existing or potential sites for particular activities. An application developed by the University of South Carolina uses a gravity model that incorporates distance decay as a component of the calculation of market potential. The aim of the model is to answer the question 'how many customers or how much in sales can I expect at this site given the existing competition?' Instead of using a market radius to select demographic variables, distance decay is calculated on the basis of the location of the proposed site and competitors, attractiveness of the proposed store (measured by floor-space), maximum market range and the type of distance decay model required.

Retail property consultants

Property consultants involved with retail site selection and development have used GIS to analyse demographic, property and geographic data in order to assess likely turnover, potential catchment, the effect of competition and planning policy for potential new sites. The catchment area (population characteristics, family income, expenditure and buying preferences), competitors and site location (accessibility, visibility) are all examined. For example, Sanderson, Townend and Gilbert, a firm of chartered surveyors in

Newcastle in the UK, used a GIS to identify sites on behalf of Global Video, who operate a chain of video rental outlets (Kirkwood, 1998). The GIS was used to integrate information about Global Video and their competitors with demographic and property data using postcode sector boundaries, road network mapping and retail Goad Plans. Possible outlet locations are identified and the GIS is used to select relevant postcode sectors in the vicinity of each site and demographic data for each sector are scrutinised to see if they match the requirements of the customer profile for a successful store. If the match is good the site is inspected. The Global Video GIS application selects sites using the following criteria:

- Busy pitch (modelled using Goad Retail Plans)
- Good car access
- Proximity to complimentary retail uses such as hot food and off-licence outlets
- Acceptable distance from the competition
- Potential population catchment area.

Data used include postcode sector boundaries, small-scale raster maps and demographic data on unemployment, car ownership and social class for each postcode sector. The process involves locating a potential site on a general location map, selecting postcode sectors that a store in that location would draw from, assessing demographic data for the proposed store against the demographic model of a successful video store, visiting the site and assessing location factors (Buchanan *et al.*, 1999). The application would be enhanced if local authorities made their planning data available digitally.

Overseas case study

In the US Wilson–Kibler, a real estate company in South Carolina, commissioned the University of South Carolina to develop a GIS application that assists retail site selection (Cowen *et al.*, 1999). The application measures market potential based on a prescribed range around the site. A circular buffer is drawn around the proposed site and the underlying census blocks within the circle are clipped. Demographic data contained in straddling polygons is allocated in proportion to area inside/outside the circle. This is illustrated in Figure 7.3. For each census tract selected summary statistics are presented for each census record. The user can then weigh the importance of each item of census data in order to derive a final 'score' for the site.

Figure 7.3 How census tract level demographic data are apportioned using area ratios.

Source: After Cowen *et al.*, 1999.

At CB Hillier Parker, a firm of property consultants in the UK, GIS is used for retail analysis using gravity modelling by research staff and for basic GIS mapping for presentation purposes (such as the calculation of catchment areas) by surveyors in general. An objective of CB Hillier Parker is to have basic mapping tools on every surveyor's desk in the next 3–4 years, primarily for descriptive mapping. It is expected that this will mature into more sophisticated analysis over time. Property data such as rents, yields, lease and occupier details are geo-coded and mapped onto large-scale OS mapping. Buildings can be shaded to illustrate acquisitions, disposals, rents and other descriptive data about the market. CB Hillier Parker also carry out a National Survey of Local Shopping Patterns that utilises GIS technology. In 1998, the first electoral roll survey of consumer shopping destination preferences for comparison and convenience goods was undertaken. This survey of more than one million households permits catchment areas and penetration rates of major comparison and convenience goods to be identified for trading locations in the UK. Before this survey, retail analysts had to rely on proxy measures of local retail market size using modelled catchment population estimates, shop counts and retail floor-space totals as indicators of market size differences. The electoral roll survey allows forecasts of local high street growth trends based on consumer spending flows. The impact of development activity, population growth and other socioeconomic factors on spending levels and sales in individual retail markets can be simulated. Retail rental forecasting models can be based on actual rather than proxy measures of market activity. The survey:

- Identifies the size and nature of catchment areas for retail goods.
- Identifies size and profile of shopping populations for individual or grouped trading locations.

- Identifies the market shares achieved by trading locations competing for trade in any catchment area.
- Identifies and predicts growth trends at the local level.
- Ranks trading locations by size, growth profile and other characteristics.
- Forecasts sales and occupational costs.
- Predicts the impact of development on shopping patterns.

GIS is also used at CB Hillier Parker to assist with the design, development, layout and retail mix of shopping centres. Analysis focuses on the flows of shoppers between origins and destinations at the postcode sector level. GIS is used to extend geo-demographic area profiling using drive-time isochrones and other trade area definitions. Data from the Survey of Local Shopping Patterns described above can be combined with other area data sets, describing income, lifestyle and expenditure, for example, to define catchment areas and other shopping population profiles for individual or grouped trading locations.

Use of GIS at Jones Lang Lasalle (JLL) was prompted by demand for retail site location analysis on behalf of specific clients. This application-led introduction of GIS can be justified financially in terms of the benefit that it offers the client in meeting the objectives of a project brief. The firm is currently evaluating the potential benefits of mapping its agency database and this was described in Chapter 4. The familiar problems of inconsistencies with in-house data are being encountered but already the pilot project is illustrating how useful even simple thematic mapping of data stored in the agency database can be. This is a more corporate approach to the introduction of GIS in an organisation and the benefits that it can offer are sometimes harder to identify and value. In terms of software and data products in use at JLL, desktop mapping is used and digital Goad Plans are available across the firm. More advanced retail mapping is used within the Information Centre. Promap is used for the generation of location plans that typically appear in reports from JLL. Also, MapPoint, which includes OS street level data for the UK, is used to generate customised plans and maps for insertion in reports and web pages. MapPoint is also capable of performing basic point and thematic mapping. MapInfo is used as a mapping package for the input of postcode zones, point data and other thematic data sets but it is not used for analysis of these data.

CACI supplies both software and data to JLL and JLL has been a user of CACI's Insite system since 1988. A customised system was built to reflect the specific needs of JLL in terms offering value added services to retailers, which was focused primarily on location and catchment area analysis. For example, using the drive-time application it is possible to evaluate specific sites for clients in terms of:

- Catchment size
- Overlapping catchments

- 'Gaps' in the market
- 'Cannibalisation' for retailers
- New locations for expansion
- Opportunities for disposal properties.

The Market-Scan application is used to determine the location of hotspots and for thematic mapping and includes 20-, 30- and 60-minute drive-times for each of the 9,000 postcode sectors in the UK, an example of which is shown in Figure 7.4.

Using CACI's Retail Footprint application primary, secondary and tertiary catchments for some 2,300 retail locations in the UK, ranging from regional and suburban shopping centres to out-of-town retail parks can be identified. Retail Footprint is based upon a gravity model, which reflects both a centre's accessibility and attractiveness (in terms of retail provision – comparison, household and food shopping) compared with neighbouring retail locations. The model defines in geographic terms the area from which a retail centre attracts its trade. During the 1990s, JLL continued to expand its use of GIS from CACI with a retail focus by adding the Paycheck data set and the Impact Module. The latter enables the potential impact of new retail development on an existing retail location, and the impact of new competitors on existing sites, to be modelled. For example, a major (fictitious) regional shopping centre, totalling some 500,000 square feet is to be built in Neston, which has an existing retail catchment of approximately 258,000 people. It is estimated that the proposed development will

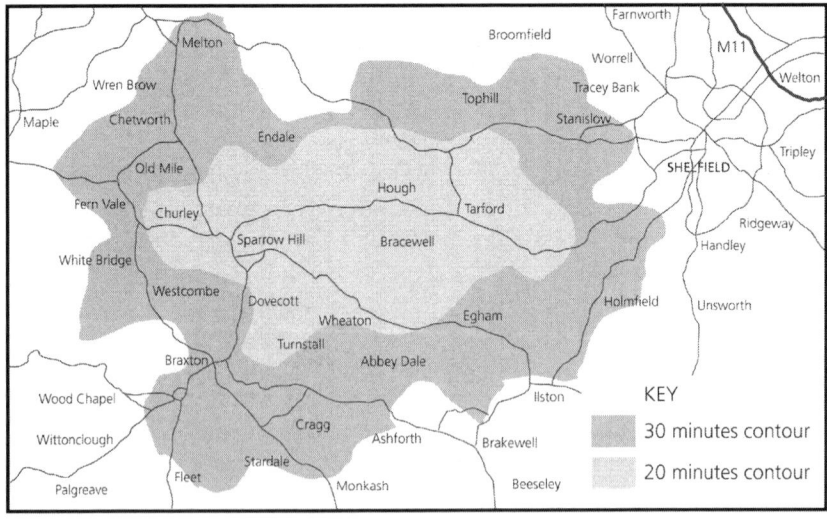

Figure 7.4 Drive-times based on postcode sectors.

increase the attractiveness of Neston as a retail location by 150 per cent. Inputting these figures into the Impact Model it is estimated that the impact of an additional 500,000 square feet of retail space in Neston will increase the retail catchment to 402,700 people. Figures 7.5 and 7.6 show how the size and shape of the primary, secondary and tertiary retail catchments have changed, underpinning this increase in population.

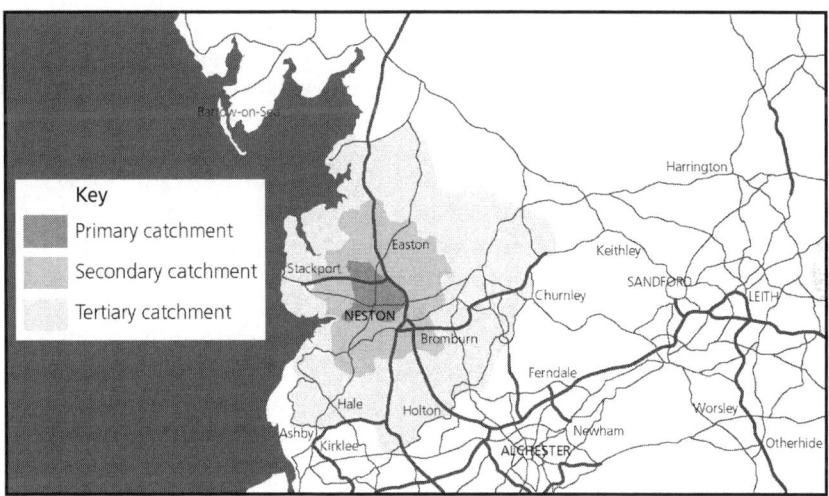

Figure 7.5 Retail catchment areas around Neston (1).

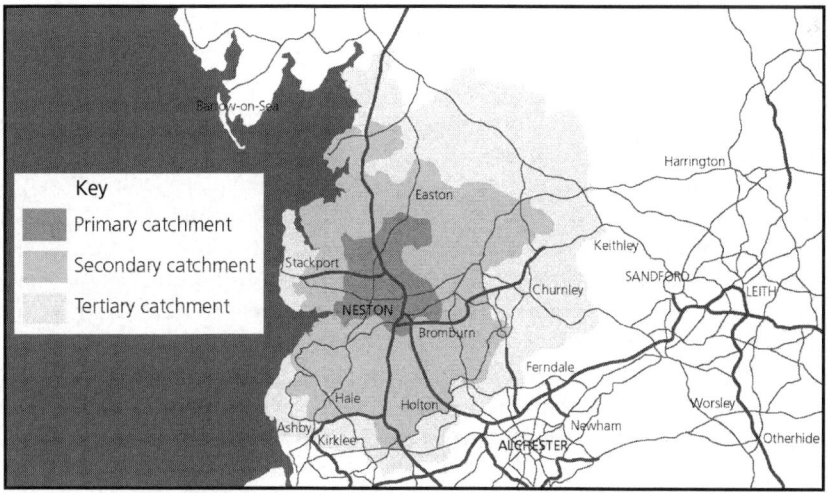

Figure 7.6 Retail catchment areas around Neston (2).

European Insite from CACI contains many of the analytical capabilities of the UK products but its main advantage as far as JLL is concerned is that it is geographically seamless and thus allows cross-border drive-times, socio-economic analysis and overlapping catchments (cannibalism) to be calculated. In the absence of this product, JLL would have to rely on each country performing the analyses, which would invariably stop at the borders!

JLL therefore uses GIS to perform retail location and relocation analysis, catchment and impact analysis, store performance measurement, customer segmentation, thematic mapping of 'hot spots', location targeting and profiling, modelling 'what if' scenarios and trend analysis. GIS is used in the retail department of JLL for analysis on behalf of a range of clients in the UK and in Europe including Costco, BAA McArthur Glen and Starbucks. A typical application of GIS for retail analysis is the location mapping of coffee houses for a particular chain in relation to the location of competitors.

Finally, JLL have developed a 'Locations Database' which is accessible over the firm's intranet. The database provides access to annually updated population, employment, business, accessibility and property data for regions, towns and urban areas across the UK. The database has a map front-end and uses MultiMap to illustrate locations once they have been selected. A useful next step would be to extend the scale of the database by allowing users to zoom in to particular urban areas, call up the large-scale mapping associated with the agency database and thus continue enquiries at the individual property level. This would go some way to realising JLL's vision of a corporate, GIS that integrates systems and databases in order to promote knowledge sharing. Future plans are to incorporate metropolitan maps for UK and Europe, develop web-based mapping capabilities, a pan European GIS system and to expand the property-related applications piloted in Leeds are described in Chapter 4.

The Investment Property Databank (IPD) and property consultants Donaldsons used GIS to analyse various geo-referenced data in order to identify potential sites for a new shopping centre. Various vector-based point and polygon data, including population density, market volatility, past investment performance, 10-, 20- and 30-mile drive-times around urban areas, infrastructure, lifestyle profiles and rental growth (shown in Figure 7.7), were converted to raster data and normalised. This was achieved by taking the surface values and reclassifying the values of each surface so that they relate to a 1–10 scale shown in Figure 7.8 for rental growth figures. The normalised surfaces were then combined by simply adding their values at each point using 'Map Calculator', a standard GIS tool described in Part I of this book. Figure 7.9 shows the result of using the Map Calculator function to add up the scores for each of the normalised variables and displays the areas according to the aggregate scores for optimum locations (1 – poor, 7 – very good). The result is a single composite map created by interpolating a series of surfaces from various vector-based point and polygon data.

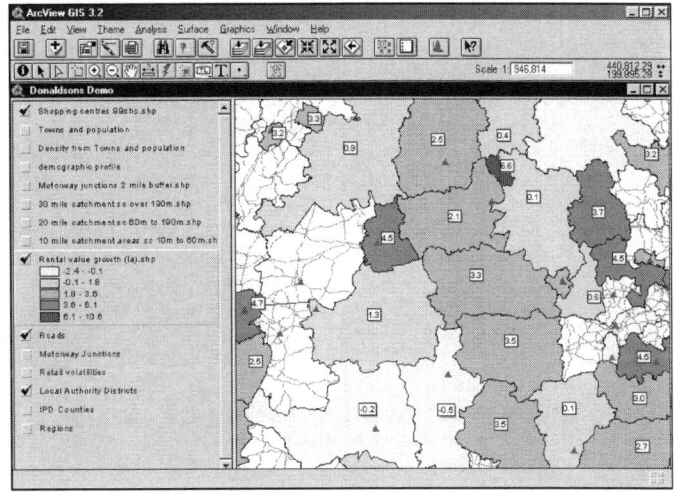

Figure 7.7 Rental growth by local authority area.

Source: IPD, used with permission.

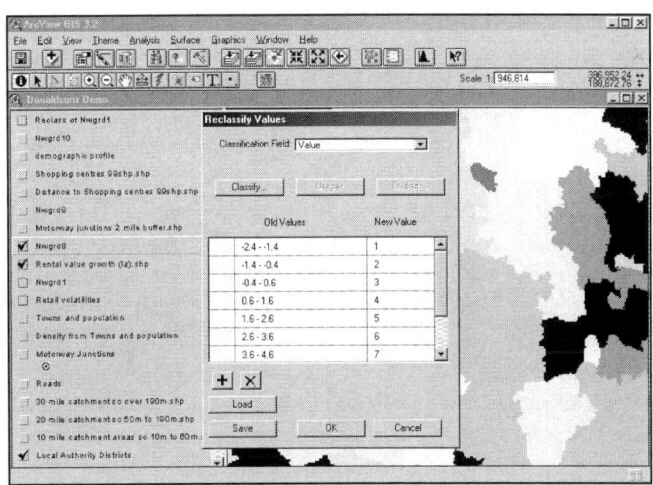

Figure 7.8 Normalisation of rental growth figures.

Source: IPD, used with permission.

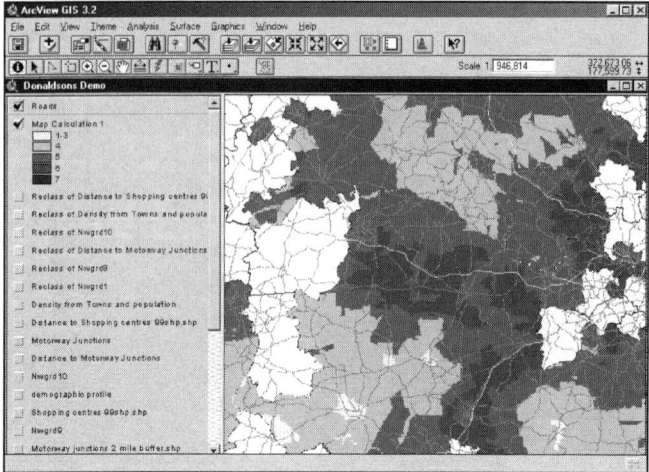

Figure 7.9 Composite map showing a ranking of suitable locations for a new shopping centre.

Source: IPD, used with permission.

Overseas case study

In the US 'appraisers are often asked to estimate the market value of retail centres that are not at 'stabilised occupancy' (90–95 per cent). Appraisers need to estimate the time for retail absorption for a geographical area in which the centre is located and for the subject retail centre itself, in other words, how long will it take for the centre to reach stabilised occupancy? GIS can be used to perform the absorption forecast (Smith and Webb, 1997). Traditionally an absorption trend analysis is undertaken by examining past absorption trends in the market and projecting the historical average over, say, the last 5–10 years. This historical data is usually available from larger national real estate brokers or private real estate information services. Data might include rents, vacancy rates and space absorption over the past few years. GIS can help forecast absorption by providing a geographical population forecast. It does this as follows; take existing retail space and divide it by the existing population estimate to calculate the current ratio of retail space per person. The ratio would vary due to population income, shopping habits and other factors that determine trade areas and retail sales. It will also be affected by the supply of new space. It is also possible to determine 'occupied' retail space per person by adjusting the retail space figure by the vacancy rate and dividing this (lower) area by the population estimate. Then look at the population forecast for, say, five years' time and multiply the number of new people by the 'occupied' retail space per person ratio to arrive at an estimate of new space needed over the next five years. From this the average annual absorption over the next five years can be calculated by dividing the figure by five (years).

Office location planning

Why do office occupiers pay more for certain locations? How do values of office premises vary across a city? Can geographical analysis of property values aid office location planning? These are the sort of questions that businesses must ask in order to optimise the use of property assets. Empirical evidence suggests that office occupiers regard accessibility to customers, clients and complementary business activities as the fundamental elements of a property decision, over and above the physical and legal characteristics of the property.

A study in Bristol used a geographical analysis technique called distance mapping to locate those office properties that were within specified access times and distances of features regarded as urban attractions. The study suggested that a geographical analysis of property values can identify movements in value and shifts in relative accessibility and concluded that this type of analysis may have an important application in office location planning (Wyatt, 1999). The results demonstrated the importance of using a communications network when calculating distances and travel times across an urban area. The route network should be based on actual transport routes and incorporate realistic impedance measures for traversal along the network. Only by developing a realistic network model of communications in an urban area can an accurate picture of how locational influences affect property value be created. The network distance analysis can be enhanced by creating more realistic impedances for traversal along different roads and certain stretches of road. Two influences in particular were found to affect the accessibility calculated using the network model. These were the configuration of the route network and the impedance for traversal along the routes. This has obvious implications for transport planning and its consequent effect on property values – the design of route networks in urban areas should seek to optimise ease of movement of vehicles and pedestrians. Accessibility, in terms of customers, clients and complementary business activities is the key determinant of the location decision for many office activities, especially the financial and professional services occupiers. This increases the demand for more accessible sites, which have traditionally been in the city centre.

As well as retail clients GIS has been used for office and industrial location analysis. Companies that are looking to relocate may require a comparison of drive-times and public transport accessibility between their existing and proposed locations. Jones Lang Lasalle (JLL) use GIS to assess labour availability in terms of the size of the workforce, economic activity, occupation profile and qualifications of the workforce. The pursuit of lower operating costs – namely property and labour costs – is an important trigger in an organisation's relocation plans. The differential in property costs (rent and rates) between Central London and provincial office markets has been a major push factor. However, organisational factors such as the need to expand and opportunities for consolidation of premises and staff have risen in importance as the primary triggers for relocation decisions. Organisations today have to ensure that their property portfolios add value to their activities and provide them with the opportunity to compete effectively in ever more demanding markets. In one particular case – it was the need to consolidate premises and staff which was the trigger for the relocation of a headquarters operation out of Central London. Following the identification and acquisition of suitable properties in and around the M25, JLL was asked as part of the relocation process to provide a detailed journey time analysis by road and public transport from staff homes to:

(i)   existing HQ location in Central London;
(ii) new location in and around the M25; and
(iii) the differential in journey time between (i) and (ii).

First, using various journey plan databases, journey time differentials were ascertained. Staff homes (using postcodes) were imported and plotted on a base map and classified according to salary group and differential in journey time. Second, 60-, 90-, 120-minute drive-times were calculated around the new site, reducing the default drive-time speeds by 50 per cent to reflect congestion on the M25 during peak hours. This exercise provided an indication of journey times by road from homes to the new location. The resultant map and accompanying database provided the client with a base on which to make initial relocating staff decisions, based on journey time to work. For any case, which was considered borderline, a replica journey was undertaken either by public transport or road from individual homes.

Another use to which GIS has been put on behalf of office occupiers is an analysis of new market potential. The CACI Market-Scan application has pre-defined drive-time catchments around all postcode sectors in the UK. Data sets, which JLL has underpinning these pre-defined catchments, include population, age-structure, population projections, occupation, social class, car and home ownership, economic activity, neighbourhood classification, retail expenditure, household income, business data and education/qualifications. Market-Scan is a useful tool for 'hot spot' mapping in terms of identifying locations that offer the best opportunities based on the analysis of the underlying data sets. For example, JLL were instructed by a financial services company to find approximately 4,000 square feet of office space in twenty different locations across the UK. The client wished to sell household insurance, household loans and mortgages. As a first step in identifying suitable locations, MarketScan was used to construct 20-minute drive-times to illustrate those locations with a critical mass in terms of number of households. It was then possible to sieve and rank the 'hot spot' locations using other key variables and identify available office properties in each location. Similarly, a bank client asked for a geo-demographic analysis of locations throughout the UK in order to identify those that contain high population density within a short drive-time. This analysis was then used to support location decisions for home insurance outlets. For industrial occupiers GIS has been used to help distribution companies locate their depots. This has involved transport network analysis.

Businesses are increasingly looking internationally for suitable premises for branch networks, production and manufacturing facilities, and distribution outlets. Grimshaw (2000) notes that leisure operators use GIS for site selection. Analysis of potential sites at the local level, in terms of the location of outlets with respect to local demand and competitor type, size and location, may help quantify market share in a region. By undertaking detailed analysis of customers of existing outlets and concentrating on socio-economic characteristics, frequency of visits, spending per head for example, a typical customer profile can be drawn up and input into a GIS. This profile can be compared with national statistical information relating to population levels, socio-economic characteristics, housing types and travel isochrones. The profile can then be used to identify specific locations for site acquisition. This process is increasingly used by major leisure operators such as Bass and Whitbread. But it is not just the private sector that requires this type of location planning. Development agencies that wish to attract inward investment for, say, regeneration must consider the location requirements (such as labour, market catchment and accessibility) of relocating firms in order to target their investment. The International Location Advisory Team at management consultants Ernst & Young use GIS to help corporate clients who wish to consider relocation options for their businesses. GIS contributes to the location analysis by screening locations prior to more detailed analysis. Property cost and land availability are the starting points. Screening then includes labour market factors (unemployment, age of workforce), drive-times, rail-times and airport locations. These data are classified (using natural breaks, as defined in Part I of the book, for example) and displayed on a map. The analysis then focuses on between

10 and 20 local authority districts after initial screening. These districts can be grouped together into more meaningful labour market areas.

The research department at FPDSavills acts as a GIS consultancy to other departments within the company. In this way GIS development is driven by departmental demand. The following sections illustrate how GIS is used for office location planning at FPDSavills.

Catchment area definition and analysis

Catchment areas are defined using journey times (peak and off peak drive-times, walk times or simple distance radii) and are used to identify areas of exclusive market catchment. The extent of a catchment area is dependent on property type but once a catchment area has been defined then profiling of data within it helps identify areas of strong market potential or 'hot spots' as well as evaluate competition and performance of the subject property (see Figure 7.10).

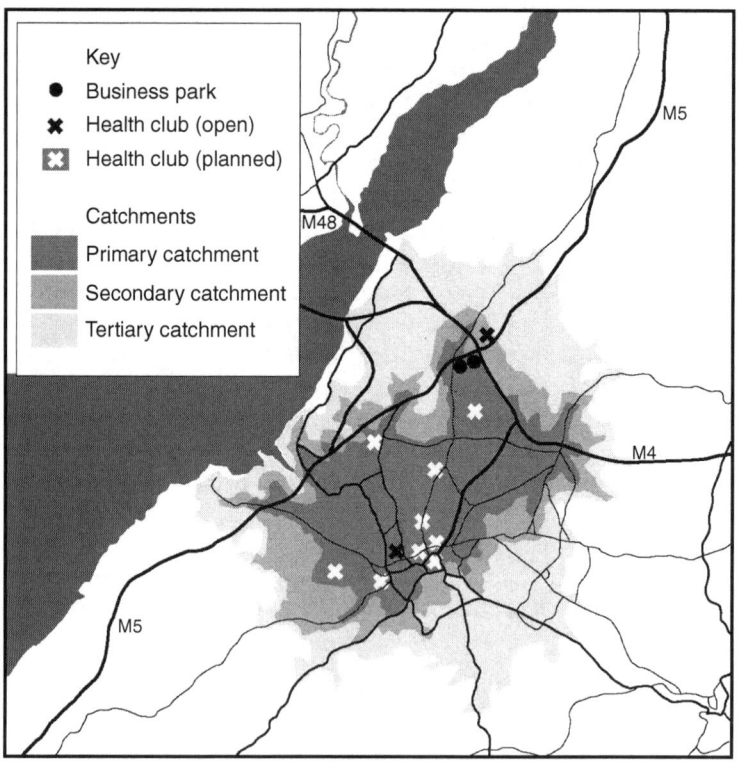

Figure 7.10 Mapping the competition, catchment and key client targets.

Typical analysis of a catchment area provides a preliminary view of potential demand and supply within a specified catchment area. Drive-times are also calculated for competitor locations so that catchment areas may be defined and overlaid with one another to identify areas of exclusive coverage and low competition. Demographic data within the catchment areas can also be examined, including census data and neighbourhood population profiles.

Customer profiling, investment performance measurement and client specific reports

Customer profiling takes the form of standard demographic census mapping and reports and CACI demographic 'ACORN' (A Classification Of Residential Neighbourhoods) profile reports. Investment performance measurement typically involves investment demand ranking. For example, a client may be interested in constructing a residential property investment portfolio and wishes to receive advice on the demographic profile of areas in which potential investment opportunities are located. Scores are allocated to postcodes using demographic data (young people for renting, for example) then data points for residential investment opportunities can be overlaid. This helps identify which investments are in high scoring locations.

An example of the sort of reports that are produced for specific client requirements is the 'pre-site check' for valuation purposes. Council tax bandings, demographic and other data are examined prior to a valuation as an indication of whether a site visit is required. Increasingly, investment surveyors require geographical analysis of property market data to support their advice and agents require high quality presentation of data together with market research to underpin their advice.

As well as subscribing to CACI data and services, FPDSavills are starting to geo-reference their own property data such as annual office turnover (take-up and supply). Because FPDSavills have a strong residential property focus, Land Registry data are incorporated by postcode sector, together with council tax data by local authority district and demographic data by enumeration district. The EGi London Office Database provides workforce data. Other data include planning and development information from suppliers like Glenigan and various property-specific data such as the location of golf clubs and health and leisure facilities. There are areal interpolation problems that arise when integrating these data sets and Part III discusses these.

Office relocation analysis

This section describes a study undertaken at FPDSavills to assess the potential impact of office relocation on morning peak rush hour travel-to-work times for staff at three office locations. For the purposes of this example, these locations have been anonymised.

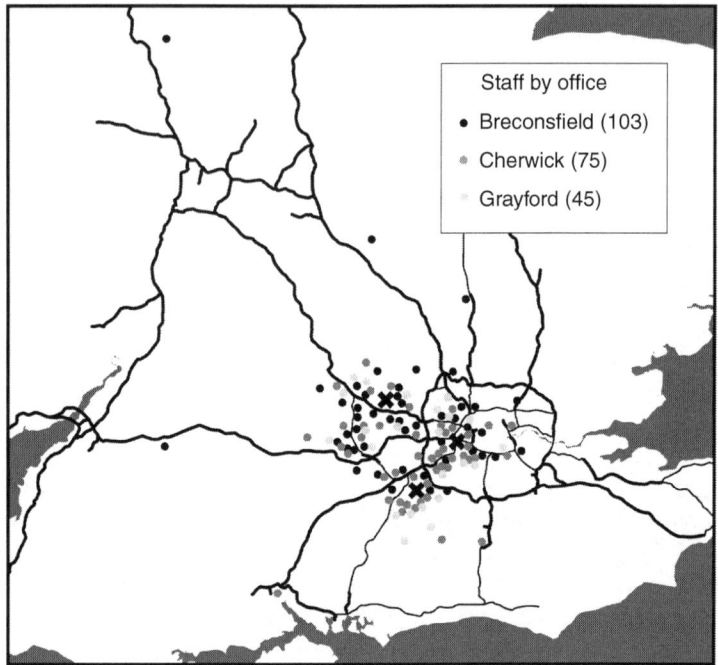

Figure 7.11 Office location (X) and staff residences.

The first stage of the analysis was to define the overall geographical extent within which the preferred relocation would lie. This was taken to be the area from which the three existing offices can be reached. Two relocation sites were chosen and the study sought to calculate the changes to staff travel times as a result of relocation to each of these sites. The three current office locations are Cherwick, Grayford and Breconsfield. Travel-to-work drive-times were generated to represent the time it takes the staff to travel to work during the peak morning rush hour. This includes modelling areas of extreme congestion within the western M25 area and tributary motorways.

The home addresses of the 223 employees were plotted (Figure 7.11) based on full postcodes. The shading of the points represented the office to which employees currently travel to work and the numbers provided in the legend of the map indicate the number of employees in each office. The crosses indicate the location of the current offices. Figure 7.12 provides a closer view and excludes from the map those employees living outside the local area (outliers).

Figures 7.13, 7.14 and 7.15 show 20-, 40- and 60-minute drive-time bands to each of the existing offices during the peak morning rush hour.

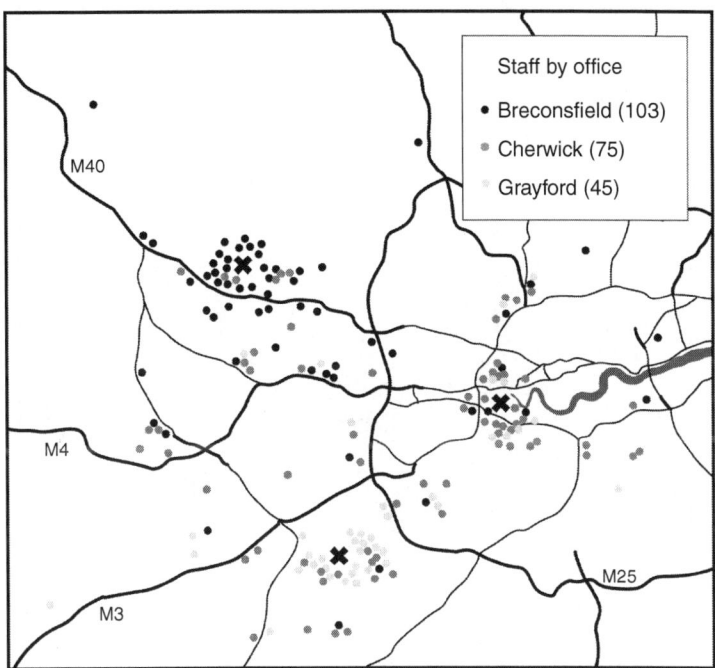

Figure 7.12 Office location and staff residences (excluding employees not commuting from within the local area).

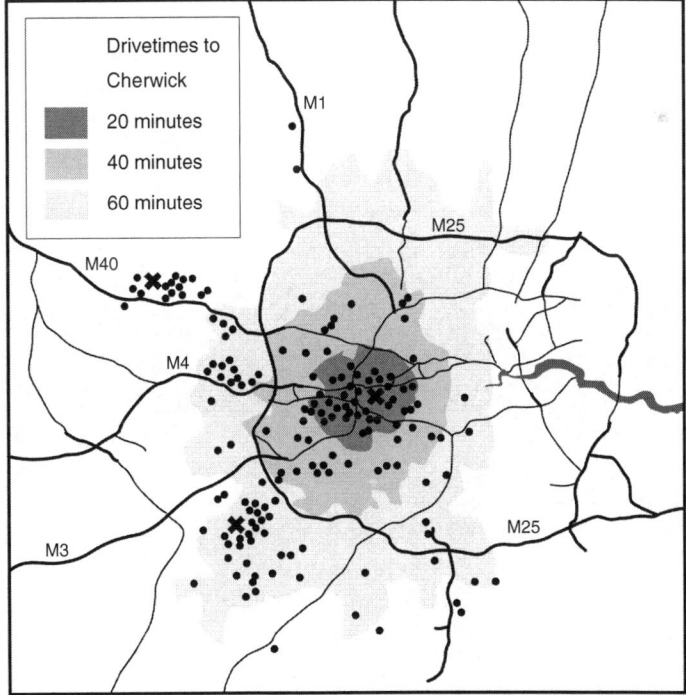

Figure 7.13 The 20-, 40- and 60-minute peak drive-times to the Cherwick office.

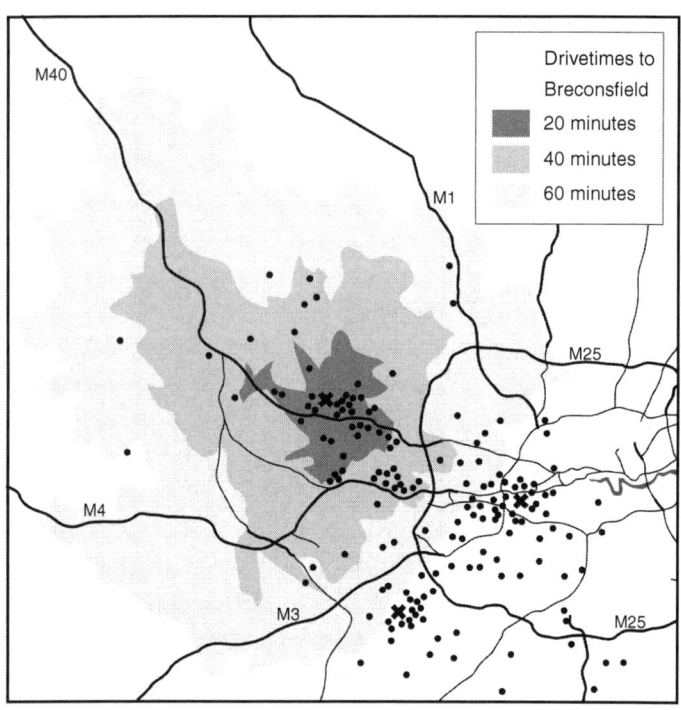

Figure 7.14 The 20-, 40- and 60-minute peak drive-times to the Breconsfield office.

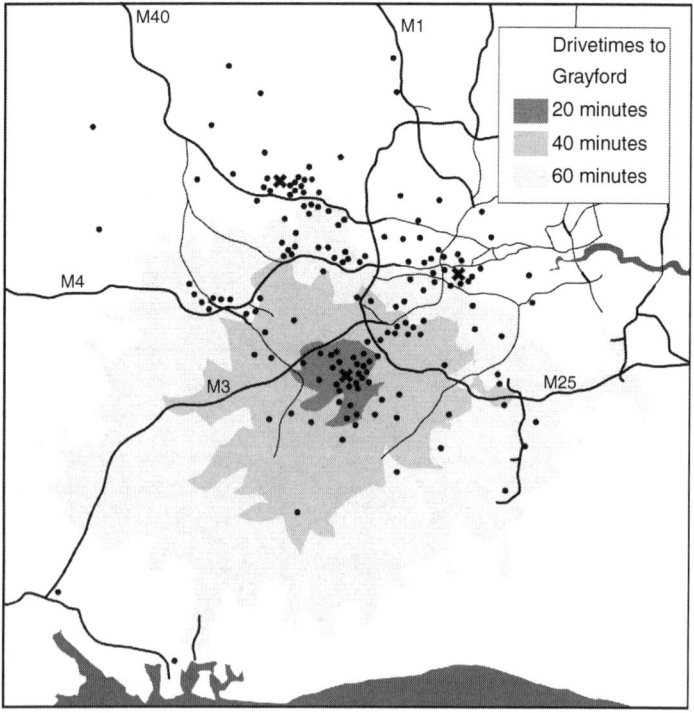

Figure 7.15 The 20-, 40- and 60-minute peak drive-times to the Grayford office.

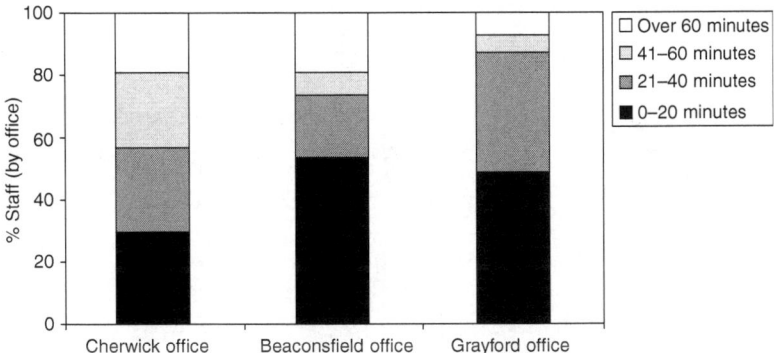

Figure 7.16 Existing morning peak travel-to-work times.

The locations of staff residences have also been superimposed onto the map. Figure 7.16 summarises the percentages of staff located within the drive-time bands of their respective offices. Breconsfield and Grayford both had approximately 50 per cent of their staff within 20 minutes morning peak drive of the office. This compared to 31 per cent for Cherwick employees. The 41–60-minute drive-time band to the Cherwick office comprised the largest proportion of Cherwick employees at 31 per cent. Breconsfield and Cherwick had over 18 per cent of their office staff spending more than 60 minutes travelling to work. Only 7 per cent of Grayford employees lived more than an hour away from their current office.

Using the peak drive-times a composite catchment map was created (Figure 7.17). The shaded areas show the locations from which each office can be reached within 60 minutes. The dark shaded areas in the centre of the map is where all three existing offices can be reached within 60 minutes. This has been identified as the preferred area for relocation. The two relocation sites indicated by crosses on the map are close to or within this preferred relocation area, in Sloughton and Blockworth. Figures 7.18 and 7.19 show 20-, 40- and 60-minute peak drive-times for the two relocation sites. Again, the locations of staff residences have been plotted on the map.

Assuming that all staff are relocated to a single office, Figure 7.20 shows what the new morning peak travel-to-work times for each proposed location would be. Overall travel-to-work time profiles for the two proposed offices were very similar. Both new sites would increase average travel time by 14–16 minutes, with three-quarters of staff travelling for between 21 and 60 minutes to get to work. Figure 7.21 breaks down these changes in travel to work to show figures as an average time spent travelling to work. The calculation of average travel to work per person per office is based on 10-minute bandings up to 90 minutes. Those staff travelling for 90 minutes

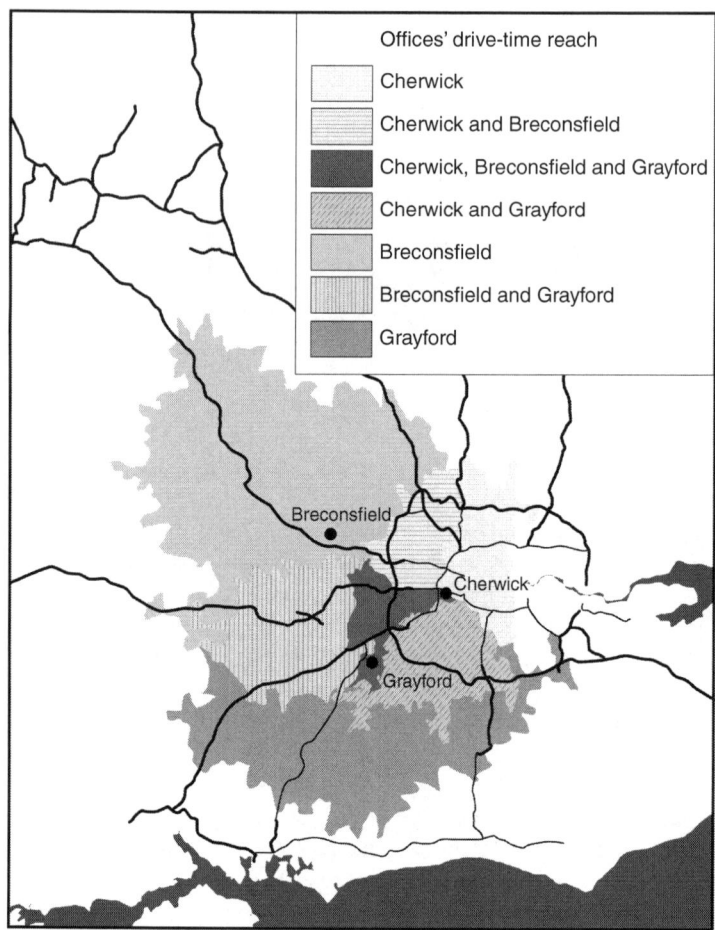

Figure 7.17 Overlap between the three main offices of 1-hour peak drive-times.

or more are averaged at 120 minutes. Tables 7.1 and 7.2 show the changes between bands of travel time in greater detail. The travel bands in the row (reading across) represent the current travel to work time. The travel bands in the column (reading down) represent the new travel-to-work time of the new site location. For example, in the top section of the table (Sloughton) in the first row is a Figure of 58. This is in the row 0–20 minutes and the column 21–40 minutes. This means that 58 people currently travelling for 0–20 minutes to work would spend 21–40 minutes travelling if the new office was located at the site in Sloughton. The figures highlighted in bold are staff who have remained within the same travel-to-work time band. The figures below the bold are those whose travel-to-work time has

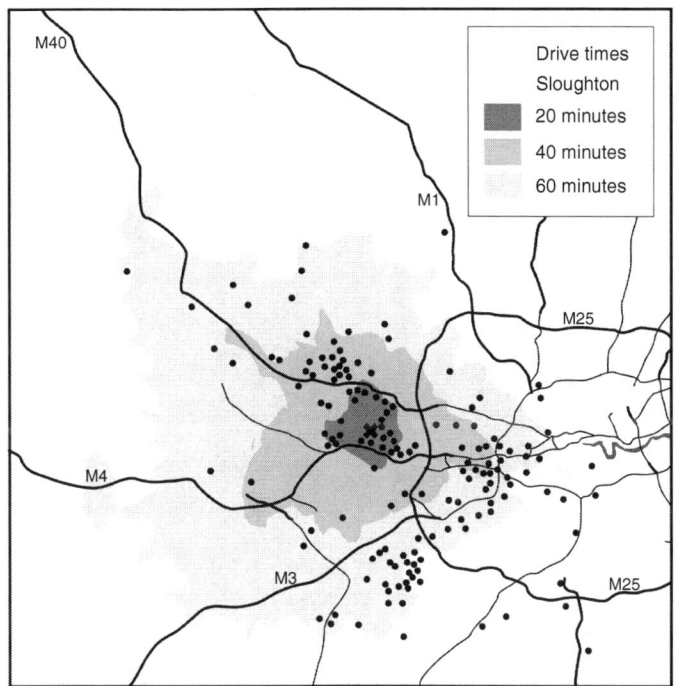

Figure 7.18 The 20-, 40- and 60-minute peak drive-times to the proposed Sloughton site.

Figure 7.19 The 20-, 40- and 60-minute peak drive-times to the proposed Blockworth site.

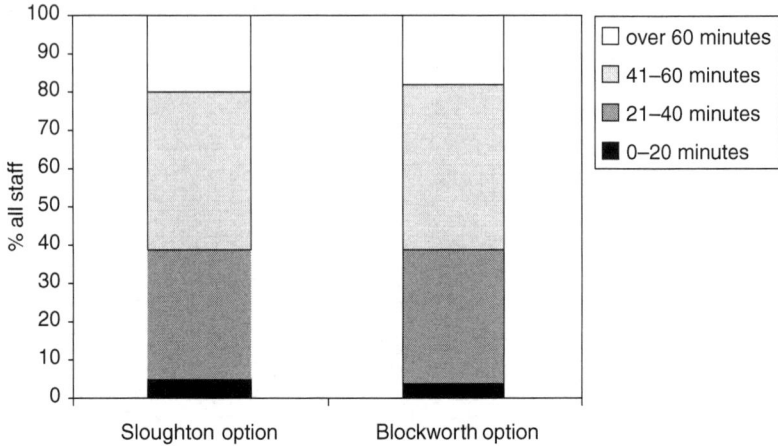

Figure 7.20 Morning peak travel-to-work times for proposed offices.

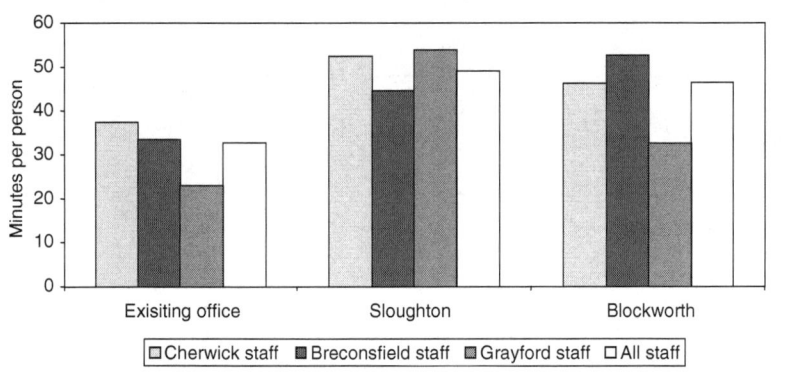

Figure 7.21 Average travel-to-work time.

Note
Travel time over 90 minutes is averaged at 120 minutes.

decreased and those above the bold have increased. Therefore the larger the numbers to the top right represents a more negative impact on the travel-to-work time and the larger the numbers to the bottom left the more positive the impact.

The impact on travel-to-work time is as follows. Average travel-to-work time for all staff is currently 33 minutes. Shortest average travel-to-work time is for staff commuting to the Grayford office with 23 minutes. Relocation to Sloughton site would mean an average travel-to-work time for all staff of 49 minutes. The Grayford staff's average travel time would incur the largest increase (31 minutes), Breconsfield staff's average travel time would increase by 11 minutes while Cherwick's would increase

Table 7.1 Drive-time matrix of all staff to each site option within drive-time bands – numbers

Current travel time bands (minutes)	0–20	21–40	41–60	Over 60	Grand total
Sloughton					
0–20	3	58	35	3	99
21–40	7	10	29	11	57
41–60	2	5	16	8	31
Over 60	0	1	12	23	36
Grand total	12	74	92	45	223
Blockworth					
Count of travel bands (Current)					
0–20	3	29	64	3	99
21–40	3	21	23	10	57
41–60	2	19	6	4	31
Over 60	3	6	5	22	36
Grand total	11	75	98	39	223

Table 7.2 Drive-time distribution of all staff to each site option within drive-time bands – percentages

Current travel time bands (minutes)	0–20	21–40	41–60	Over 60	Grand total
Sloughton					
0–20	1	26	16	1	44
21–40	3	4	13	5	26
41–60	1	2	7	4	14
Over 60	0	0	5	10	16
Grand total	5	33	41	20	100
Blockworth					
0–20	1	13	29	1	44
21–40	1	9	10	4	26
41–60	1	9	3	2	14
Over 60	1	3	2	10	16
Grand total	5	34	44	17	100

by 16 minutes. Of the 20 per cent who would have to travel for more than 60 minutes, 10 per cent would not have had to do so previously. Three-quarters of the staff would travel for between 21 and 60 minutes to get to work. Relocation to Blockworth site would mean an average travel-to-work time for all staff would be 49 minutes and a significant increase in average travel-to-work time for the Breconsfield office (up 19 minutes). The average travel time would increase by 9 minutes and 10 minutes for the Cherwick and Grayford offices, respectively. Of the 17 per cent who would have to travel for more than 60 minutes, only 7 per cent would not have done so previously and a majority of staff, 78 per cent, would travel for between 21 and 60 minutes to get to work.

To summarise, overall travel-to-work time profiles for the two proposed offices are very similar. Both new sites will increase average travel time by 14–16 minutes, with three-quarters of staff travelling for between 21 and 60 minutes to get to work. Blockworth would be the preferred option for the Grayford office, increasing travel time by 10 minutes compared to 31 minutes for a move to Sloughton. Sloughton would be the preferred option for the Breconsfield office with average travel time increasing by 11 minutes compared to a 19-minute average increase for a move to Blockworth. Blockworth would be the preferred option for the Cherwick office, increasing travel time by 9 minutes compared to 16 minutes for a move to Sloughton.

Overseas case study – US low-level radioactive waste disposal facility (Cargin and Dwyer, 1995)

Federal law requires that each US state be responsible for ensuring that low-level radioactive waste generated within its borders is disposed of safely. A state may do this individually, or may combine their efforts with other states and form a compact. Pennsylvania, Maryland, West Virginia and Delaware joined to form the Appalachian Compact, with Pennsylvania agreeing to site the first low-level radioactive waste disposal facility for the Compact. A GIS, designed to locate a suitable site for the facility, has become an integral part of the screening process. Screening is divided into two tasks: disqualification (determining where the facility cannot be located) followed by evaluation (determining where the facility should be located).

The disqualification process eliminates areas of the state that are not suitable for siting a low-level radioactive waste disposal facility. During the disqualification phase of the project, a total of eighteen disqualification criteria were processed, which involved the application of thirty-five individual disqualification data layers in a GIS (see Table 7.3). The disqualification data layers covered a wide range of size and complexity, from National Fish Hatcheries (four polygons and 300 acres disqualified) to Geological Stability (nearly 20,000 polygons and 423,000 acres disqualified), and Exceptional Value Wetlands (5,500 polygons and 9.9 million acres disqualified). Disqualification was applied in a three stage process, with Stage One being applied at a statewide scale (1:250,000), Stage Two at a regional scale (1:50,000), and Stage Three at a local scale (1:24,000). The staged approach was used to apply criteria that removed large areas at an early stage in the process.

Once an area was disqualified by one criterion, there was no need to look any further in those areas for any other disqualifying features. The disqualification process eliminated approximately 75 per cent of the state from consideration as potential disposal facility site.

In the evaluation process, the remaining non-disqualified portions of the state were screened as to their potential for siting the proposed facility. Fourteen additional criteria were applied, each of which designated an area as either more or less favourable for licensing. Many of the evaluation screening criteria involved buffering map features to identify less favourable areas. For example, it is less favourable to be within a mile of a Wildlife Area Boundary. Other criteria, however, such as Road Hazards, presented a much more difficult challenge. The Road Hazards evaluation involved investigating each non-disqualified/non-deferred area to determine if it was possible to travel from the nearest limited access highway to the area without passing through four intersections in any given 1-mile segment. Given the fact that this evaluation must occur on a statewide basis, and that there are over 1,500 possible exit points from the limited access highway system, this quickly became a substantial task. GIS networking capabilities were used to automate this analysis as far as possible. The process then combines these technical scores with a public values survey to factor in public rankings and develop an overall rank–sum index score for each area, with the top 100 areas moving forward into Site Selection. In Site Selection, the remaining criteria were used to narrow consideration from the 100 areas to a subset from which three potentially suitable sites were chosen.

Insurance and finance

Insurance

Insurers traditionally targeted house insurance risk to a geographical extent that is determined by the first half of the postcode. This equates to approximately 10,000 properties. In other words the risk that a single property was estimated to be subject to was shared between 10,000 'similar' properties in the locality. With the aid of GIS, insurers are increasingly able to focus on the best risks by using more sophisticated analysis to target risk at a more refined geographical level. Insurers can combine data from the British Geological Survey and other data providers to group properties that are believed to be subject to similar risk by using the first digit of the second half of the postcode. This equates to approximately 2,000 properties. Consequently, properties in locations that are 'high risk', from subsidence

Table 7.3 Disqualifying criteria

Masking facilities
Active faults
Geologic stability
Slope
Carbonate lithology
River floodplains
Coastal floodplains
Important wetland
Dam inundation
Public water supply
Surface water intake
Wildlife Area Boundaries
State forests and game lands
Watersheds
Oil and gas areas
Agricultural land
Mines
Protected area boundaries

---

or flooding for example, are becoming increasingly difficult to insure and premiums are high in comparison to those that are considered low risk.

The UK international reinsurance broker Benfield Grieg Ltd specialises in identifying and quantifying risks associated with catastrophic hazards in the UK – essentially windstorms (such as the 1987 storm) and coastal flooding. Clients of Benfield Grieg usually hold portfolios of property, boats and aircraft, all of which have to be insured. The reinsurer's role is to spread the accumulation of exposure of individual insurance companies, by purchasing reinsurance in the major international markets, such as Lloyds of London. Prior to 1993, reinsurance was relatively cheap and easy to find and the UK was not considered a catastrophic risk area. In addition there was a 'relative non-sophistication' in the reinsurance business, with the technology to do meaningful analysis via GIS not being widely available. The 1987 storm in the UK was considered a one-off freak, however more storms in the 1990s indicated a change in the established weather patterns. The increase in global natural catastrophies changed perceptions and meant that it was necessary to have a scientific understanding of risk and an appreciation of the financial amounts involved.

The introduction of a GIS has played a pivotal role in spreading the risks and, in some cases, reducing premiums in the wake of the troubles that reinsurers suffered in the early 1990s. The GIS allows layers of sometimes complex data to be presented on a map, allowing rapid understanding of what the data means. GIS permits the integration of scientific data with insurance data to give decision support for financial decision-making. The system enables the results from the data analysis to be depicted as maps, giving a 'healthcheck' that the underlying model makes sense. This visualisation of the assessment of clients' exposure to catastrophic perils, and associated levels of probability, helps them decide how much reinsurance protection to buy. Data sets used include digital terrain data, weather information and internal data such as flood damage claims.

Benfield Greig Ltd has six GIS specialists in its 31-strong Research and Development (R&D) team, which also includes statisticians, econometricians, meteorologists and geophysicists. The team provides services to clients, typically looking to insure blocks of 500,000 houses, around the world. It started its GIS programme in 1992, using desktop mapping systems. However these lacked the power and functionality of a full GIS, which it began using in 1994, although desktop mapping is still used for visualisation purposes, such as thematic mapping. The company developed a flood modelling GIS application to determine the financial impact of environmental catastrophies on insurance companies. The real strength of a GIS is that it can read many different computer formats, manage huge data sets and carry out spatial analysis. This linking of locations on maps with data held on various databases has permitted the development of four modelling applications; flood, wind, terrorism and earthquakes. Most of these models are grid-based, usually wind and flood models are produced in 100×100 metre cells (giving 60 million cells for the UK coastline). This allows the production of different models for different types of land surface, as trees, hills and rough ground will all affect flood and wind flow in different ways. Simulations of extremes can be run in order to see what happens in an area, the data used often being based on historical patterns.

Overseas case study

The US Aon Corporation is an international company that specialises in insurance brokerage and risk management products and services. The organisation has offices in more than 120 countries and Aon UK has developed an Insurance Portfolio Management System, which enables insurers and reinsurers to establish the risks associated with a particular peril. The GIS-based system enables the exposure to risk (coastal and river flooding, freezes, crime, wind storm damage, subsidence, for example) to be assessed. Clients can assess exposure to risk for their own portfolio and claims by examining maps, and tables and graphs are used to give a statistical view of the data. It is claimed that the system ensures that insurance rates more accurately reflect the risks involved and improved response times to process requests for insurance and claims.

In the past insurers and reinsurers relied on claims' history to calculate their exposure. In terms of the geographical nature of their portfolio they had little understanding of what perils they were exposed to. The use of GIS enables a whole range of data sets to be incorporated into a variety of models that allow the actuaries to analyse and model the exposure to risk more accurately. Data sets used include addresses and postcodes, height data, soil and geology, weather, socio-economic statistics, crime statistics and internal data such as previous claims and portfolio details.

The Insurance Portfolio Management System consists of five modules:

1   Coastal and river flood modelling – calculates property damage scales based on type, age, location and socio-economic group of occupants. Two databases of over 1.7 million commercial properties, combined with the Department for the Environment, Farming and Rural Affairs/Institute of Hydrology river flooding model, show which properties are at risk from 1-in-100 year floods.

2   Freeze modelling – assesses freeze damage probability in different regions. Known losses in the past fifteen years are used to associate predicted claim frequency with actual loss. This allows the prediction of an overall market- or company-specific loss for a given set of temperatures.

3   Crime modelling – correlates subsets of census information with police statistics at the postcode level and is used to model domestic theft. Integrating this model with claims experience helps produce a scientific model to assess crime risk and to price insurance accordingly.

4   Wind modelling – integrates hourly weather data to generate wind speed contours and a 3D surface of wind activity. Sixty-five individual events have been identified from wind activity over the last twenty-five years, which can be used in the model to simulate storms of varying intensities and these help derive loss estimates based on damage factors.

5   Subsidence modelling – utilises a geo-hazard determined from a formula that takes into account soil type percentages, property type and age. This geo-hazard together with other types of subsidence risk (mining, natural cavities, erosion of silts and fine sands, cambering) allows the risk to be estimated. The model also includes an analysis of soil moisture deficit for different time periods.

Finance

Henderson Global Investors is an international investment management company. Its main business areas are active and fixed interest asset management, property services, private capital, listed investment trusts, retail unit trusts and offshore funds. It also provides portfolio management and private client services to individuals and charities in the UK. Henderson enjoys a strong reputation for its global property expertise. GIS has been used for property investment research at Henderson for the last five years and it is regarded as a source of competitive advantage. There are four areas in which GIS is used to assist property investment decision-making.

(a) Mapping and presentation: Much of Henderson's property investment data is geo-coded. GIS is used to map the locations of property investments within individual funds in order to illustrate the geographical spread of assets at local, regional and national levels. GIS is a useful visualisation tool in this regard. Henderson uses small-scale raster mapping as a base map for locating individual investment properties.

(b) Regional analysis of property investment performance: The performance measures of property investments (total return, income and capital returns, rental growth) are monitored geographically to identify geographical patterns or time-series trends. This is done at a regional level and is particularly useful for office and industrial investment property because geographical patterns are more discernible at this scale than they are for retail property, where more asset-level or micro-scale analysis is appropriate.

(c) Demographic analysis (retail catchments and drive-times): Demographic analysis consists of drive-time calculations in order to evaluate customer and labour potential of urban areas and regions. Road mapping is used together with drive-time software. Henderson is seeking to enhance this sort of analysis by including data derived from retail store loyalty cards. This will provide customer-centred information (which is better than retail store surveys but not as good as the CB Hillier Parker electoral roll survey described in the 'Retail property consultants' section because loyalty cards do not capture new markets and are biased towards loyalty card users).

(d) Asset-level performance measurement: Property investment research needs to shift its focus from the portfolio to the asset level because fund management decisions inevitably distil to property specifics. Henderson is finding limitations in the use of GIS at the property level. The key drawback with geographical analysis at this scale is the lack of data points on which to base reliable statistical analysis. For example, looking at the effect that an out-of-town retail development may be having on city centre retail activity is difficult without sufficient data. The lack of data points makes the results from any surface interpolation questionable. Ideally it should be possible to analyse property investment activity at the asset level, particularly with regard to retail property. The real power of GIS for property analysis lies with its ability to reveal trends and patterns at the property level but there is still some way to go before this type of analysis is reliable.

Henderson has produced rent maps at a regional scale for office and industrial property (at this scale retail rent maps are of little use). The use of GIS to map accessibility against rent is interesting but more revealing would be to map accessibility against turnover as this is a more refined measure of profitability. This type of analysis is discussed in Chapter 8. It would also be interesting to look at the way in which planning decisions affect retail performance at a micro level. At the moment local authorities primarily use GIS for data integration and management but clearly use of GIS for forward planning will be very powerful (e.g. location of housing land having regard to say flood plains and previously developed land). Local authorities need to look at the implications of their planning decisions and planning policy. As can be appreciated from Chapter 5, the stage that local authorities are at with regard to GIS implementation is fairly typical across the property profession – the full potential of GIS has not yet been realised.

Overseas case study – US real estate advisors (Harder, 1997)

SSR Realty Advisors, Inc., a subsidiary of The Metropolitan Life Insurance Company, is a commercial real estate investment firm that manages assets worth $2.6 billion. The firm's performance is measured largely by reference to the returns on the investments that it manages. At any one time SSR may be reviewing up to 300 properties. Investment acquisition, management and disposal decisions rely on an extensive research database. On this database, research data is organised geographically. The user interface displays a map of the US with the 100 largest Metropolitan Statistical Areas (MSAs). From this interface map queries and access to specific tools is possible. The asset management tool shows a map overlaid by points that represent properties managed by SSR, coloured according to property type. Polygons representing the MSAs and points identifying the location of specific property assets are shown. After selecting a particular MSA the map zooms to its geographical extent and displays further information (earthquake faults, toxic waste sites, freeways, emergency services). Once the asset manager has selected his or her name from a drop-down list, the display is filtered to show only those properties under his or her management. Properties can be shaded according to their performance against a national benchmark. Further buttons on the interface allow analysis of lease terms, cash flow and property performance. Each dot (representing individual properties) can be selected to reveal all property information, photographs and floor plans. Regional economic data and office property vacancies can be accessed by clicking on the library button.

Summary

The use of GIS for retail market analysis has been one of the major application areas of GIS in recent years, particularly in the private sector. The nature of these applications differs from the local government-based property management systems described in Chapter 5. The retail applications tend to be more project-specific both in terms of geography and property decision. They also tend to involve a greater degree of analysis rather than data management. Much of this analysis relates to demographics, customer data, store location, competitor locations and the route network. Geographical analysis and modelling can be quite sophisticated.

The use of GIS for financial services is still in its infancy but potentially this is a very significant application area. In this chapter we have concentrated on areas where this market overlaps with property, namely property investment decisions and property insurance risk analysis.

Retail and financial services applications are well served with geo-demographic data, software and service providers but these data and service providers are currently quite fragmented. Over the next few years many of them will undoubtedly move away from providing standard data products towards a data integration service, pulling data together from various suppliers. This sort of service will rely heavily on the widespread adoption of the NLPG, described in Chapter 4. Also, more and more users are starting to geo-code their own data. They are encountering problems in the process and therefore initiatives such as NLIS and NLPG will help here. Some of the problems relating to the price and availability of data are discussed in Part III.

It should be clear by now that the chapters within this part of the book have progressively introduced application areas with an increased level of sophistication of analysis. Comprehensive land and property information management systems have given way to more project-specific applications that often integrate disparate data sets to help inform strategic decision-making. Perhaps unsurprisingly, there are more data management applications of GIS than there are data analysis applications.

References

Buchanan, H., Fairbairn, D., Taylor, G., Stevenson, P. and Wall, J., 1999, The use of geographical information systems in site selection, RICS Research, RICS, London.

Cargin, J. and Dwyer, J., 1995, Pennsylvania's Low-Level Radioactive Waste Disposal Facility Siting Project: Process Summary, ESRI User Conference, www.esri.com/library/userconf/proc95

Clark, M., 1991, Developing spatial decision support systems for retail planning and analysis, AGI Conference Proceedings, Birmingham.

Cowen, D., Jensen, J., Shirley, W., Zhou, Y. and Remington, K., 1999, Commercial real estate GIS site evaluation models: interfaces to ArcView, ESRI User Conference Proceedings, www.esri.com

Grimshaw, D., 2000, *Bringing Geographical Information Systems into Business* (2nd edition), John Wiley & Sons, New York.

Harder, C., 1997, ArcView GIS means business, Environmental Systems Research Institute, Redlands California.

Hornby, R., 1991, Successful application of GIS at Nationwide Anglia, AGI Conference, 1.2.1–1.2.6.

Goodwin, T., 1997, Fighting the supermarket sweep, Mapping Awareness, June, 16–18.

Kirkwood, J., 1998, GIS insight on site, Estates Gazette, 9847, November 21, 130–131.

Longley, P. and Clarke, G. (eds), 1996, *GIS for Business and Service Planning*, Wiley, Chichester, UK.

Walker, J., 2000, Developing a network optimisation model for a major retail brand, AGI Conference, London.

Wicks, M., 1995, Increasing management effectiveness – beyond the spreadsheet, GIS for Business, Geoinformation International, Cambridge, UK, 52–54.

Wilkinson, C., 1991, Store performance – evaluation the ingredients of a successful site, AGI Conference, 1.1.1–1.1.5.

Smith, C. and Webb, J., 1997, Using GIS to improve estimates of future retail space demand, *The Appraisal Journal*, October, 337–341.

Wyatt, P., 1999, Geographical analysis in property valuation, Research Report, Royal Institution of Chartered Surveyors, London.

8 Property market analysis

Introduction

This final chapter in Part II describes how GIS is being used to examine characteristics of property market activity as an aid to decision-making. It may seem odd that an industry, which produces, manages, occupies and trades in a good that is inherently geographical has taken so long to take advantage of GIS technology. After all, every property is unique, if only in terms of its geographical location, so geography plays a central part in most property decisions. The spatial dimension of property is a distinguishing characteristic that has contributed to the creation of a separate field of study. However before the development of GIS software, it was difficult to accurately measure the explicit impact of location in property models. As Rodriguez *et al.* (1995) state, GIS technology provides users with the ability to enhance property analysis.

The incorporation of geography in property decision-making varies depending on the process. Developers know only too well the importance of choosing the right location for a new shopping centre or a distribution park. We have already seen how planners and certain business sectors use GIS to assist their location decision-making in Chapters 6 and 7. The wider profession is now beginning to realise the benefits of and rewards from examining the geography of property decisions more carefully. This chapter begins by describing the pioneering work of property data suppliers such as PMA, Property Intelligence and IPD in incorporating mapping into their products and services. The chapter then turns to more advanced uses of GIS in property market analysis and research.

Visualisation of property data

Simple bespoke maps help illustrate the geography of the built environment in a way that text description, numerical or statistical analysis fails to impart. Showing the location of competitor businesses, sizes of catchment areas, proximity to other urban land uses, for example, helps inform property decision-making.

Property market analysis

PMA is a property data consultancy whose main product PROMIS (Property Market Information Service) is a compilation of data on property market conditions, take-up and availability, current and potential supply, market prospects and transaction details. It provides Internet access to reports on the office, industrial, town centre retail and out-of-town retail markets in over 100 towns and cities. The service also includes information on the local economy, town, employment and unemployment, demographics and catchment analysis. Allied to PROMIS is PMAs development database, which includes data from all stages of the development process (application, permission, construction and completions). For office and industrial uses historic construction/development pipeline data are also recorded to enable PMA to create a full completion and stock series back to 1975. These data were geo-coded by local authority district, but since 1995 this has been undertaken on a property-by-property basis. Take-up and availability data are collected from surveyors and tend to relate to urban centres as defined by surveyors rather than standard local authority districts. Transaction data are collected from various published sources and stored on a building-by-building basis.

Within PROMIS key development schemes and transaction details are illustrated on topographic maps. Town centre properties are illustrated on custom raster maps, retail maps are based on simplified Goad Maps whilst out-of-town schemes are presented using vector maps. PROMIS classifies property as industrial, retail, out-of-town retail or office and maps are included for each property type showing the location and transport infrastructure for the urban area under investigation. For example, Figure 8.1 is a national scale map illustrating 2.25- and 4.5-hour drive-times for industrial property around Luton. Key industrial locations are highlighted on a map (Figure 8.2) of the hinterland of the selected urban area, together with the main transport routes. Associated attribute data record the address, vintage and key occupiers for each location. A similar scale map shows the locations of key industrial developments, and attribute data provide further details such as address, developer, completion date, size and planning use class. Another map shows the locations of key industrial deals and attribute data record the address, size, occupier, date of transaction, rent and other lease details.

Retail PROMIS includes maps showing the primary catchment area (similar to that shown in Figure 8.3) and larger scale town centre maps, such as Figure 8.4, identify retail provision (including key stores, shopping centres and developments underway) and market conditions (including key deals and zone A rents). Another map, shown in Figure 8.5, indicates the location of competing retail centres. Out-of-town Retail PROMIS includes regional maps that show the locations of retail warehousing (attribute data record the address, retail occupiers, their respective trading activity and

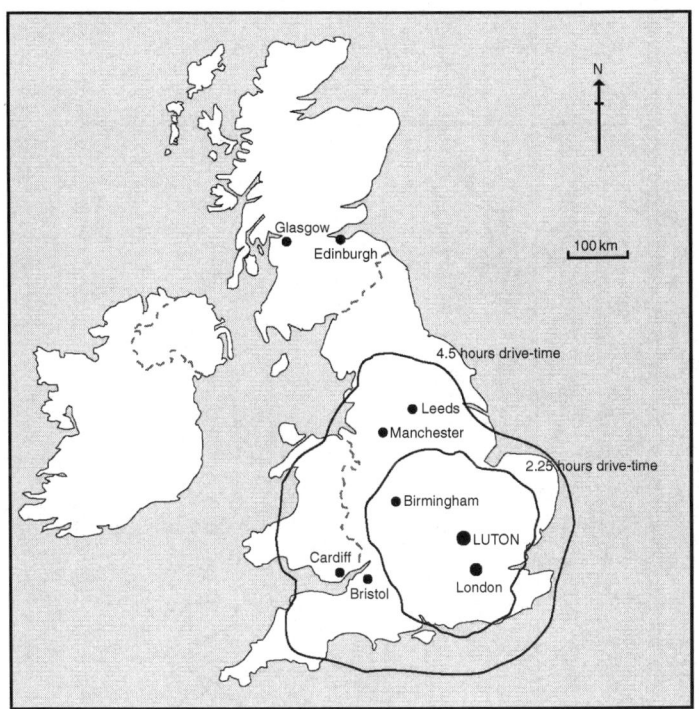

Figure 8.1 Luton drive-time map.

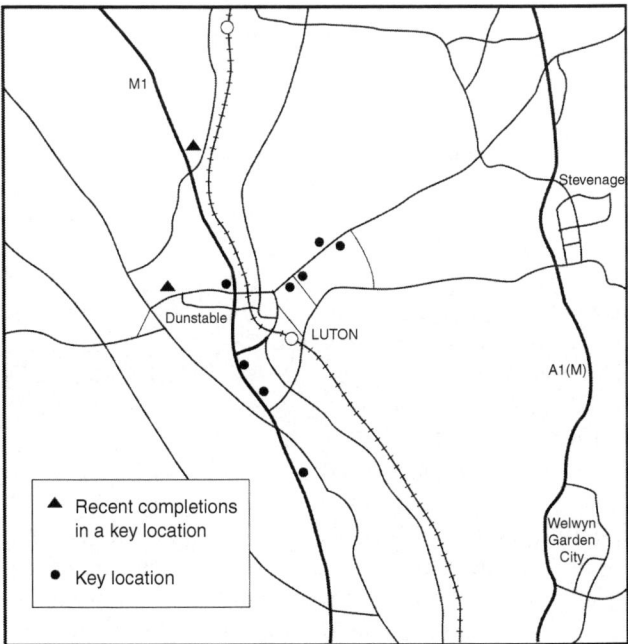

Figure 8.2 Luton key industrial locations.

Figure 8.3 Primary catchment area around Neston.

floor area), superstores (attribute data record the address and retailer name) and out-of-town retail developments in progress (attribute data record the scheme name, developer, size, development stage and other comments).

Office PROMIS includes maps that illustrate the location of business and science parks in and around the chosen urban area as well as the town centre and out-of-town development pipeline. Attribute data record the name, address and size, completion date (where applicable) of the schemes. Further maps show the location of key town centre and out-of-town deals and availability. Attribute data record the address, occupier, size, date of transaction, rent and other lease details.

EuroPROMIS follows a similar format to the UK service, covering 50 retail markets and 40 office markets in 16 countries. Like PROMIS, EuroPROMIS identifies and maps key property information. Town centre records are illustrated on custom raster maps of the key streets, whilst out-of-town schemes are visualised using vector maps. Data include catchment population, supply, household growth and market conditions. GIS has been fundamental in the creation of consistent data sets across all of the centres. In most cases catchments are based on administrative areas at the European

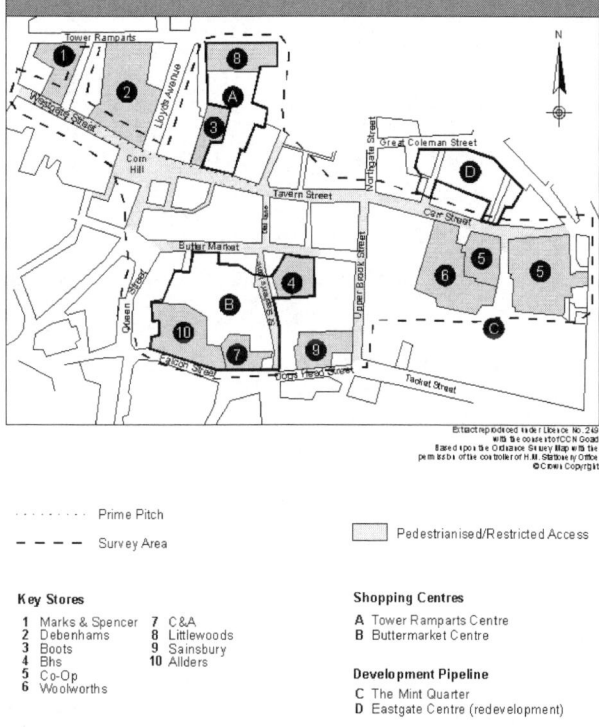

Figure 8.4 Ipswich town centre – retail provision.

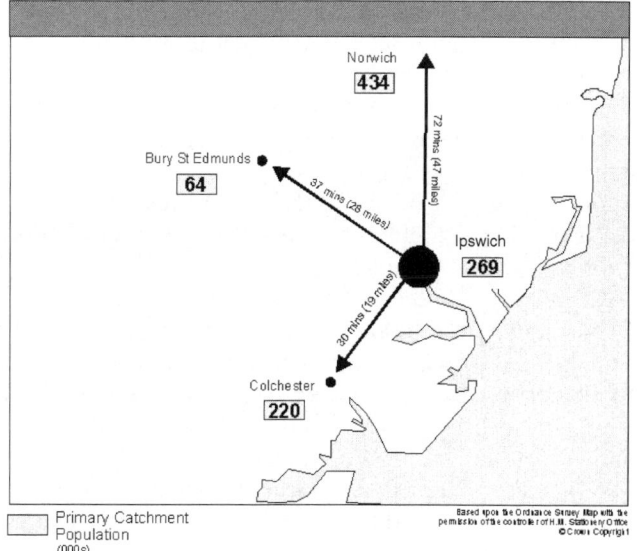

Figure 8.5 Ipswich and competing centres.

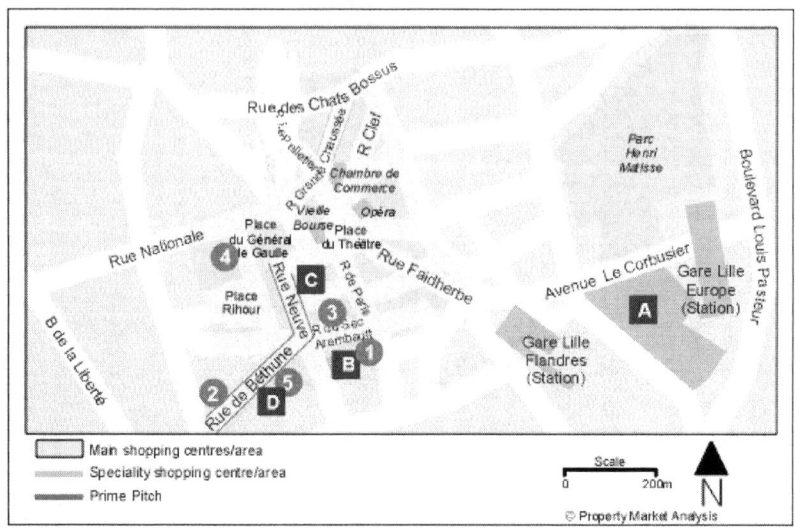

Figure 8.6 Street-based property map for Lille, France.

NUTS (Nomenclature des Unites Terratoriales Statistique) 5' level, equivalent to wards in the UK. Economic, demographic and household information has then been collected from official national sources for the defined catchments whilst property demand and supply variables for the same area have been obtained by collating property specific information. Geo-references for each property have been captured to identify whether it falls within a defined catchment. Simple street and regional maps of the urban centres throughout Europe, such as that illustrated in Figure 8.6 for Lille in France, are used to show the location of key city centre retail stores, development projects, out-of-town and suburban shopping centres and retail parks. Associated attribute data provide further details such as the occupier name(s), year opened, size and other details regarding the physical nature and ownership of the properties.

Information, collected through PROMIS at the local level and, wherever possible, on a building-by-building basis, also supports PMA's other key services. Those which use GIS include:

- Retail consultancy: On bespoke projects shopping surveys are carried out to identify the primary catchment. Based on this area, demographic and population profiles are produced from Enumeration District level data; potential demand and competing supply profiles are produced from individual building information.
- Office consultancy: In addition to incorporating mapping within bespoke projects, by holding demand and supply databases on a building by building basis, GIS is used to compare town centre markets against evolving

out-of-town office markets and to analyse office markets across regions and sub-regions (M25 west vs Central London, for example).

- Central London Office Forecasting Service: As well as providing forecasts of rents, yields and values. The service also provides submarket analysis. The basis of the service is EGi's London Office Database, which has been enhanced by providing geo-codes for every building. This enables user-defined sub-markets to be created and analysed in comparison with the rest of the market. Figure 8.7 provides an example of rent analysis in the city of London.

Focus

Focus is an information service that is widely used in the property industry for:

- Collecting market transaction data for comparable evidence for valuations.
- Finding company details and property ownership details.

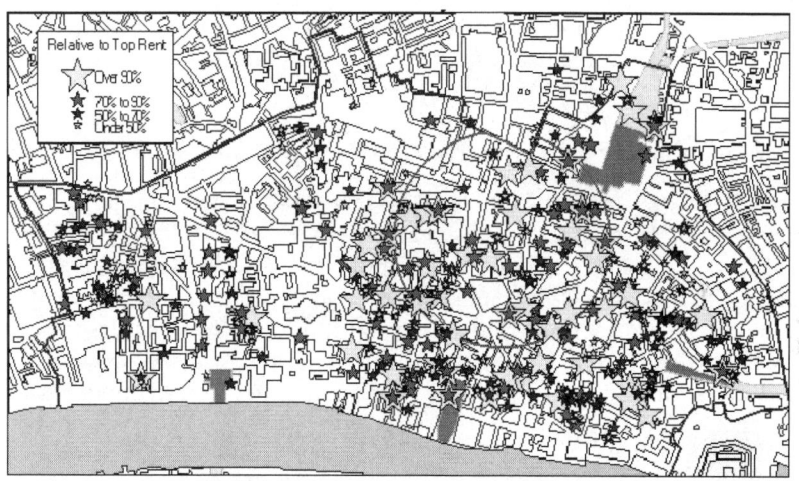

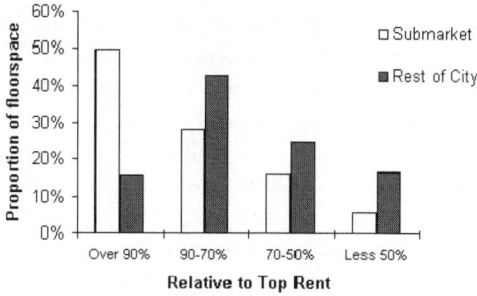

Figure 8.7 Distribution of deals by rental levels.

- Examining demographic and socio-economic data for urban areas.
- Viewing location and relocation intentions of retailers and office occupiers.
- Assessing office space availability in London.
- Accessing rateable values, auction details and results.
- Monitoring the development pipeline.

FOCUS services are Internet-based and are available as separate databases including news, property databank, town profiles, rating lists, retailer requirements, relocation leads, office availability, local authority plans, property auction guide and a London office occupier database. For Central London these services have been integrated so that the various data can be viewed simultaneously for a particular property. The Central London database uses a BS7666 compliant gazetteer and information held by clients can be integrated with the FOCUS data. Data can be viewed by floor, building or street and photographs and maps are an integral part of the system. The digital mapping is taken from the OS 1:10,000 scale map base but larger scale mapping can be incorporated. The mapping module allows the user to print, zoom and pan around the map and there is also a facility to select buildings via the map and view attribute data.

Thematic mapping of property data

Perhaps the most straightforward way that GIS can assist property market analysis is by helping to visualise patterns over space in key variables such as values, rents, yields, vacant property or land use. Thematic mapping provides a window on property market data that conventional statistical analysis cannot provide. The examples in the following sections describe how GIS has helped reveal trends in property market data that have assisted residential and commercial property investment decision-making.

Her Majesty's Land Registry

The Land Registry publishes residential price information aggregated to postcode sector level. Consideration is being given to extending this to unit postcode level. The Residential Property Price Report, published each quarter by the Land Registry since 1995, lists average prices and sales volumes in the residential property market for England and Wales. The data are classified by property type and by local authority district. A breakdown of the average sale prices of old and new properties by property type (detached, semi-detached, terraced and flat-maisonette) is also incorporated. The data are based on standard statistical regions (administrative and postal) and can be easily related to maps or incorporated into a GIS. Figures 8.8–8.12 show the average house prices for regions in England and Wales for the third

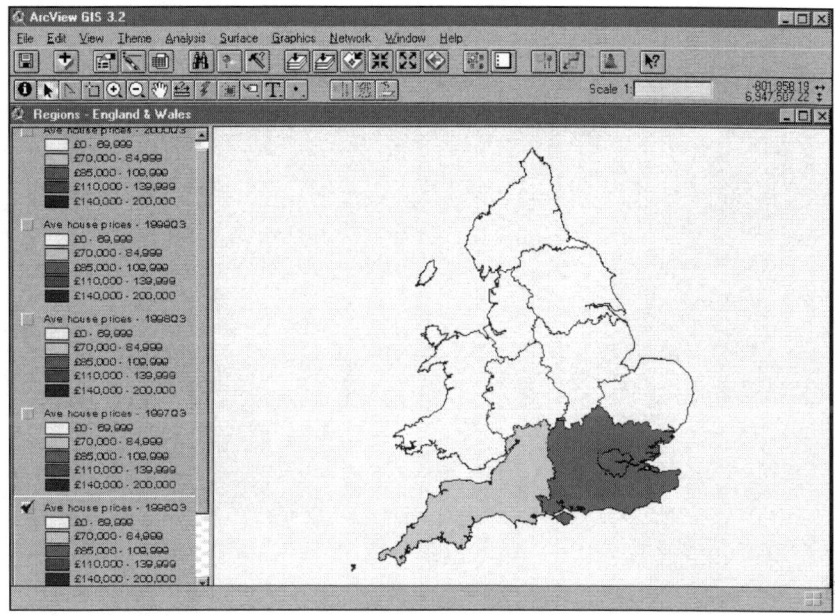

Figure 8.8 Average house prices in the third quarter of 1996.

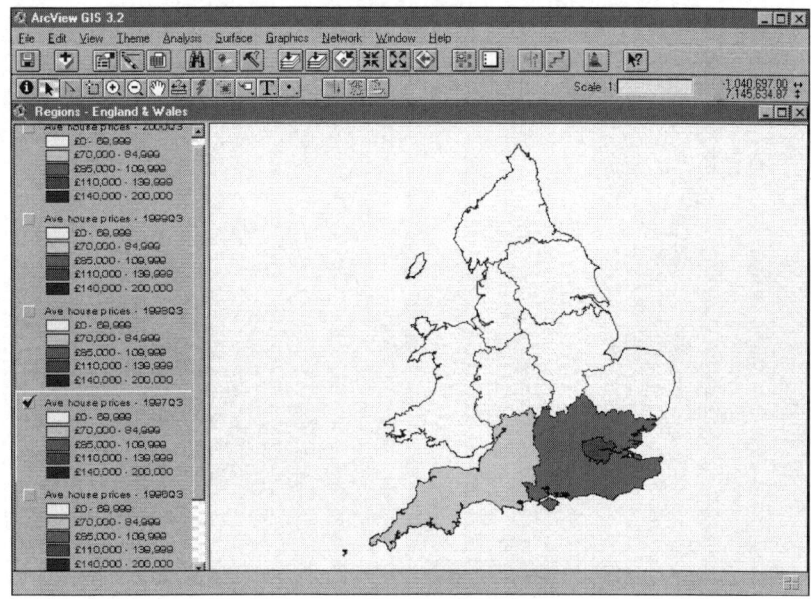

Figure 8.9 Average house prices in the third quarter of 1997.

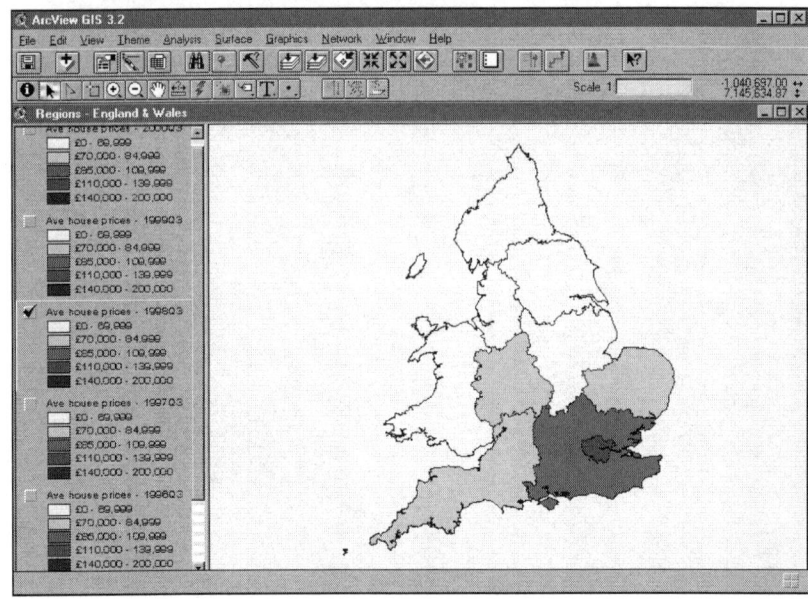

Figure 8.10 Average house prices in the third quarter of 1998.

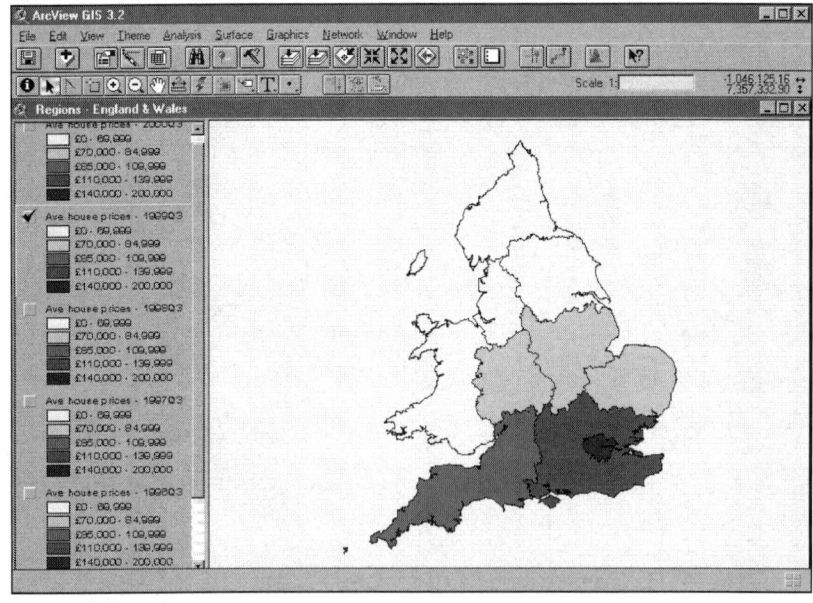

Figure 8.11 Average house prices in the third quarter of 1999.

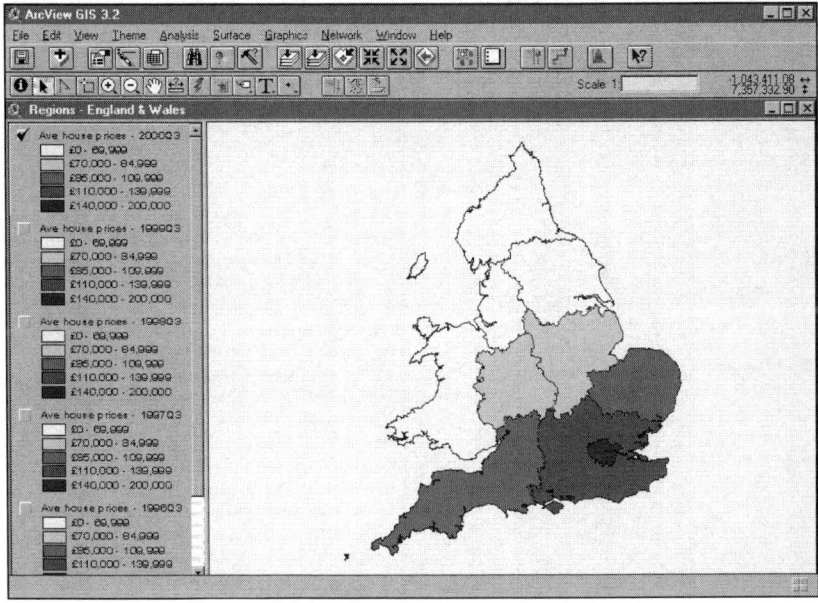

Figure 8.12 Average house prices in the third quarter of 2000.

quarter of each year from 1996 to 2000. Over this period the UK experienced significant growth in average house prices, but the way in which this growth not only differed geographically but also 'filtered' out from London and the south-east over time is clear from this sequence of maps. At a more refined geographical scale, that of local authority district, Figure 8.13 shows average house prices in the third quarter of 2000 for some of the London Boroughs. The band of high price boroughs stretching from Camden to Elmbridge is plain to see.

Investment property databank (IPD)

Investors succeed in locations that improve, not those locations, which are currently the best: the best location already commands a premium price. GIS can help investors identify favourable locations by displaying past trends and forecasting future patterns. GIS is ideal for examining the geographical patterns and trends in property performance. Property data can easily be classified according to standard or user-defined geographical boundaries. Such geographical segmentation is difficult to accomplish without a GIS (Rodriguez *et al.*, 1995).

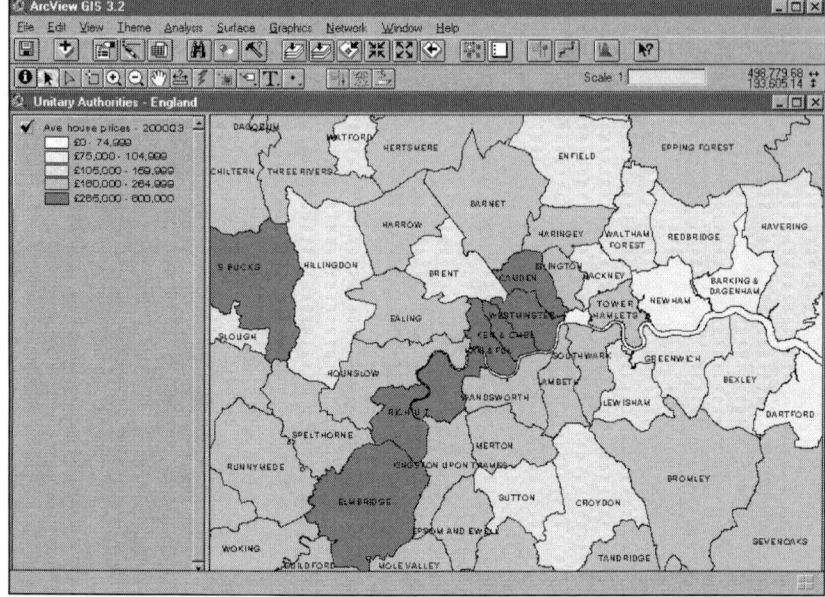

Figure 8.13 Average house prices in London in the third quarter of 2000.

Prior to the introduction of GIS at IPD none of the company's publications contained even basic illustrative maps, let alone maps showing actual trends across different geographical areas. As the 'Visualisation of property data' section describes, other data suppliers such as PMA were using maps in publications or in promotional documents. Therefore IPD (with the wealth of property data at its disposal) also decided to do so. All IPD data can be mapped including total return, capital growth, income return, rental growth, yield impact and portfolio structure. Maps used include 1:10,000 street maps for major urban areas, 1:10,000 map of London postal districts and local authority districts across England and a Wales, 1:50,000 road network, district and county boundaries and a 1:500,000 UK map.

Use of GIS at IPD can be divided into two areas: mapping of property data using standard geographical areas and mapping of property data using user-defined geographical areas. These are now considered in the following sections.

Mapping of property data using standard geographical areas

The GIS acts as a geographical front-end management tool used to query the properties in the databank. Fund managers and property researchers may use the graduated symbol function within the GIS to portray their properties according to capital value, rents received, total returns, rental growth, etc. A fund's property performance statistics may be mapped at regional, county, local authority district, postcode (illustrated in Figure 8.14) and street level (illustrated in Figure 8.15). Funds can interrogate attribute data, label and shade data points on a map according to any of the attribute data in the database. Properties may be queried on-screen, providing full locational, descriptive and property performance details including street number, street name, town, postcode, co-ordinates, property type, capital value, total return, rental value growth and other variables. It is possible to compare investment properties with sub-market averages, link to sale and purchase details, digital maps and access those data on-screen.

GIS visualisation and display techniques add a further dimension to property research. The ease with which the data held in tabular form can be linked to a map means that locational patterns can be analysed geographically. For example, Figure 8.16 shows the density of retail property investments in the databank at local authority level. The density is simply calculated by dividing the total capital value for each local authority district by the area in square kilometres in each district. Because many of the local authority districts have insufficient samples for a total return to be reported, the maps are also useful for illustrating the location of property

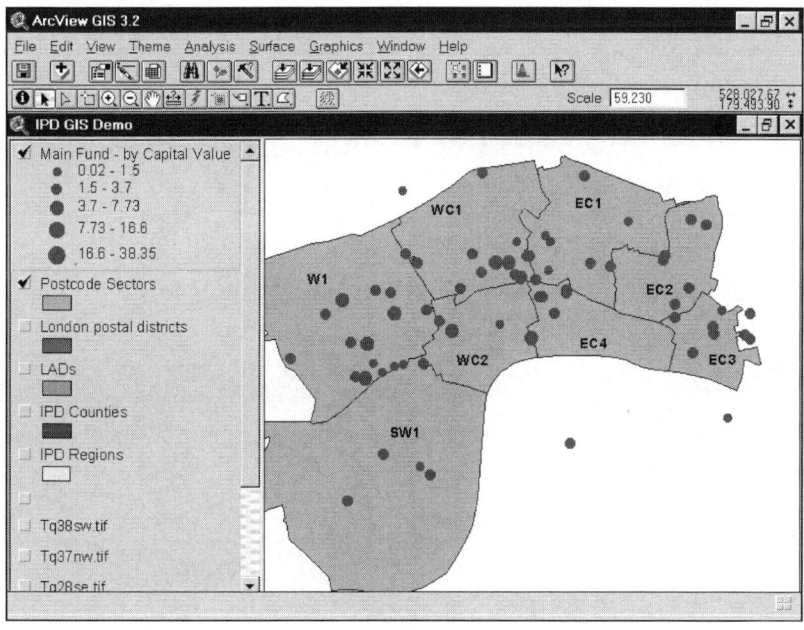

Figure 8.14 Property investments in Central London classified by capital value.
Source: IPD, used with permission.

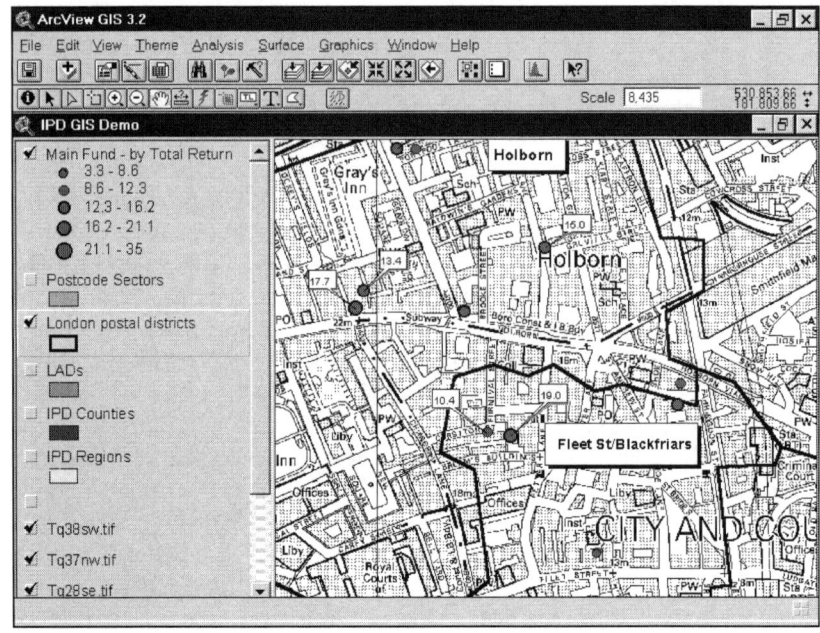

Figure 8.15 Property investments in Central London classified by total return.
Source: IPD, used with permission.

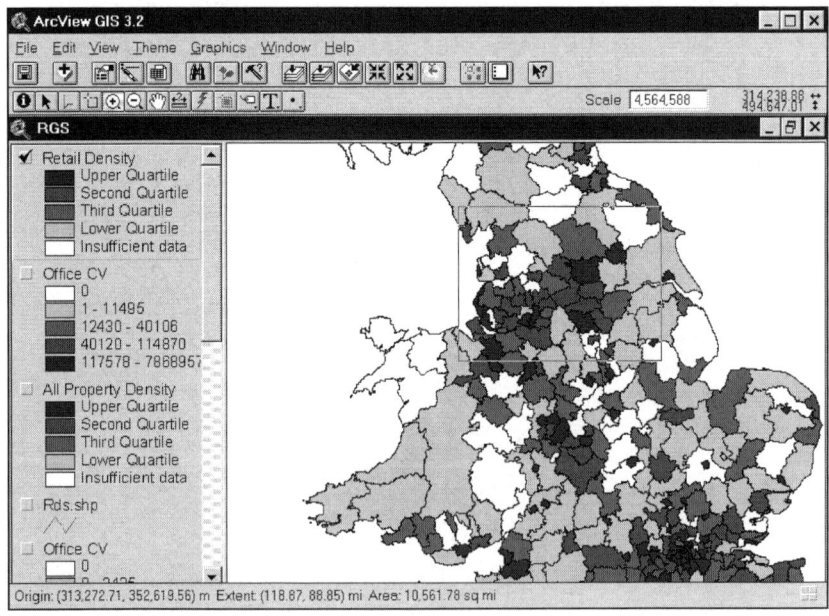

Figure 8.16 Density of retail property investments in the IPD databank at local authority level.
Source: IPD, used with permission.

investment activity by property type. Fund managers can examine the location and spread of investment properties at the local level using these density maps. It is also possible to examine trends in investment performance
over time, classified by standard geographical areas. This is illustrated in
the sequence of figures from Figure 8.17 to 8.19, which show how the IPD
Index of Total Return on property investments in south-east England
changed geographically between 1996 and 1998.

Mapping of property data using user-defined geographical areas

At IPD, custom areas are created by merging standard geographical areas
(such as local authority districts), by tracing boundaries around city centre
core and secondary areas using street maps (see Figure 8.20) or by creating
buffers around motorway sections, for example. At a larger geographical
scale bespoke geographical areas can be created by a fund manager to
measure the performance of, say, retail property investments along Oxford
Street. Caution is needed during this process because geographical boundaries can be manipulated to present required statistics. This, and similar
analysis issues, will be discussed in Part III.

Because all of the properties in the IPD are geo-coded this means that
maps can be produced showing the location of retail, office, industrial and

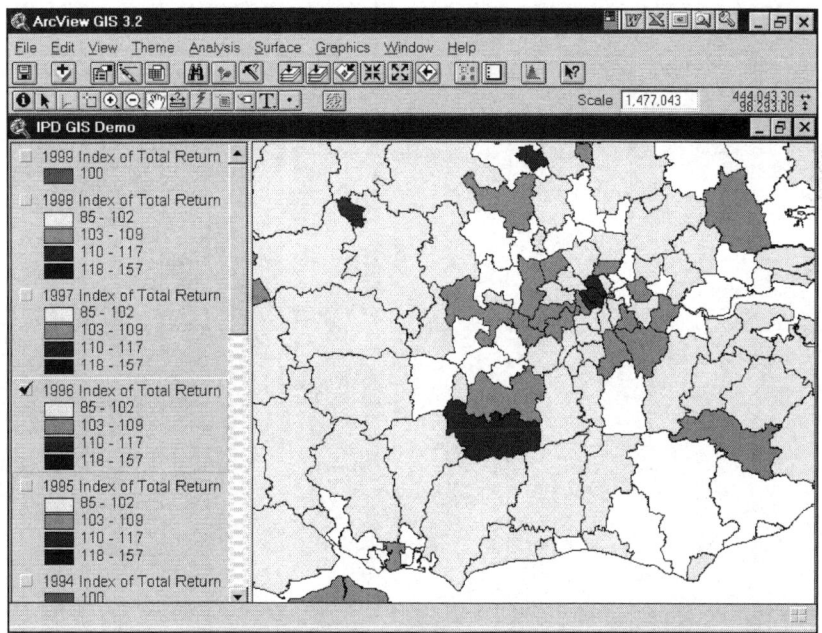

Figure 8.17 Total return on property investments in south-east England in 1996.
Source: IPD, used with permission.

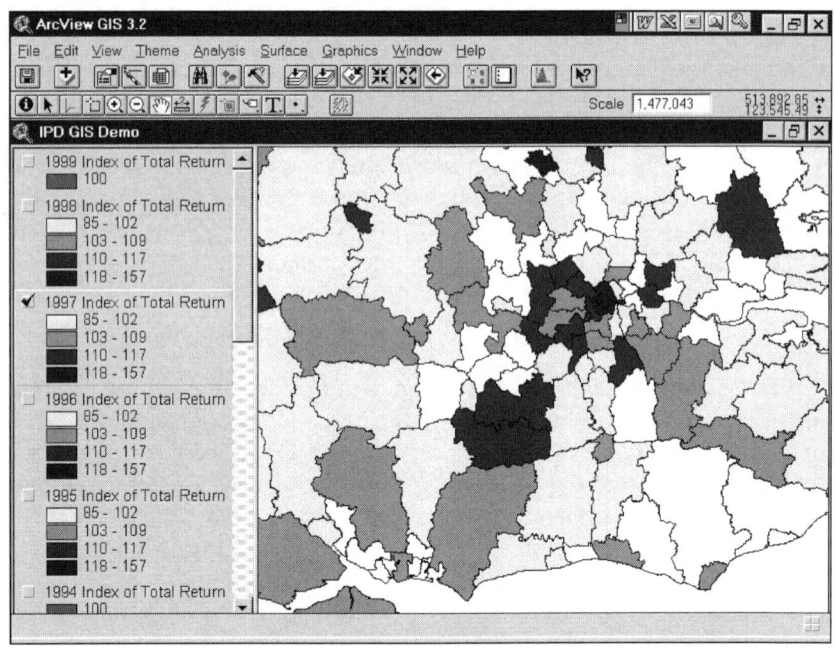

Figure 8.18 Total return on property investments in south-east England in 1997.
Source: IPD, used with permission.

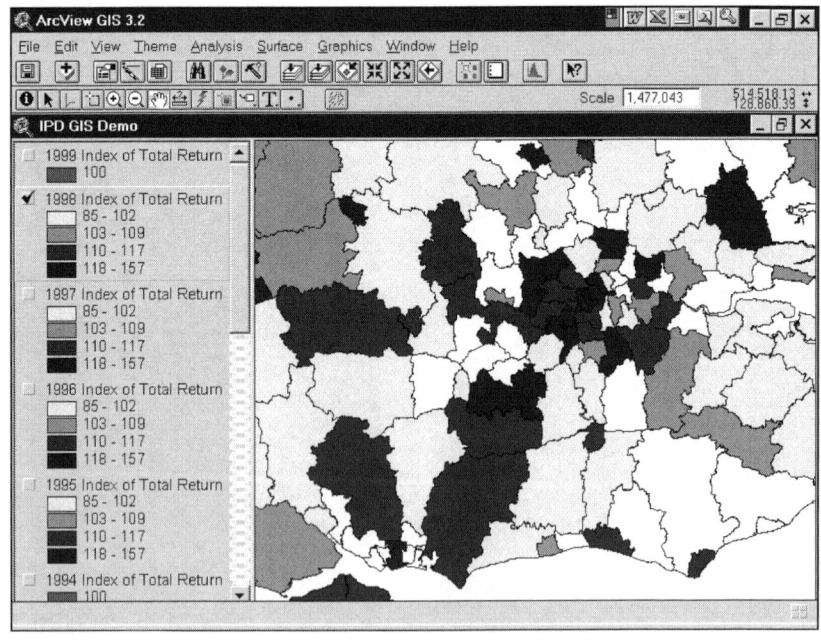

Figure 8.19 Total return on property investments in south-east England in 1998.
Source: IPD, used with permission.

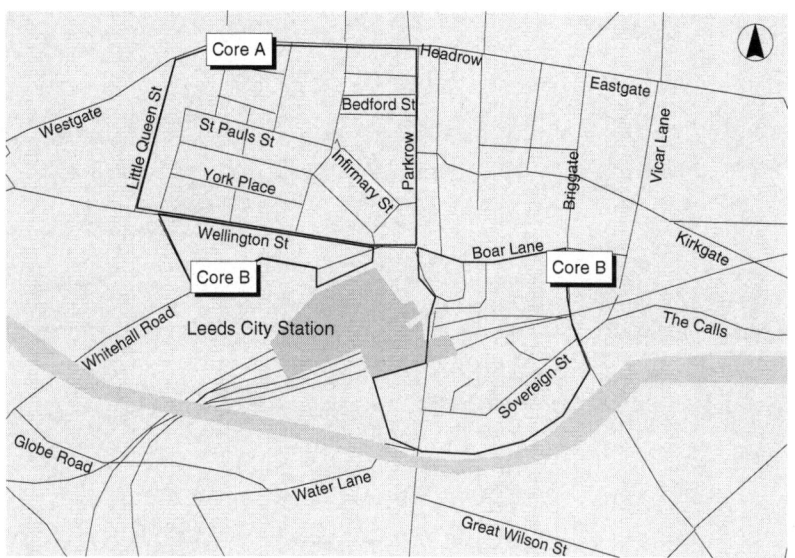

Figure 8.20 Core areas in Leeds city centre.

other property investment classes prior to the delineation of custom geographical areas. This prevents areas being defined that contain too few data points for statistically significant analysis. Prior to the use of GIS in this way such a task was very time-consuming but the ability to disaggregate markets to user-defined geographical areas is central to the benefits of developing a GIS at IPD. These benefits are illustrated in the two examples that follow.

MOTORWAY CORRIDORS

The first involves the delineation of motorway corridor markets for office property. The concept of motorway corridors is popular with investors and commercial property agents but it is difficult to examine the location and extent of these markets without the use of a GIS. By using GIS to create motorway corridor areas, or buffer areas around the motorways, it is possible to analyse the performance of various motorway and sub-regional locations and conclude, for example, that M25 offices (as defined by a 1.5-mile border/buffer around the motorway) have outperformed M4 offices.

A number of motorway corridors have been created, illustrated in Figure 8.21, and offices falling within these areas have been analysed to compare market share and investment performance. Of key importance is whether or not these areas have separate investment performance character-istics. For example do some motorway corridors perform better than others

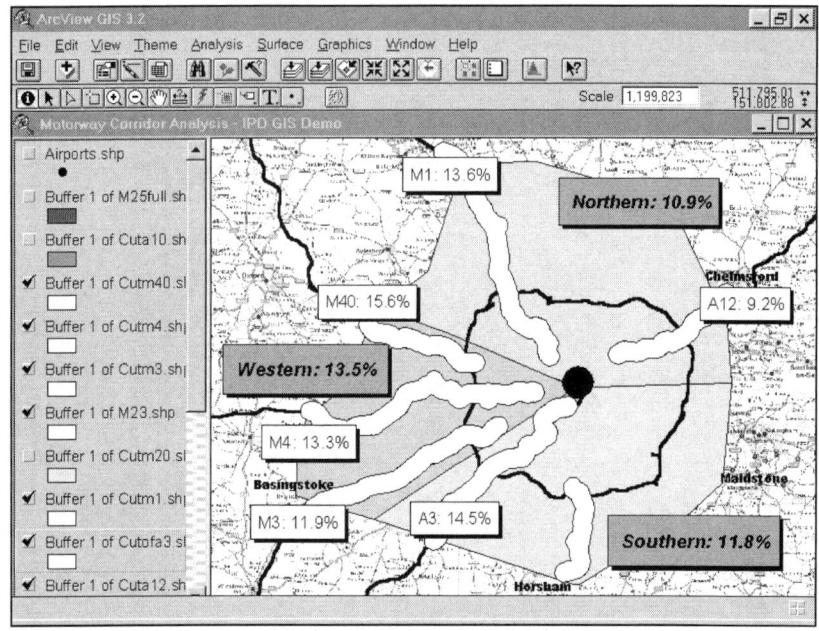

Figure 8.21 Motorway buffers around London.

Source: IPD, used with permission.

or does proximity to a motorway affect performance? These questions can be answered using performance measures from each of the delineated corridors but it would also be possible to use performance measures to delineate the areas in the first place. Within one-and-a-half miles of the M25 there are 104 offices recorded in the IPD database with a capital value of some £630 million. On average every square mile of the corridor space contains £1.8 million worth of office stock. The comparatively lower density of office stock than that of the M3, M4 and M40 corridors owes much to the fact that many of the M25 offices are located in the west. In terms of the major radial routes leading out of the capital and in terms of both number and capital value, IPD offices are most concentrated along the M4 corridor, averaging £7.3 million for every square mile. The M40 corridor leading towards Oxford has, on average, almost £5 million of stock per square mile. Further research could test proximity to motorway junction as an influence on property performance.

This type of analysis has been commissioned by Lambert Smith Hampton, a firm of property consultants, for their annual Office Market Review of the M4 Corridor. The report describes letting and investment market activity in the major office markets along the M4 motorway between London and Bristol.

BRISTOL OFFICES

The second example illustrates the use of GIS to examine the location and investment performance of office property in Bristol. GIS is used to define and create prime, secondary and tertiary areas of office space and the long- and short-term structure and performance of office portfolios in those different city centre locations is compared.

GIS analysis has shown that the edge-of-city office space on the north fringe of Bristol accounts for a quarter of the total office stock in the whole south-west region of the UK. In particular, urban development on the north fringe is a clear rival to central Bristol, accounting for almost as much stock by capital value as the combined 'waterfront' and 'core' areas of central Bristol. Despite containing half the number of properties than that of central Bristol, the northern fringe is approximately 60 per cent the size of central Bristol office market measured by end year 1998 capital value. Taken together, northern fringe and central Bristol offices now account for almost two-thirds of the entire regional office market.

These two examples have demonstrated the application of GIS in analysing the location and the spatial variation in office markets in the UK. The motorway corridor analysis has shown that proximity to motorways has helped property performance over the long term but that short-term effects can put individual locations or corridors at a disadvantage when compared to the wider market. The provincial office analysis demonstrated the role of GIS in analysing sub-markets in Bristol. Research commissioned by property consultants Jones Lang Lasalle has developed this idea with a further six provincial office centres selected for comparative analysis. The role of GIS in disaggregating the property market is clear. It represents a tool with which to define corridor markets more accurately and flexibly, by altering buffer distances along motorways for example. Most property data are aggregated into regions, which are large, not classified by economic function and which contain a mix of disparate sub-areas. The introduction of GIS at IPD allows for greater clarity and transparency in defining and analysing property sub-markets. Conventionally defined standard regions may no longer be important in defining and analysing property markets.

As is the case with many GIS applications in property, retail clients have driven the use of GIS for more sophisticated property market analysis. This includes the measurement of performance at transport nodes and the evaluation of the impact of infrastructure and other development projects on investment performance. Funds can view property investment performance at a local level for user-defined geographical areas. Attribute data can then be analysed at the property level and results can be presented in the form of tables, charts and thematic maps to show, for example, movements in performance. The IPD GIS offers the ability to measure investment performance against market trends geographically using either standard of user-defined geographical areas. This aids the geographical management of investment portfolio data.

Occupier Property Databank (OPD), a sub-division of IPD is in the process of establishing a geo-coded radio mast sites database, which currently contains over 20,000 sites. The database allows for geographical rental comparisons. For example, the highest rents for rooftop sites are paid in London while Wales and the south-west are regions of the country where the telecom operators are paying the lowest rents. Within the regions there is a wide variation of average rents paid. Spatial rental patterns are clearly evident – decreasing rents with increasing distance from Central London can be visualised. GIS analysis can focus on particular areas of the country. For example, examining the pattern of average rents in Greater London it is apparent that large parts of Central London have some of the highest average rents, with the city of London, city of Westminster and the Royal Borough of Kensington and Chelsea all averaging over £8,000. A north/south pattern is also apparent in that those districts immediately south of the River Thames have lower average rents than those just to the north of the Thames. This would be an expected pattern as the focus of activity in terms of shopping and business is very much north of the river.

Property market analysis

Beyond thematic mapping lies more sophisticated analysis of property market data using the more advanced functionality of GIS. This section describes the type of geographical analysis that is possible using GIS and that is now being undertaken in support of property decision-making. Intelligent Space Partnership (ISP) is a planning consultancy that specialises in pedestrian modelling. They have developed sophisticated spatial modelling tools within a GIS framework that can forecast likely pedestrian flows within new developments. This is of particular relevance to the property market as pedestrians are critical to the economic success and functional performance of retail development schemes.

Visualising pedestrian flows

Surveys of pedestrian flows provide key evidence that helps planners and developers understand how well an existing site is being used and what its development potential might be. ISP use two main methods by which to collect pedestrian data:

(a) pedestrian flow surveys
(b) pedestrian route choice surveys.

Below are examples of how each of these methods of data collection and visualisation using GIS provides the evidence to aid the decision-making process.

Pedestrian flow surveys

Survey evidence is collected to represent the pattern of flows around a property development site and to help calibrate a model of flows within a development. Usually data have to be collected in the field as existing data are
often not suitable, but the focus of analysis on discrete geographical areas
makes data collection self-contained. Survey evidence of pedestrian flows
was used in a Pedestrian Safety study of St Giles Circus in Central London
undertaken by ISP. Camden Council is involved in a major urban redevelopment scheme to create a public plaza at the foot of the Centre Point
Tower at St Giles Circus. This site currently has the worst accident record
of all road intersections in Camden. The council wished to assess the impact
of the scheme on pedestrian flows and accident risk in the area. By understanding why pedestrian flows were at such risk in the existing scheme, the
council wished to ensure that future plans would remedy the current problems. Pedestrian flow was measured and mapped using GIS to show the
spatial pattern of movement (Figure 8.22). The levels of movement vary
considerably within this local area, which can be easily seen when represented in this manner. The width of available pedestrian space is key to
the ability of a street to support pedestrian movement. It is therefore important to identify what the constraints on pedestrian flow are for different
pavement capacities. The relationship between pavement width and pedestrian flow acts as a measure of how crowded pavements are in terms of

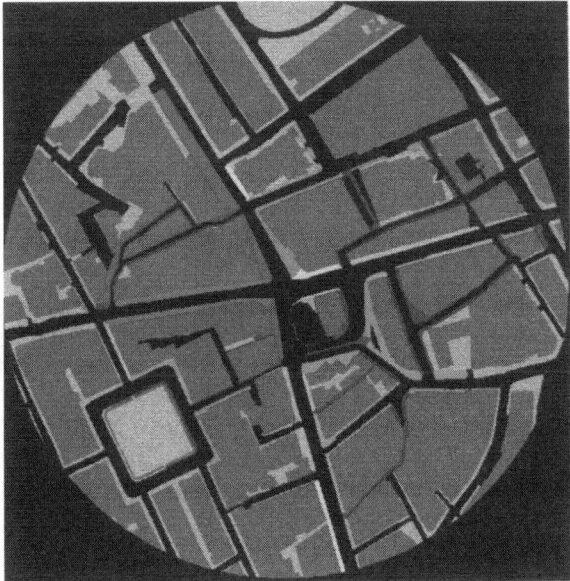

Figure 8.22 Average movement on each pavement.

Source: ISP, used with permission.

pedestrian flow per metre of available pavement. The highest level of crowding is on the tiny pavement underneath the Centre Point tower.

Pedestrian route choice surveys

To provide a more detailed explanation of how pedestrians move around an area, their route choices can be analysed. This can help identify street sections where high numbers of pedestrians cross the road. If these locations are not supported by signalled crossings, a heightened risk of accidents is present. Because the location of crossings in relation to pedestrian flows is of vital importance to road safety, some roads can be more dangerous to cross even though there is less traffic. For the St Giles Circus study, CCTV data were used to plot the road crossing locations. A GIS was used to map the intensity of use of formal road crossings and the level of informal ('red man' or away from formal facilities) crossings. When these data are overlaid with the location of accidents, a strong relationship between the pattern of road crossings and the pattern of accidents as a whole can be seen. The majority of accidents occurred outside designated crossing facilities but concentrated on paths of informal crossings or where visibility is impaired.

This kind of survey method was also used to evaluate Transport for London's traffic reform proposals for the Shoreditch Triangle Area of London. ISP conducted a survey of the routes pedestrians take when crossing each of the main roads within the area. Key findings of the analysis were that pedestrian crossing patterns are determined by 'desire lines' of visibility and informal crossings occur where crossing facilities are not co-located with desire lines. Using this analysis, recommendations were made to improve road safety, by locating the new crossings on desire lines and thereby reducing the level of informal crossing. Transport for London adopted six of the seven recommendations in their revised plan for the scheme.

Pedestrian route choice analysis can also be used to assist public space design. As part of its expansion, Cambridge University commissioned a master-plan for the development of five new buildings on its Sidgwick site. One of the aims of the plan was to improve the public realm on the site by creating legible and well-used pedestrian movement routes, enjoyable public spaces and by supporting ease of cycle access. Intelligent Space undertook an observation survey of pedestrian and cycle movements from each of the main entrances, A, B, C, D and F in Figure 8.23. Each of the individual movement traces was input into a GIS, shown in Figure 8.24. The results identified one area where an informal route exists on the site that is not supported by paved movement space. This issue is important for development of the site, as a mismatch between natural movement and landscaping is a common problem for public space design. A more significant finding on natural movement patterns is that the legendary culture of strict adherence to formal paved routes in Cambridge is reflected in the observation data. People do not walk on the grass, even in dry conditions. This is of particular

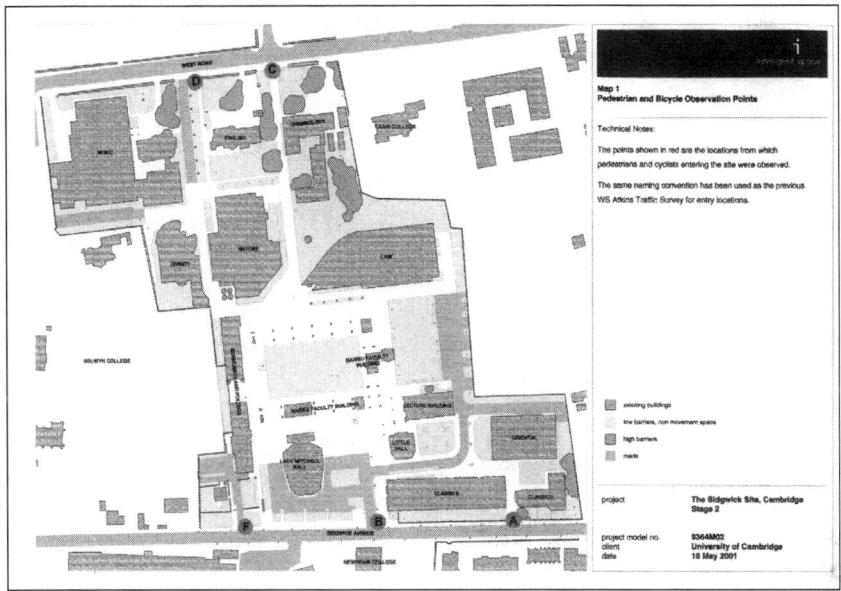

Figure 8.23 Pedestrian and bicycle observation points.
Source: ISP, used with permission.

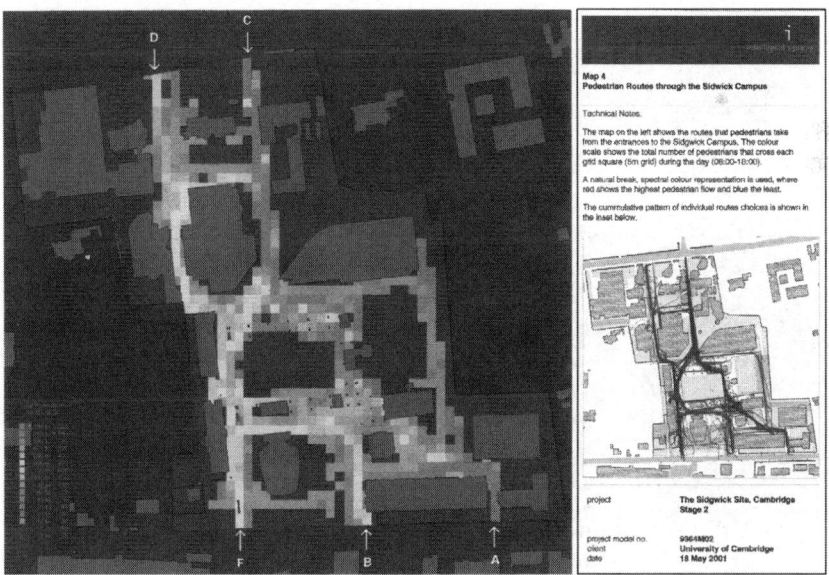

Figure 8.24 Pedestrian routes through the Sidgwick Campus.
Source: ISP, used with permission.

interest to the Sidgwick site because of the location of a grass area in the middle of the key 'desire line' between the north and south of the site as emphasised visibility analysis. In previous studies, pedestrians have been observed taking extreme risks in order to cross roads on a desire line rather than at a formal crossing nearby, but the grass area does effectively stop people walking through the main route at present. Although pedestrians seem to have learned the established rules about where it is allowable to walk, the culture of cycle parking is anarchic in comparison. Cyclists ignore many empty cycle racks in some parts of site and prefer to chain bikes on railings or lean them against walls in certain key concentrations on site, Figure 8.25. There are two strategies that can be used to bring more order to the cycle parking. Either the cyclists have to be taught to accept parking in undesirable locations, or the bike racks need to be relocated to better reflect the demands of users. The first strategy is to treat the issue as a management problem and the second strategy is to treat the issue as a design problem. As a design problem, there are a number of key characteristics of the areas where parking is concentrated that can be used to guide future provision. The observation study suggests that people want to park their cycles

- close to movement routes
- in highly visible locations
- near building entrances.

Figure 8.25 Cycle routes through the Sidgwick site.

Source: ISP, used with permission.

The main issue regarding the provision of cycle parking on the master-plan is, therefore, not increasing the overall levels of cycle parking provision on the site (although this may be desirable) but ensuring that there is sufficient parking provision adjacent to the main movement routes and in locations that are highly visible from these routes. Without this it is likely that there will continue to be a large amount of informal parking remaining on the site even after completion of the master-plan design.

'Desire lines' and visibility analysis

Spatial analysis techniques are used to measure the factors that influence pedestrian movement, such as visibility, accessibility, pedestrian capacity and the location of key attractors. Techniques employed include Visibility Graph Analysis (VGA) (which captures the 'desire lines' of visibility within the street grid) and shortest path analysis. With VGA, at each point on a grid of observations throughout the pedestrian movement space, a visual field is calculated by checking which other points can be seen. This gives an area that is visible to a pedestrian standing at this point. In the example shown in Figure 8.26, all points in dotted light grey can be seen from the highlighted point. The same calculation is performed for each point on the grid over the whole study area. The pattern of visibility can be represented within GIS using an equal range spectral colour scale, from dotted light grey

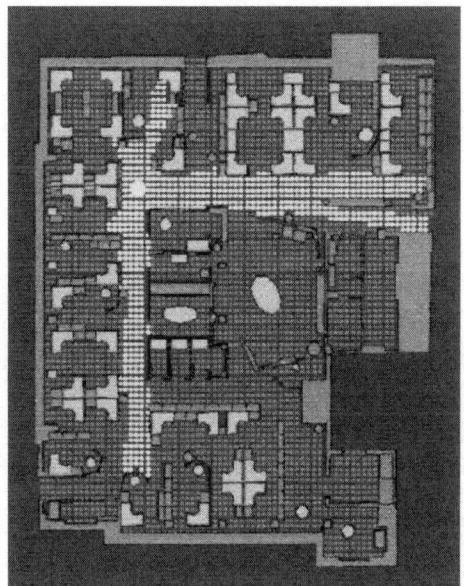

Figure 8.26 An example of observation points for VGA.

Source: ISP, used with permission.

(highest visibility) to dotted dark grey (lowest), as in Figure 8.27. Spatial analysis techniques can also be used to calculate 'shortest path' or accessibility measures for pedestrian movement systems. Shortest path analysis calculates how much of the pedestrian area is accessible from each grid point using a simple journey of four changes in direction. The smaller the area of the retail centre reached from any point, the less accessible the location is. One example of this kind of analysis is shown in Figure 8.28.

Natural surveillance analysis

Analysing visibility for pedestrians with GIS can also be used to evaluate way-finding and 'natural surveillance' levels in new developments. This provides developers with an evaluation of how easy it will be for the users of a building or site to find their way around, and how well surveyed the public spaces are for safety purposes. The location of building entrances is an important aspect of the safety of the public realm because they provide a degree of 'casual surveillance' of the site by the users themselves. The spatial characteristics of a site may be analysed for 'casual surveillance', visibility and the integration of movement routes using GIS.

Figure 8.29 shows casual surveillance from building entrances in the Sidgwick site of Cambridge University. Areas in medium grey are those where

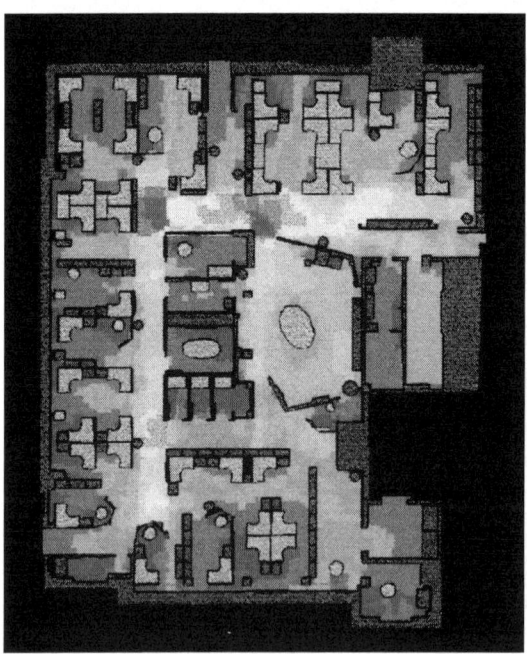

Figure 8.27 An example of the visibility measure inside a building.
Source: ISP, used with permission.

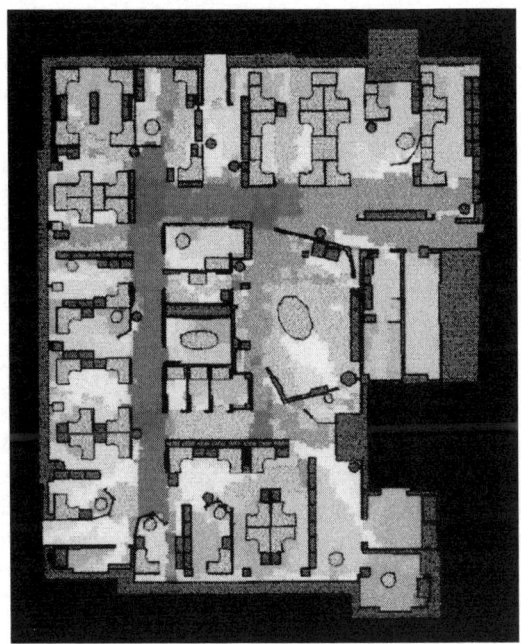

Figure 8.28 An example of accessibility analysis inside a building.
Source: ISP, used with permission.

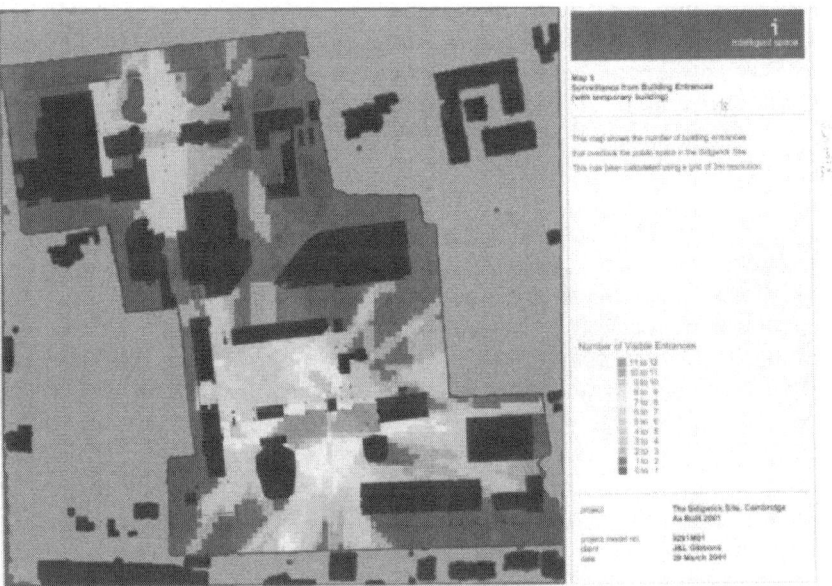

Figure 8.29 Surveillance from building entrances.
Source: ISP, used with permission.

a passing pedestrian would only have surveillance from either one building entrance, or none at all. The levels of natural surveillance in open sites with free-standing buildings (such as this) are low compared to traditional streets, where many more entrances and windows are typically visible. Spatial analysis techniques such as these can be used to help improve natural surveillance when the site is developed further.

Pedestrian modelling

For the purposes of pedestrian modelling, all the variables from the analyses outlined above are linked together within a GIS. This allows a statistical model to be created showing which variables are most influential in determining pedestrian flows. Once the model has been calibrated, the spatial analysis is re-run for the design scheme and the model is applied to the new development. In this way, the calibrated model is used to forecast flows and visibility for planned development. Applications include retail developments, pedestrianisation schemes, shopping centre development, re-routing transport network and building design. In the case of retail schemes, rents, vacancy rates and other key variables can be analysed together with pedestrian flows and visibility. This will help the developer optimise the design of the scheme so as to maximise revenue. It is also possible to model the location of traders within the retail scheme in order to optimise layout. One particular study evaluated the redevelopment of the retail centre in large UK town. The purpose of the study was to provide evidence of the likely effects that planned changes to the centre would have on pedestrian flows. Pedestrian flow data, including what pedestrians can see and where they can go, was collected for the retail area. These pedestrian movements can then be modelled to predict flows in the proposed retail centre using Visibility Graph Analysis and shortest path analysis. Pedestrian models of shopping centres are designed to predict 'passing trade' or circulation within the centre, clearly of vital importance to retailers. A model was created that predicted 90 per cent of the observed pedestrian flows around the centre. The significant variables were visibility, area of retail floor-space for each unit and an attraction factor based on national averages of customers as a percentage of entrants to the retail unit (normalised for unit area and number of entrances). The results of the study showed that some of the design schemes being considered for the redevelopment of the centre would alter visibility detrimentally but floor-space and attraction positively. When the new values of these variables were input into the model the pattern of modelled pedestrian movement was not significantly changed. This type of analysis can help developers and investors optimise the design of retail schemes so as to maximise pedestrian flows and thus maximise potential rental revenue, particularly if turnover rents are charged.

Property research

In the past, property research has been constrained by inadequate technology and unavailable data. But in recent years significant advances have been made in both these areas. To an extent they are closely related – as technology improves the ability to collect and handle large volumes of property data becomes more achievable. This is evident as many surveying firms computerise their property files and commercial data suppliers collect, integrate and market increasing volumes of property data. Furthermore, in the public sector, the OS, Land Registry, VO, the ODPM, utilities and local authorities are custodians of substantial property data sets, which are increasingly held in digital form. The following sections looks at some of the ways that property research has harnessed GIS technology.

Property value maps

A value map displays geographical variations in property values. They are rare in this country but well known abroad where they are often used for taxation purposes. Research that uses value maps is also rare although interest in the geographical representation of property values has increased since GIS have been widely available. Potential uses of value maps include:

- planning and development (land and property use allocation, compulsory purchase and compensation);
- taxation, in Switzerland and Denmark the public can check their assessments by viewing value maps and transaction details;
- identification of areas of latent value for investment and development purposes (e.g. vacant or under-utilised land in high value areas);
- policy evaluation, for example, show changes in value over time, perhaps due to planning blight, road improvement, green belt and other planning designations;
- valuation, examine the geographical distribution of values as an aid to valuation for example.

Value maps were first used in Britain by Anstey (1949) who produced value contour maps to show how property development would alter the pattern of land values in the Barbican area of London. Figures 8.30 and 8.31 illustrate the two maps that were produced before and after major redevelopment of the Barbican area, and the 'isovals' or value contours, which join sites of equal value, show the transformation in value distribution for this area.

Figures 8.32 and 8.33 are two examples of value maps that were produced by planning departments of local authorities before the advent of GIS. These paper-based value maps related to one land use only and a specific point in time. Data collection and map production were time-consuming and expensive and they were too general to be of use for valuation purposes

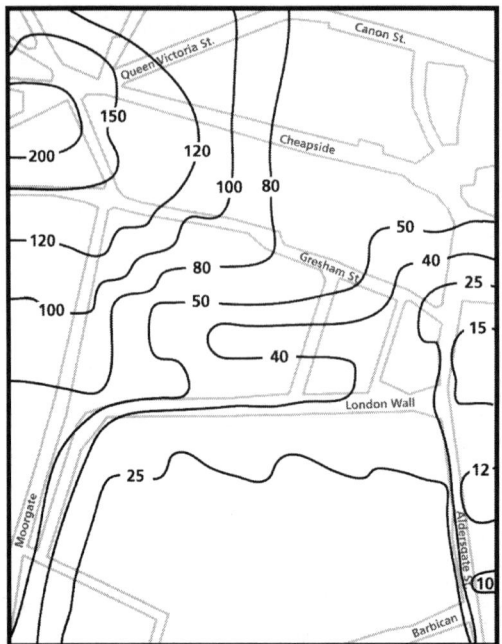

Figure 8.30 Barbican, overlaid by isovals of 1939 land values (pounds per squarefeet).

Source: After Anstey, 1965.

Figure 8.31 Barbican, overlaid by isovals of 1969 land values (pounds per squarefeet).

Source: After Anstey, 1965.

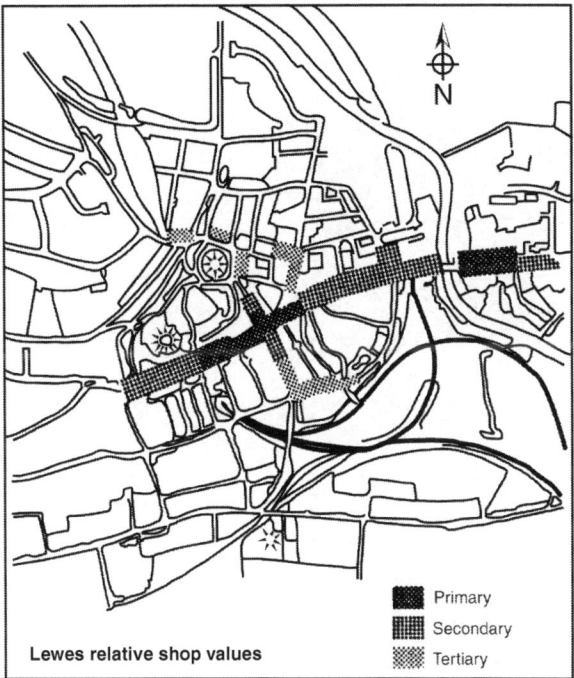

Figure 8.32 Relative shop values in Lewes, east Sussex.
Source: Howes, 1980.

where individual property details are required. Yet, despite these draw-backs, Howes (1980) believed that value maps were useful property deci-sion aids and he investigated different methods of displaying property value data geographically.

GIS are able to address some of the problems inherent with traditional value maps by producing them efficiently and as part of a wider suite of data analysis techniques. Howes acknowledged the limitation of a value map with regard to its information content but the functionality of GIS means that a value map need not be a static display but can form part of a more comprehensive analytical process. Digital maps allow frequent updat-ing, quicker production time and greater analytical potential. Problems in the past related to lack of staff resources, the production of expensive paper maps and the complex calculations involved. A GIS can manage the data input, calculations and presentation of the results in a graphical format. The spatial analysis of property data enhances the valuer's understanding of locational influences on property value.

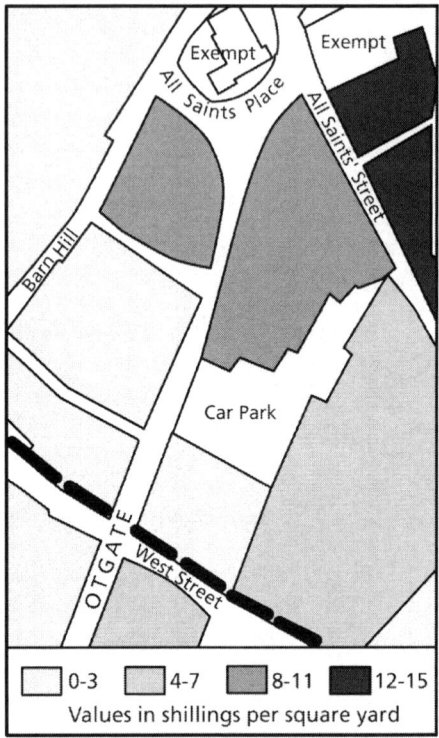

Figure 8.33 Rateable values per street block.
Source: Howes, 1980.

Property valuation

The value of property reflects its capacity to fulfil a function. With regard to commercial property, functional qualities may include (with examples given in brackets):

- locational influences (accessibility to the market-place, proximity to suppliers of raw materials and important nodes such as railway stations, car parks and open spaces);
- physical attributes (size, shape, age and condition);
- legal factors (lease terms and restrictive covenants);
- planning and economic factors (planning constraints, permitted use and potential for change of use).

Property valuation is the process of identifying and quantifying these 'value factors', illustrated in Figure 8.34. The result should be an informed

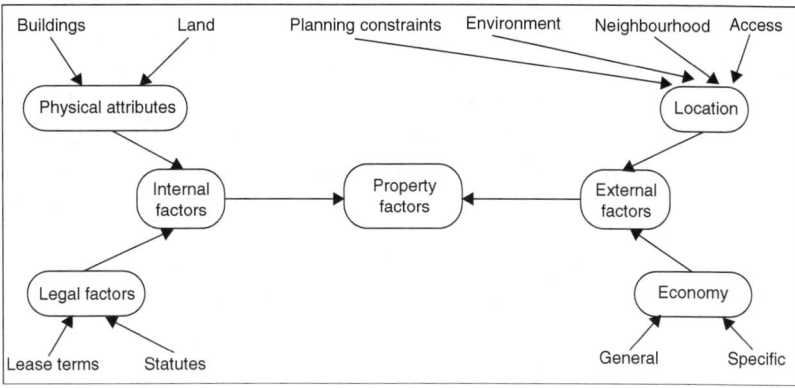

Figure 8.34 A taxonomy of factors that influence property value.

opinion of value based on an assessment of those factors considered relevant to the value of the subject property. The prevalent method of valuation is the comparison technique. Popular with valuers and favoured by the courts, the comparison method involves the analysis of properties similar to the one being valued in order to isolate and quantify individual influences on value. Many reconciliation procedures have been developed to quantify physical and legal influences on value: depreciation allowances for obsolescence and age, adjustments for abnormal rent review patterns and the payment of premiums. However, such established techniques do not exist with regard to geographical value factors. Valuers often rely on their local knowledge to quantify differences in value caused by a property's proximity to competitors, potential customers and suppliers over the infrastructure network. Developments in GIS provide an opportunity to investigate locational influences on value more objectively. Geographical analysis of comparable evidence is implicit when valuers use their local knowledge and experience to financially adjust the values of properties that are in different locations. Properties may have physical and legal characteristics in common and these can be identified and reconciled but spatially no two properties are the same. Sudden changes in value can occur over a short distance in the case of retail property, perhaps due to dead frontage, pedestrian flow or a busy street.

Despite widespread recognition that the location of a property is a primary influence on value, research that attempts to explicitly measure that influence at the intra-urban level is lacking. To date, research has concentrated on the description and display of spatial influences on property value rather than analysis. This is due to two reasons: a lack of data on which to base research and the absence of a suitable technology capable of undertaking complex geographical analysis.

With regard to data availability, the collection and analysis of evidence of transactions provides the valuer with the data required to identify and quantify value factors. Substantial effort is expended collecting evidence. Given that more value is attached to a product that has taken time and effort to collect, an attitude of 'data secrecy' has developed in England and Wales (RICS, 1994). Such an attitude is not evident in many other countries where transaction details are in the public domain. Data secrecy is inefficient and leads valuers to rely on secondary and incomplete information. It produces a conflict between the need for confidentiality of client information and the need for a comprehensive database for valuation purposes. Consequently valuers are cautious in the selection of comparable evidence, choosing only those comparable properties that are near to the subject property and from the recent past. It is left to the skill of the valuer to quantify physical and spatial differences between properties because an objective methodology relies upon a greater quantity of data. Often, the most reliable comparable evidence is to be found close to the subject property in the recent past, however, such evidence is rare and valuers must resort to evidence from further afield in time and space where adjustments for differences between comparable properties and the subject become difficult. With wider data access then not only does the quantity of comparable evidence increase, which directly aids valuation, but potential value patterns among properties can be explored.

With regard to technology, a more sophisticated analysis of geographical influences on value can be undertaken using a GIS. Dixon (1992) points out that surveyors base values on locational criteria yet the potential of GIS for such an operation remains largely undiscovered. The use of GIS offers the ability to identify the locational factors that affect property value in the same way that physical and legal factors, such as differences in age or lease structure, are quantified by valuers. By using a GIS, locational factors such as access to customers, raw materials and suppliers, links with other businesses and the workforce catchment area can be measured and analysed. This is significant because if we are able to draw out the locational factors that affect property value then valuers will no longer be restricted to searching for comparables that are near the subject property. Rather, valuers will be able to search for comparables that are influenced by similar locational factors regardless of their proximity to the subject property.

In the US, the Appraisal Institute predicted that 'GIS is expected to have a strong impact on traditional location analysis just as computerised discounted cash flow modelling has dramatically influenced financial analysis in the past two decades' (Appraisal Institute, 1992). In the UK, Waters (1995) suggests that surveyors should harness new IT developments to help with the core of their expertise, namely location. He argues that valuation depends on comparison, comparison requires access to transaction data and a key element of transaction data is location.

Overseas case study

In Switzerland a GIS-based valuation application was developed, which integrates GIS concepts with valuation parameters (Din, 1995). A method for obtaining an aggregate score for locational influences on residential property value was developed. The system acts as a decision support tool by calculating a 'geo' index obtained from large-scale digital maps of Geneva and Zurich, which are overlaid with environmental themes such as distance to the commercial centre, noise sources, traffic volume and schools, public amenities and shopping centres. It acts as a Decision Support System (DSS) for property professionals and purchasers. Empirical studies established weights for each environmental factor by mailing around 800 homeowners (weights may be specific to an urban area, however). A simple geo-index weights variables for a particular location (determined by inputting the address).

Geographical analysis of property values

Research that has sought to assess the determinants of property value has either ignored detailed location analysis or dealt with it only in a broad sense – for example, by undertaking the study in an area small enough to assume that its locational value characteristics are constant (Adair and McGreal, 1986, 1987), or by dividing the study area into zones for which certain assumptions can be made such as 'a high value area' or 'the central business district'. Previous studies have recognised the significant effect that location has on value yet they have concentrated on identifying and quantifying the effects of physical attributes on value, such as the number of habitable rooms or the presence/absence of certain amenities. The development of GIS means that computerised locational analysis is now possible. Accessibility is an important influence on the location of retail property and GIS may be used to investigate the effect of accessibility on retail property value at the individual property level.

Scott (1988) comments that valuers infer a substantial amount of information about a property from its location. This inference is based on local knowledge and experience. Valuers divide urban areas into homogeneous 'sub-locations' within which the properties are assumed to be locationally similar. The definition of such value areas is intrinsic to many valuation models that apply statistical analysis, such as multiple regression analysis. However, the definition of value areas is discriminatory as well as predictive and therefore there is an element of inevitability that their ability to predict values will be high. Consequently, little is learnt about why properties in different locations attract different values. Regression results are also suspect due to multicollinearity, spatial auto-correlation can occur in geographical analysis and the correlation between two variables changes substantially when different spatial aggregations of the same data are used (Norcliffe, 1977). This leads to the main difference between valuation models that apply homogeneous value areas and proximity variables: homogeneous value areas only discriminate between zones of different value whereas improvements in computerised spatial analysis techniques allow a more rigorous approach to the inclusion of spatial value factors.

Isolating the impact of different physical and locational characteristics on property price formation at a given location has been the subject of many studies. This type of research has various practical applications in the property industry such as tax assessment, market research, land use planning and site selection. Multiple regression techniques for property value formation, the hedonic price model, are based on the assumption that the total price equals the sum of its parts – some error notwithstanding. That is, the sum of the market adjustment factors of all relevant structural and locational characteristics of the dwelling is seen as an estimate for its market value. The location is usually treated as a bundle of variables indicating accessibility (general and specific), neighbourhood factors (physical, service infrastructure, social, aesthetics and other), density, local government and negative disturbances. Alternatively location may be treated as a separate, usually discrete variable. Hedonic analysis of large data sets has been widespread since the seminal paper by Rosen in 1974 but early hedonic price models were not spatial – location was considered without the scattered effect of objects situated elsewhere. During the 1990s, the locational element of property value has been taken into consideration more explicitly using GIS and spatial econometrics in order to capture locational differences and spatial interactions. This has been done in the main by creating indicators as proxy variables for locational value.

Rodriguez *et al.* (1995) examined how GIS can improve regression models for residential property. Specifically they showed how GIS can provide a superior location variable relative to the commonly used straight-line distance assumption. They concluded that GIS can facilitate the creation of many types of variables that can be useful in property market analysis. GIS-created variables can be neighbourhood or regional characteristics that would be otherwise difficult or time-consuming to create. For example, GIS can easily generate a variable that identifies all properties adjacent to a property meeting a particular set of criteria. This type of identification would be difficult without a spatial database. Rodriguez *et al.* (1995) also point out that distance variables are also easily produced with GIS. For example, a GIS can provide more accurate distance measures by using the actual transport network rather than straight-line distance and by assigning various impedances to traversal along sections of the network. Use of spatial area characteristics usually means assuming zero distance between units within the geographical area. A GIS allows other assumptions such as a uniform spread of units throughout the area. More importantly, according to Clapp *et al.* (1997), GIS can provide spatial statistical analysis capability such as tests and corrections for spatial autocorrelation. A GIS also performs geo-coding and allows visual inspection of spatial linkages and attractive presentation of results (Clapp *et al.*, 1997).

To date the predominant use of GIS and geographical analysis in hedonic modelling of property values has been limited to the display of resultant values or residual errors and the measurement of Euclidean distances for proximity variables. For example, Des Rosiers and Theriault (1992) used GIS to display the results of a regression analysis. Transactions and features that influence sale prices (major work places, neighbourhood shopping centres and welfare housing) were mapped. This visually established spatial relationships among natural features, transport and communications, activities and services, socio-economic data and regression value predictions. A GIS was used to generate 'isovalue' curves and a 3D representation of the value distribution. In another study Longley *et al.* (1994) used GIS to display deviations between predicted house values and council tax assessments. They suggested that a GIS provides a suitable means of devising property assessments based on comparisons with the asking prices of 'similar' properties in a given area. In a mass appraisal study by McClusky *et al.* (1997) the application of GIS was limited to a visual analysis of the pattern of predicted values. Location was handled in the hedonic model by specifying 'ward group' as a variable by which comparables are selected and their values adjusted. Bible and Cheng-Ho (1996) used a GIS to generate distance variables (to work, schools, shopping malls, etc.) in a study of apartment rents.

Recently, more sophisticated use of geographical analysis has been evident. For example, Rodriguez *et al.* (1995) used a shortest path algorithm to calculate the route along the road network between each property and the central area. Gallimore *et al.* (1996) are more cautious in their view of measuring the influence of accessibility on residential property value. Accessibility to employment and shopping centres, leisure and educational facilities, exposure to neighbourhood and environmental factors exert themselves in a complex interaction when considered as influences on value. Numerical measurement of such factors, even where possible (distance to nearest school for example) may not reflect qualitative effects such as type of journey. Also, multicollinearity problems may arise. Gallimore *et al.* (1996) suggested that location value can be modelled as a 3D surface. They measured the variance between actual selling prices and prices predicted by a Multiple Regression Analysis (MRA) model, which has no location variable. The predicted values will clearly overvalue in some places and undervalue in others. The ratios of over- and undervaluation can be plotted to identify positive and negative value influences, the impact of which can be determined by measuring the distance from the influence to the property. Alternatively, the ratios can be used as z-co-ordinates and a GIS used to interpolate a 3D surface from which it is possible to estimate a location influence value for any point (property) on the surface. When these values are used to adjust the locationally blind MRA predictions, significant improvement in predictive power is observed.

Wyatt (1997) sought to correlate 'locational' values of shops with an 'accessibility' index. The aim of the study was to determine areas of high and low locational value, calculate the positions of maximum and minimum accessibility using network analysis and test for correlation between the two variables. The network-based accessibility index provided a measure of locational value at the individual property level. The methodology used expert system heuristics to select comparable properties from a database by asking a series of questions about the subject property. The values of the comparables were adjusted to account for all physical differences between them and the subject property. Value maps displayed values that have been reconciled for all differences except those attributable to location, an example of which is illustrated in Figure 8.35. They are therefore known as 'locational values'. A network model was devised to test for correlation between the accessible locations and high value areas displayed on the value maps. The network was constructed using a GIS and comprised road centre-lines and pedestrian routes. Movement along the network was influenced by link impedance, which refers to the cost associated with traversing the network in either direction and was based on connectivity, impedance, demand, barriers, turns,

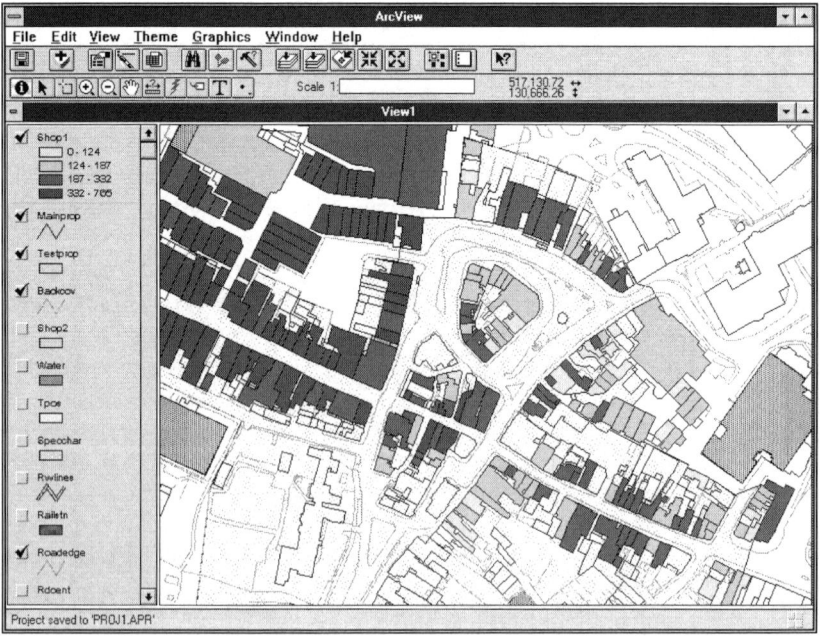

Figure 8.35 GIS-based value map of the central retail area of Horsham in west Sussex.

centres and capacity. The accessibility index value of a property on the network is an aggregate measure of how reachable it is from other properties. A gravity model that assumes that the effect of one location on another is directly proportional to its attractiveness and inversely proportional to its distance was used to calculate the index values. A variety of impedance and distance decay functions were used to evaluate the optimum correlation coefficient between accessibility and locational value. A relatively simple log–linear model that distinguished vehicular and pedestrian routes performed well and there was a significant positive correlation between locational value and accessibility. Wyatt (1998) found that two influences in particular were found to affect the accessibility calculated using the network model. These were the configuration of the route network and the impedance for traversal along the routes. This has obvious implications for transport planning and its consequent effect on property values.

Figueroa (1998) highlighted the problems with using discrete neighbourhood variables (dummies and scores) to handle location; the treatment at boundaries and the averaging effect throughout the neighbourhood distort reality. Instead a Location Value Response Surface (LVRS) can be constructed. This is a 'location-blind' regression model where the residuals are used as a proxy for location value. The spatial distribution of these 'location' residuals may be mapped using several methods but an interpolation grid is regarded as the best because it is able to show dramatic changes in value. An irregular interpolation grid means that the grid points are interpolated from the exact locations of the residuals rather than a spatially weighted average between regular grid points. Figueroa (1997) constructed a regression model using more than 6,500 time-adjusted residential sales over a period of two years. Seven variables describing physical attributes of the properties (including size) were used to construct the model. The model provides adjustment factors for each variable and performance statistics. The residuals (difference between predicted value from model and the actual time-adjusted sale price) were calculated by adjusting each sale to reconcile all physical differences so that differences in predicted prices would only be due to location. The residuals were geo-coded and an interpolation grid constructed by calculating the average of the residual values of neighbouring properties within a 500-metre radius weighted by the inverse of the distance between the subject and neighbouring property. These distances were also used to construct connectivity matrices needed for the calculation of Moran's Coefficient (MC) of spatial autocorrelation. The grid-estimated location values are then fed back into the location-blind regression model. The improved performance that resulted was significant. But, as with Gallimore *et al.*'s study above, this is inevitable because a location-blind pricing model will have a residual that can partly be explained by location. Simply interpolating a grid from these residuals and feeding these back into the model will inevitably improve its predictive power but does not explain the causes of locational variation in the predicted prices. It would be more useful to compare the interpolated location values with a hypothetical explanation for the variation observed (see Wyatt, 1997 op. cit.).

Desyllas (1998) undertook a similar study but used a regression model rather than an expert system to adjust for non-locational factors. Desyllas studied office rents in Berlin between 1991 and 1997. MRA was used to derive a residual figure for the amount of rent not explained by non-locational factors, what Desyllas terms 'location rent'. These residuals were plotted using a GIS and non-random patterns emerged. To model and predict location rent, a spatial variable that correlates with the pattern of residuals is required. Desyllas (1998) suggests that 'one approach to finding an independent variable is to model the street system as a network and calculate accessibility values based on the relationship between individual streets and the configuration of the system as a whole'. Desyllas' suggestion for a location variable is based on the hypothesis that the spatial pattern of rents in an urban area correlated with the spatial pattern of streets. As Desyllas states 'when many individual firms make complex location decisions based on accessibility to specific places important to their business, the pattern of demand that emerges mirrors the general configurational structure of the street grid'.

In summary GIS opens up new areas of statistical analysis and provides tools for identifying possible causes of errors in statistical modelling of property data such as spatial autocorrelation. Results from several studies of hedonic regression models show that prediction errors (measured in terms of Rsq) are dramatically reduced by including spatial autoregression. Studies suggest that hedonic models should test for and probably correct for spatial autocorrelation. 'GIS removes the most important impediment to this advance in econometric technique by allowing measurement of relationships among spatial units' (Clapp *et al.*, 1997).

Summary

Central to most built environment decisions are issues regarding proximity, neighbourhoods, accessibility, separation and complementarity. Distance mapping, network and gravity modelling represent sophisticated techniques for handling locational influences on property value. In the future planning and transport policy, congestion levels and developments in ICT will alter the current pattern of accessibility at an increasing rate. The GIS analysis techniques described in this chapter may reveal relationships between and trends in land and property value over a wider urban area. This has obvious benefits for valuation and location planning.

There is a well-used statistic in the GIS community that states that over 80 per cent of the data used by organisatons is geographically referenced. However, this in itself is no justification for using a GIS to store, manage, analyse or display these data. What is more important is whether your current and future business activities require those data to be geographically referenced in order to be of use. Implementing a GIS to handle property information depends on what the aim is. Some initial questions might be; how can GIS technology help an existing function or change a way a goal is achieved, what data will be needed and how will they be made ready for input into a GIS, what data can be shared, what results are required and who should be responsible for maintenance of the system and the data? It is necessary to evaluate all implementation options, understand how the organisation uses data and what the business functions will be. The effort expended depends on the size of the project and investment in GIS must be based on a sound business case. The extra sophistication provided by GIS must be warranted.

Despite significant barriers to IT development in the property industry in the form of restricted data access, confidentiality and secrecy there have been many innovative applications of emerging computer technologies. Feenan and Dixon (1992) believe that the public sector has produced excellent work with GIS for property decision-making and management. However, until a comprehensive property information system is established such work will remain peripheral to the land and property professions. The development of the NLIS and NLPG initiatives is watched with interest.

References

Adair, A. and McGreal, S., 1986, The direct comparison method of valuation and statistical variability, *Journal of Valuation*, 5, 41–47.

Adair, A. and McGreal, S., 1987, The application multiple regression analysis in property valuation, *Journal of Valuation*, 6, 57–67.

Anstey, B., 1949, Value contour maps: a new tool for planners. *Planning Outlook*. Iss 2, 29–39.

Anstey, B., 1965, A study of certain changes in land values in the London area in the period 1950–64, land values: being the proceedings of a colloquium held in London on March 13 and 14, 1965 under the auspices of the Acton Society Trust, ed. Hall, P., London, Sweet and Maxwell.

Appraisal Institute, 1992, *The Appraisal of Real Estate*. Chicago, Appraisal Institute.

Bible, D. and Cheng-Ho, H., 1996, Applications of geographic information systems for the analysis of apartment rents, *Journal of Real Estate Research*, 12(1), 79–88.

Clapp, J., Rodriguez, M. and Thrall, G., 1997, How GIS can put urban economic analysis on the map, *Journal of Housing Economics*, 6, 368–386.

Des Rosiers, F. and Theriault, M., 1992, Integrating geographic information systems to hedonic price modelling: an application to the Quebec region, *Property Tax Journal*, 11(1), 29–57.

Desyllas, J., 1998, When downtown moves: isolating, representing and modelling the location variable in office rents, Cutting Edge, RICS, Leicester.

Din, A., 1995, Construction of a geo-index for real estate values in the framework of geographic information systems. International Real Estate Society, Stockholm, Sweden.

Dixon, T., 1992, IT applications in property. *Mapping Awareness & GIS in Europe*, 6(2), 49–51.

Fairbairn, D.J., 1993, Property-based GIS: data supply and conflict. In Matter, P.M. (ed.) *Geographical Information Handling – Research and Applications*. London, John Wiley & Sons, pp. 261–271.

Feenan, R. and Dixon, T., 1992, *Information and Technology Applications in Commercial Property*, Macmillan, Basingstoke, UK.

Gallimore, P., Fletcher, M. and Carter, M., 1996, Modelling the influence of location on value, *Journal of Property Valuation and Investment*, 14(1), 6–19.

Figueroa, R., 1998, *GIS Supports Property Reassessment in Regina*, Geo Information Systems, Hune, 32–36.

Howes, C., 1980, *Value Maps: Aspects of Land and Property Values*. Geo Books, Norwich.

Longley, P., Higgs, G. and Martin, D., 1994, The predictive use of GIS to model property valuations, *International Journal of Geographical Information Systems*, 8(2), 217–235.

McClusky, W., Deddis, W., Mannis, A., McBurney, D. and Borst, R., 1997, Interactive application of computer-assisted mass appraisal and geographic information systems, *Journal of Property Valuation and Investment*, 15(5), 448–465.

Norcliffe, G.B., 1977, *Inferential Statistics for Geographers*, London, Hutchinson & Co.

RICS, 1994, The Mallinson Report: Report of the President's Working Party on Commercial Property Valuations. London, Royal Institution of Chartered Surveyors.

Rodriguez, M., Sirmans, C. and Marks, A., 1995, Using geographic information systems to improve real estate analysis, *Journal of Real Estate Research*, 10(2), 163–173.

Scott, I., 1988, A knowledge based approach to the computer-assisted mortgage valuation of residential property, unpublished PhD thesis, University of Glamorgan.

Waters, R., 1995, *Chartered Surveyor Monthly*.

Wyatt, P., 1997, The development of a GIS-based property information system for real estate valuation, *International Journal of Geographic Information Science*, 11(5), 435–445.

Wyatt, P., 1998, The use of digital mapping and geographical analysis in property valuation: a case study, RICS Research Report, RICS, London.

Part III

GIS issues in land and property management

9 Information management

Introduction

A GIS provides a means of managing information digitally and in a geographical context. Many property management decisions require consideration of location, geography and space. The technology is well suited to this, but constructing a GIS-based property information system that records the complex interests in land and property is a daunting task. The volume of data is often enormous and the cost of establishing and maintaining a database will be prohibitive unless a means of utilising existing resources can be developed. Historical property data can be hard to trace due to the poor quality and maintenance of some records. Property interests are heterogeneous, if only due to their unique locations in space. Add to this a plethora of physical and legal factors that typically characterise property interests and it is no surprise that many property information systems are developed for specific applications and it is only later that the potential benefits of linking data sets and databases becomes apparent. Yet to collect, store and manage information in digital form is clearly more efficient than any other means. Consequently the property industry has witnessed the digitisation of substantial amounts of property data with the aid of increasingly sophisticated technology.

The management of land and property information is not just about computer systems. There are human, information and commercial issues to be considered. Even when a project has been clearly specified and system and data requirements outlined, access to certain information may not be possible due to confidentiality constraints or legislative barriers. Certain land and property information is commercially and personally sensitive and must be handled accordingly. The laws of copyright and the data protection regulations must be adhered to. The information that the project requires may need to be integrated so that analysis can occur, but the data formats may be incompatible. Cost, time and requisite skills are all issues that must be considered when implementing an information management policy.

A number of information management issues have arisen during the implementation of GIS applications for the built environment and these are considered in this chapter. They can be categorised as follows:

(a) Data quality and liability

- Measuring data quality
- Errors in land and property data input, analysis and display
- Liability for data integrity and misuse of data.

(b) Data access

- Data ownership and intellectual property rights
- Data protection and privacy
- Confidentiality.

(c) Data standards

- Standards for geographical data
- Standards for describing data
- Standards for land and property data.

A key advantage of information management is the synergy arising from the pooling of information. A particular benefit of GIS is the ability to integrate data from a wide range of sources using location as a common point of reference. The process starts with the definition of items of interest, progresses to the organisation of those items and their representation in the database and ends with the communication of data to others. If managed properly the process of data integration will alleviate or neutralise problems of data duplication, reduce data limitations and help to reduce data capture costs. It will also improve the information base as a whole and allow users to combine data sets in ways that were previously unavailable to them.

Although an integrated data resource is likely to provide significant benefits, merging sets of geographical information in a meaningful way is a complex process. A wide range of factors can influence success and the effective management of resources is critical. Factors that need to be considered are:

- Political, institutional and organisational: the significance of these factors often far outweighs the technical obstacles to GIS-based data management. Leadership, enthusiasm and co-operation between individuals, departments, organisations and governments are vital if geographical data resources are to be utilised effectively. A willingness to co-operate results in the development of standard procedures and increases the ability to integrate and exchange data.
- Social and cultural: often historical reasons for secrecy can play a major part in retarding the data dissemination process.

- Financial: although the unit price of computer hardware has dropped significantly in real terms over recent years, this is offset by the increased processing resource needed to support modern operating systems. GIS software is often expensive and the costs of data – particularly map data – have remained stubbornly high. Without cost-effective data sharing policies in place within and between organisations, the cost of data purchase, update and management can prove to be too high in some cases.
- Technical: although the technology itself is not perceived as a major barrier, the availability of skilled staff has been a problem.
- Data-driven: somebody wishing to merge data sets from different sources faces difficulties because different applications tend to represent information in different ways. Classification tends to be carried out with a specific objective in mind and it may be that the classification scheme adopted by one organisation is inappropriate for another.

Data purchased from different suppliers are often incompatible and may require manipulation before they are useable on in-house software. This has been a significant barrier to the uptake of GIS in the property industry. Even in-house there are problems regarding standard property referencing procedures, between surveyors and analysts, for example, often because the former fail to recognise the value that a comprehensive information system can offer. A 'critical mass' of data is often required before recognition is achieved.

Data quality and liability

As GIS usage has become widespread and the technology has moved from university departments to government and corporate environments, the need to consider the implications of bringing together diverse data resources has become more pressing. Issues of data quality include error, uncertainty, scale, resolution and precision in geographical data. All of these aspects affect the ways in which information can be used and interpreted. There is a need to consider:

- What data quality is.
- Ways of measuring and reporting data quality.
- The types of error that might occur in land and property data.
- The documentation of geographical information.

What is data quality?

What is meant by the term 'quality'? In general terms words like 'good' or 'excellent' are used to describe the presence of quality, but we need a much more precise definition if we are to talk about high and low levels of quality in terms of geographical information. Generally, in order to assess the quality of something we need to have a purpose in mind for it. The quality of a geographical database will depend largely on the uses that it is to be put to.

A useful definition of data quality is 'fitness for purpose'. The London Underground map provides a good example of how something can be 'high quality' and 'low quality' at the same time. While it is high quality for the user who is deciding on the best route from A to B using the tube, it is of a much lower quality if one wishes to know how far it is from A to B, or how long the journey might take. The application to which the data are put will determine whether or not they are suitable for the intended purpose and so data quality is also application dependent.

We must always ask ourselves 'Are the data fit for the intended purpose?' The answer depends on the type of analysis we wish to undertake. The only way that this question can be answered is by knowing something about the data sets that we are using as well as the task that we plan to undertake. Having defined the task and set the parameters of the project, attention should be turned to the data themselves; how accurate are they, when and how were they collected, how often will they be updated in the future? Poor data quality is just as likely to jeopardise a GIS project as hardware or software problems. We need to consider:

- How well the digital structures that we have created represent the real world?
- How well the different algorithms that we use to operate on these data sets compute the true value of products?
- What happens to errors in the data when we perform a sequence of GIS operations?

Measures of data quality

The development of standards for data description in the GIS field (see, for example, Federal Geographic Data Committee, 1994a) have led to the recognition of the following quality parameters for geographical information: accuracy, lineage, completeness, consistency and currency. These are considered below.

Accuracy: of position, topical attributes and temporality.

Accuracy can be defined as *the closeness of results, computations or estimates to true values (or values accepted to be true)* (Chrisman, 1991). Since spatial data are usually generalisations of the real world, it is often difficult to identify a 'true' value and we work instead with values that are *accepted* to be true. For example, in measuring the accuracy of a contour in a digital database, we might compare it to the contour as drawn on the source map.

Although the accuracy of a spatial database will certainly impact upon the quality of the products generated from it, the nature of this relationship is often complex and difficult to assess. So, for example, the accuracy of a slope, aspect or watershed computed from a DEM is not easily related to the accuracy of the elevations in the model itself.

In the property sector, accuracy is dependent to some extent on application. A key consideration is whether accuracy can be guaranteed by some assurance of authenticity. If this cannot be given, the market will soon determine whether information is reliable or not and will bid a price accordingly. For example, it is reported that 6,000 square metres of office space were let for £150 per square metre on 1 April 1998, that the lease was for a term of ten years with a rent review in year five and there was an initial rent-free period of six months. All of these facts are important to a property valuer. There is no room for inaccuracy. For example, a rent review clause that states that the rent can be increased or decreased in line with open market rents is substantially different from a clause that states that the rent may only be increased at review. Facts must be distinguished from opinion.

Precision, which is clearly distinct from accuracy, can be defined as the degree of detail or refinement present in reporting a measurement. Precision is not the same as accuracy – high precision in a measurement does not guarantee that the measurement is accurate! GIS tends to work with high levels of precision, often much higher than the accuracy of the data itself. All spatial data are somewhat inaccurate but they are generally represented in the computer with high levels of precision, which can often be misleading.

Since all spatial data are inaccurate to some degree, three important questions are:

- How can we measure accuracy?
- How can we track the way in which errors are transmitted through GIS operations?
- How can we ensure that users do not ascribe greater accuracy to data than they deserve?

Lineage: details of the source and history of the data and a description of the set of processes that are used to produce a dataset.
Lineage is a record of the data sources and of the operations which created the database. This would include how the data were digitised and from what sources, when they were collected and by whom, and what steps were used to process them. The documentation of lineage is a very useful tool in assessing the quality of data. An experienced GIS user can quickly assess how appropriately the data were developed, and whether any unreliable methods have been used in constructing them.

Lineage information provides vital information to the user when determining fitness for purpose. Lineage information is often a component of metadata (which we can define succinctly here as 'data about other data') and standards for producing metadata are described in the section on 'Data standards'. As an example, the date on which the data were reported and dates of important facts reported in the data such as a lease commencement date should be included in any description about those data.

Completeness: of data coverage, classification and verification.
Completeness is affected by rules of selection, generalisation and scale. In this context it refers to geographical completeness (national coverage as opposed to London only) and sector completeness. A property database that is complete should cover all sectors of the market, including residential, office, industrial, retail and leisure premises. More specifically, completeness concerns the degree to which a database conforms to its specifications. Are all the records that should be logged present? For example, there are 21 million land parcels in England and Wales. Of these, about 17 million are logged in the databases of HMLR and 15 million or so of these are digitally registered.

Consistency: longitudinal (or temporal) and cross-sectional (or geographical) consistency.
Ideally, property researchers want consistent data that are disaggregated to the individual property level, are consistent for the geographical extent to which they relate and are consistent over the time period for which they have been collected. Analysis of property data often takes the form of cross-sectional comparisons, between two or more urban areas for example, at a particular point in time, or longitudinal or time series analysis of a specific urban area over a period of time.

Currency: up-to-date and timely data.
The final element of data quality is how up-to-date the data are. Generally, old data are less useful than new data (unless you are interested in historical analysis). Also, combining data sets from different time periods can cause problems. Consider the example of an analyst who wishes to model the relationship between socio-economic characteristics and property prices. His comparisons require access to property and census information, but the only census data available to him were collected in 1991 while property information has been collected for 1998. His analysis of the data will certainly produce some results, but their validity may be called into question because of the seven-year gap between the collection dates of the inputs.

Currency can be measured in a number of different ways:

- Date of publication
- Date of digital capture
- Date of survey.

Generally, the date of survey is the most useful of these measures. This represents the time that the data were collected and there may be substantial delays between capture and publication.

Location details of people and businesses change frequently. For GIS to be effective, locational accuracy needs to be kept up-to-date at the property level (McKeon, 2001). For many property professionals, currency of data is seen as a major issue. If a surveyor is acting for a client on a rent review, then up-to-date, comparable evidence is required to support the opinion of rental value.

There are some recurring themes regarding the strengths and weaknesses surrounding the quality of land and property data in the UK (Smith and Wyatt, 1996):

Strengths:
- Property indices, compiled by the IPD and property consultants, are generally of good quality; but such indices are slow to appear.
- The timeliness and accessibility of on-line sources is good.
- The VO Rating List has good coverage and is readily accessible.
- UK land and property data are perceived to be of higher quality than those that are available in most overseas markets (possibly excepting the US).

Weaknesses:
- Public and private sector cross-sectional data are often incomplete and inconsistent.
- Longitudinal or time series data often have a short history with infrequent observations and discontinuities.
- Data are slow to be published compared with other investment markets.
- There is a heavy reliance on samples and proxies.
- Data are often expensive to obtain.
- Geographically, property data suffer from inconsistent boundary definitions.
- Information on stock, such as quantity (floor-space), quality, occupancy levels, vacancy rates and occupier activity is poor.
- Property transaction-based measures are weak (e.g. rents, yields and capital values).
- There is a lack of measures for specialist property types and secondary property.
- Largely due to the above factors, published statistics on property market performance sometimes report conflicting results when the underlying data purport to measure the same thing.

Table 9.1 classifies property data sources according to their coverage. The ideal is a 'comprehensive' data set, which can be disaggregated if required to the level of its most fundamental components. Many datasets have been

Table 9.1 Aggregated and disaggregated property data (Society of Property Researchers, 1996)

Data coverage	Description	Examples	Comments
Comprehensive	Data relating to features of the property market in total, or wholly representative of the sub set of properties being studied	Total stock figures; property GDP' measure; national rateable value	Very few examples of truly comprehensive data sources. Highly aggregated data is of limited use; ability to disaggregate is critical if maximum benefit is to be obtained
Sample	Data relating to groups of properties which are subsets of the total population being studied	Most sector, region and locally based data; indices; market survey results	Most available data sources fall into this category. 'All property' indices are based on samples andare only a proxy for unavailable comprehensive data
Specific	Data specific to a particular property	Floor area, lease terms, rent paid	Detailed data complete form except where such data represents an extreme value and hence provides a benchmark for the wider market (e.g. the 'top' rental value in a town center; largest letting, etc.)

aggregated by location and/or market sector so their utility to many property advisors is diminished. Whilst it is accepted that 'information overload' may result if data are inappropriately chosen or mishandled, in general the more data that can be assimilated at the finest level of detail (building or sub-building level) the more robust is the analysis.

Errors in land and property data

A GIS is capable of generating an enormous amount of map data, much of which may be of questionable quality. Being able to distinguish data that are unreliable, inaccurate, or unfit for the intended purpose from those that are important, usable and reliable forms a major part of any GIS project. The user needs to make sure data are accurate, precise, up-to-date and well maintained. We can define error as *'the quantity measured when assessing the degree of closeness of observations to the truth* or as *the deviation of a measurement from a true value or value that is accepted as being true'*

(Chrisman, 1991; Rybaczuk, 1993). There are three broad categories of error in spatial information:

(a) Positional errors – errors in the location of an object.
(b) Attribute errors – errors in the set of information associated with an object.
(c) Temporal errors – errors in the time of observation of measurements or time of update of digital records.

These three broad types can be subdivided into conceptual and measurement errors. Conceptual errors are errors in the communication process, leading to misinterpretation of meaning or mistakes in encoding. For example, what is meant by 'house' or by 'area'? Measurement errors are errors in the recording of information due to operator mistakes or equipment failure.

Errors can appear in three different ways, categorised as gross error, systematic error and random error. Gross errors are normally caused by severe equipment failure or serious mistakes in the measurement process. They are often large and are usually easy to spot because they are normally very different from other measurements. Systematic error (also known as bias) is a uniform shift of the values in the data. Here, a systematic pattern arises in the errors that are occurring. If, for example, a survey instrument was badly calibrated, all the distance measurements with that instrument might be 1 metre out. There is some consistency in the error pattern that is produced. Finally, random errors occur because of the imperfections that always exist in instruments and processes of data interpretation. No measurement can ever be 100 per cent accurate and no operator can ever be one hundred per cent reliable. Random errors are usually small, but unavoidable.

Some of the ways that errors enter geographical data handling and analysis are described in the following sections.

Errors in data input

Typical data input errors include (after McKeon, 2001):

• Abbreviations; for example, male (m), female (f), c/o
• Data entry accidents; wrong field or incorrect spelling
• Duplicate or incomplete records
• Geo-referencing inconsistencies; postcodes, grid references, addresses
• Name conventions; Robert Smith Ltd or R Smith Ltd, Kensington High Street or High Street Kensington
• Different standard formats; for dates, units of measurement, etc.

Errors in data analysis

AGGREGATION

Classifying data into a small number of large groups will hide patterns of variation within each group. In some data sets, such as the British census, data are deliberately subjected to random change at high levels of detail to preserve anonymity whilst the correct totals are preserved at aggregated levels. The analyst who seeks to aggregate data should study the data distribution before committing it to a classification. This will reveal whether the data are normally distributed, skewed and where any outliers may lie. It is also important to note that 'relationships observed at a particular level of aggregation do not necessarily hold for the individual observations' (Martin, 1996). This problem is known as the 'ecological fallacy' and is illustrated below where the mean house price of a unit postcode is £400,000 but this aggregate value is actually only found in two cases and is totally unrepresentative of the large value in the middle of the list (Martin, 1996).

£300,000
£400,000
£400,000
£1,200,000
£300,000
£300,000
£100,000
£200,000

One of the key spatial analysis problems arising in the aggregation of data is the Modifiable Areal Unit Problem (also known as MAUP) (Openshaw, 1984). MAUP occurs when data for one level of spatial resolution (either a set of areas or a point data set) are aggregated to another (namely a larger set of areas or from a point data set to a zoning system). The resulting transformation may change the scale and resolution of the original data. This in turn may amplify or suppress existing patterns and may even create wholly new spatial patterns, which are a direct consequence of the interaction between the data being aggregated and the properties of the zoning system used for the aggregation (Raper *et al.*, 1992).

The MAUP illustrates the difficulty of choosing one of potentially many areal units that may be defined. The problem is to decide upon a suitable level of aggregation for the display of the underlying data (Martin, 1996). A key area of research in computational geography is this problem of zone design to select the most appropriate reporting regions for geographically distributed phenomena.

AUTOCORRELATION

Spatial autocorrelation is the formal term for the 'first law of geography' (Tobler, 1979), which states that 'everything is related to everything else, but near things are more related than distant things' (Longley *et al.*, 2001). Spatial autocorrelation is an attempt to quantify the degree to which geographical features are related. It is important because a central tenet of many statistical analysis techniques is that the input data values for the variable that is being measured must be independent. However, spatial autocorrelation infers that, geographically, the values of a variable are dependent to some extent on neighbouring values. For example, house prices in a particular neighbourhood tend to have some relationship or similarity with one another.

Similarly, if data are collected for the same variable over a number of time periods, temporal autocorrelation or smoothing may result. The more that time periods are aggregated the more we hide possible temporal influences on geographical patterns. For example, aggregation over time might mask temporal trends such as the effect of new transport development on property values or land use patterns (Monmonier, 1996). 'Just as temporal periodicity may be missed if measures are made at an inappropriate temporal scale, so spatial periodicity may only be detected if the spatial scale of measurement is appropriate to the distribution of the spatial phenomenon' (Longley *et al.*, 2001). So aggregation to various scales affects time as well as space. Taking several years of data reduces the effect of chance but increases the range of possible causal agents.

It is possible to propagate error by combining the effects of spatial and temporal autocorrelation. For example, data collected from the same geographical units, such as office rents over several years, which are then averaged by postcode sector for each year represent moving averages. This combination of spatial and temporal aggregation causes a great deal of smoothing in data values. Research has shown that autocorrelation effects are reduced when data are analysed at as disaggregated a level as possible, both spatially and temporally.

AREAL INTERPOLATION

There are problems that are caused by interpolating geographical data that have been collected at one particular level of aggregation and are translated to a different level for analysis and reporting. Errors can enter results if a pattern that is based on areal units leads to conclusions based on individual properties. For example, an average income in a postcode sector of £16,000 may mask substantial differences in more discrete geographical areas or between individual properties within that postcode sector. A map that shades average income for each unit postcode might reveal markedly different results. Therefore, any such map should include a caveat that the data are only reliable at a particular level of aggregation. In the UK, house price data are only available at postcode sector level but there is a temptation to use them as a local-level indicator for targeting initiatives and investment especially when thematic mapping is so easy on a desktop GIS (Ferrari, 1999). Monmonier (1996) warns that it would be very unwise to interpolate patterns from an aggregated level to conclusions based on individual locations. Research into the effects of areal interpolation has been undertaken by Flowerdew and Openshaw (1987), who provide a useful review of the problem, and some solutions are introduced by Flowerdew *et al.* (1991), Voss *et al.* (1999) and Simpson *et al.* (2000).

ERROR PROPAGATION

We have said that errors are inherent in geographical data sets. As a result, when we combine two or more different data sets using standard GIS operations, the errors that are present in the source data sets will be carried forward through our analysis. This phenomenon is normally referred to as 'error propagation'. Error propagation is a problem, because how it works is not fully understood and an important research activity in GIS is the investigation of error behaviour as data are combined in sequences of GIS operations. For example, Li *et al.* (2000) show how the compounded effect of errors can influence the results of simulation procedures. More comprehensive treatment of this topic is beyond the scope of this book, but it is important to be aware that complex GIS overlay and data manipulation procedures are likely to cause error propagation. As successive data sets are combined and manipulated, the inaccuracies that are present in each information source will interact with one another, causing new patterns of error.

Data quality and the display of land and property data

All measurements have errors, and maps made using them will include these errors. The quality of the message given by map information depends upon the way in which information is collected and depicted. Again, fitness for purpose should be the overriding concern. In other words, is the map display appropriately accurate for the task at hand? The user is referred back to the discussion of different types of map and generalisation approaches in Chapter 2.

There are obvious physical limitations on the amount of information that can be displayed on a map without it becoming cluttered. As far as thematic mapping is concerned, it is important to consider what information needs to be displayed and what can be stored in the database. Sometimes when inappropriate colours and shading are used the map confuses rather than illuminates, particularly when trying to communicate varying densities. Simple presentation is paramount.

Choropleth maps, introduced in Chapter 2, can be used to describe geographical effects such as land use type or council tax bands. However, 'by manipulating breaks between categories of a choropleth map,... a mapmaker can often create two distinctly different spatial patterns' (Monmonier, 1996). It is important to explore cartographic alternatives to identify geographical trends, groupings and relationships. For further details, consult Robinson *et al.* (1995) or Monmonier (1996).

Choropleth maps can be overlaid with other maps in order to pick out patterns. Often, in order to make the map easier to interpret and due to the availability of aggregated data, these choropleth maps relate to areas or polygons with boundaries such as electoral wards or postcode areas. The actual size of these polygons varies according to factors normally unrelated to the data for which they are being used as a display; the size of enumeration districts depends, largely, on the number of people within each. Therefore they are much smaller in size in the dense urban areas than they are in rural areas. If they are used to display average age, for example, then the rural polygons will dominate the map because they are bigger. It is important that the resultant image does not mislead because the boundaries do not relate to the variables being displayed.

Size differences among area units such as parliamentary constituencies can radically distort map comparisons. The map of the UK in Figure 9.1 shows how the conservative party's 25 per cent share of the vote in the 2001 general election appears greater because of the high number of geographically large rural constituencies within that share. The 63 per cent labour government share is weighted towards the geographically smaller urban constituencies.

In summary, with regard to data quality and possible sources of error, the analyst must be aware of the consequences of combining data from disparate sources that may have been collected or surveyed at different scales. The same level of caution needs to be exercised as that employed by the statistician when exploring and presenting data: various classification schemes might be tried (equal interval, quartiles, natural breaks) in order to best represent the variation in the data and careful consideration taken of any outliers in the data sample and their relevance to the analysis. When looking for patterns a combination of maps, scatter-plots and correlation coefficients might be used.

In recognition of the need for information about data quality, the Association for Geographical Information (AGI) has published a set of guidelines on geographic information content and quality (AGI, 1996). This publication is a much needed introduction to the pitfalls of using and bringing together spatial information from different places. The guidelines are designed to help those, who require geographical information, determine that it is fit for its intended purpose and how to decide which aspects of quality are important given different user objectives. A summary of the recommended procedure is given in Table 9.2.

Liability for data integrity and misuse of data

There should be a clearly defined framework of responsibility for data quality and liability for errors. As countries move towards multi-purpose land and

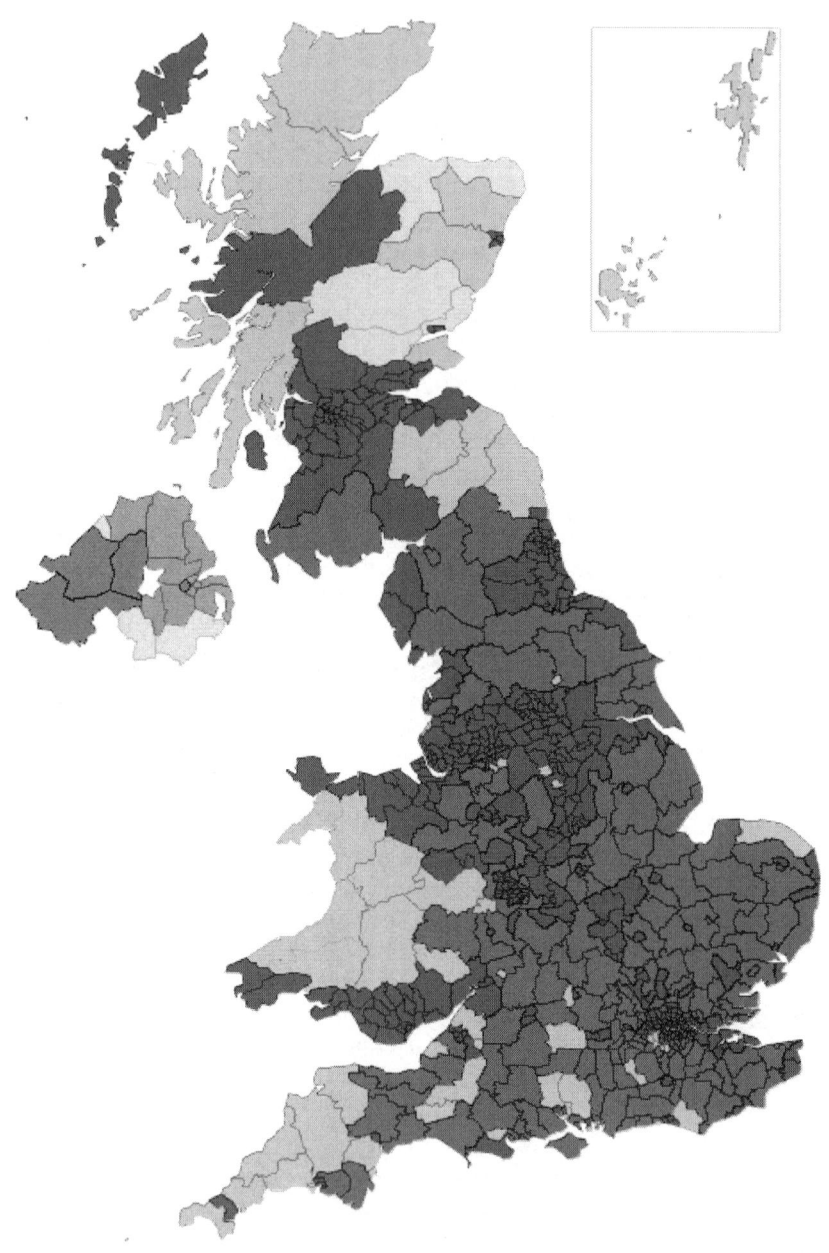

Figure 9.1 2001 general election results by parliamentary constituency.

Source: http://news.bbc.co.uk/hi/english/static/vote2001/results_constituencies/pol_map.stm, used with permission.

Table 9.2 Summary of process for assessing quality of geographical data

---

Define the task, its geographical and temporal extent and output requirements

Identify the contents of data sets that the task requires
Specify spatial attributes; feature types, scale, accuracy of positioning, topology, etc.
Specify temporal attributes; intervals, duration, etc.
Specify thematic attributes; units of measurement etc.
Other issues; availability of existing data sets, resourcing, format, restrictions on
   use, etc.

---

property information systems, more data are combined and it is harder to assign responsibility and determine liability if problems arise. If nobody can guarantee quality, this may reduce the value of data. Compulsory land registration and maintenance of the history of ownership for each land parcel may help to detect errors and omissions in a national land information system such as the UK NLIS described in Chapter 5. Another answer is to only make factual data available. Regarding the dissemination of such data, liability and responsibility for accuracy should be made explicit. This is particularly important where data from a variety of suppliers are integrated and accessed from one source. There should also be a source or link for obtaining further information about each item of data. The market knowledge behind the facts is crucial to the understanding of certain land and property data.

Overseas case study

Under plans for introducing a NLIS in the UK (described in Chapter 5), in terms of data quality, the NLIS hub manager does not accept responsibility for mistakes of fraud. Consequently, data must be validated by the supplier. A distinction may therefore need to be made between 'registered' land and property data (i.e. data collected and managed by statutory authorities such as local authorities or HMLR), and 'recorded' data, that might be provided by a commercial organisation. The latter might serve to inform rather than be relied upon legally (property values, investment yields and vacancy rates, for example). A conveyancer might be expected to rely on registered data but may have no duty to seek recorded data, although a test of reasonableness might apply. The distinction between such legal and business advice is not always easy to draw but the NLIS should not change the duty of the conveyancer, rather, it should make it easier to conduct investigations that there is an obligation to make.

Data access

Legal issues have a significant impact upon the use of geographical information. A geographical information manager must be aware of the implications of copyright and data protection legislation, understand how these may limit the use of data sets and appreciate how to avoid misusing data.

Information is increasingly being seen as a tradable commodity. Initiatives like NLIS rest upon this concept. Tradable commodities imply the existence of vested interests and rights over information. The protection of rights is important, because data are valuable and time and effort has been expended in developing them. The main reasons for legal action against those who misuse data are as follows:

(a) Money – loss of earnings through data leakage and a desire for punitive damages/cost recovery.
(b) Reputation – 'The OS is the definitive source of information about UK mapping, and misuse of our data could result in products that bring it into disrepute'.
(c) Privacy – an individual or organisation may wish to protect its data resources from outsiders.

Unfortunately, there is no single point of contact for information law in the UK. It is incomplete, diverse and in some cases contradictory. The main sources are ownership law, including copyright (data sets) and patents (software, data models etc.), and privacy law, including the Data Protection Act and Employment legislation.

After a description of the main players in and the structure of the land and property data market, the following sections present an overview of copyright, data protection, privacy and confidentiality issues for land and property information.

Data ownership and intellectual property rights

Is it possible to 'own' data? In law, this topic is covered by the Copyright, Designs and Patents Act of 1988. The law does not consider that individual data items are 'ownable'. Instead, it is the creative effort spent in selecting and recording data that is recognised and protected. The producer of a set of information has rights associated with its use, for example, the right to buy and sell information and the right to be identified as the author of a data set. Maps are specifically mentioned in the Copyright Act, as are computer files and other digital data. However, the gap between a paper map and the digital work that lies behind it is not clearly considered. Copyright law was not intended to protect rights over data, but to recognise the time, effort and creativity that the originator of a work has committed in selecting and arranging the information. It is based on the principle that you cannot protect an idea but that a final product that results from that idea is protectable.

Without copyright protection or other forms of Intellectual Property Right, suppliers of property data will be reluctant to permit the dissemination of data that has taken time and money to collect and compile. A detailed discussion of the requirements for legal protection of Intellectual Property Right concerning geographical data is beyond the scope of this book but others, Larner (1996a,b) for example, discuss these issues in depth. In general, traditional legal frameworks may require adaptation if they are to be applied to digital geographical data. An example of how the interpretation of copyright laws can have a direct impact on users is the British OS's interpretation that the transfer of map data from disk to RAM and display on screen within a computer system constitutes copying. As a result, OS has introduced a charge for data use based on how many users are manipulating the data. Such charges do not apply in the case of hard copies.

There is a balance of rights between the user, the author and the subject of a set of data. If information is to be protectable under the Copyright Act, it must be original. Originality is proved by 'sweat of the brow' – effort, money and ideas that have been spent in creating the work. Original literary and artistic works are copyrightable. In other countries, a work may have to demonstrate intellectual or artistic merit to be copyrightable. In the UK, we consider how much it costs to create data as well as 'sweat of the brow'. This is not considered in the rest of Europe or the US.

Geographic information is mentioned in the UK Copyright Act in certain ways:

- Section 4(a) refers to graphic works and photographs including maps, photogrammetry and satellite imagery under 'original artistic works'.
- Section 1(1)a refers to original literary works, which would include digital maps and text and direct digital data capture.

Rights over information are limited. In general, it is up to the prosecutor to prove that these rights are being infringed somehow. The main issues are:

- Is the work original? The effort put into the selection and presentation of information has to be your own – 'sweat of the brow'. For example, if I take an OS map and process it, the rights of the OS go through to the new work that I have created. I should have the permission of the OS both to produce and to distribute the new map and if I intend to sell it then there is a strong argument that the OS would be owed some form of royalty as a result.
- Copyright law is limited by 'good of society' arguments. Information copyright is a powerful way of restricting access to data, and so there is a clause that states that copying has to be 'substantial' to infringe

rights. The main point of argument in any legal action is usually the determination of how much damage has been caused by infringements.

- Distinguishing between 'use' and 'copying' in a digital environment is difficult. Computer network transfer and disk to disk or disk to RAM transfers are all copying activities. This implies that a data supplier could charge a copyright fee for the use of their product, since it involves copying.
- If I could establish how a piece of software works and launched a product that achieved similar results through my own endeavours, I would not be in breach of copyright unless it could be demonstrated that I had adopted an entirely similar approach. The same is applicable to data sets through 'value added' arguments.

Under UK law, the standard duration of copyright protection is the life of the author plus fifty years. For government held data sets, the length of copyright is 125 years for secret information, and fifty years for published information. The length of copyright for computer generated products is a maximum of fifty years. Updates to a computer database do not result in copyright protection for a further fifty years unless an entirely new and different product is created. However, the recent European Database Directive provides new rights for database creators within the EU, allowing a database owner to restrict rights of extraction and duplication of a database for a period of fifteen years following inception. This right can be extended for an additional fifteen years if substantial changes are made to the database. Mirchin (1997) provides a useful summary.

The willingness of a surveyor to pay for a property information service would depend on the extent to which it saved time in collecting data and any improvement in the quality of advice given to a client that can be reflected in the fee charged. With regard to public sector data, an aggressive and opportunistic pricing policy, which is designed to charge commercial rates for data despite the fact that the data were often collected at the taxpayers' expense, means that the data are not charged at the marginal cost of reproduction but at the price the market will bear (Maffini, 1990). In the UK therefore, data pricing and licensing arrangements can be a significant barrier to GIS uptake, especially in the case of access to large-scale OS map data. Copyright is also a significant barrier to the use of digital mapping and is a major cost when a consultancy wishes to pass on results of GIS analysis to clients. This will inevitably lead to reduced accessibility of geographical data for some users and particularly for a fragmented and price sensitive property industry. Also, the onward use of data may be restricted, usually through the use of license agreements or registration of the purposes for which data are to be used. A data provider issues a license to use data in particular ways:

- Copy the work
- Distribute to the public/issue copies (there are internet distribution and security implications here)

- Perform or display the data
- Broadcast/cable programmes
- Adaptation – for example, summarise data, filter/analyse.

An information manager needs to know the terms of these licenses so that he/she can ensure that usage falls within licensing restrictions. This becomes more difficult as the size of an organisation increases but intranet technology can be used to police usage through process/license management and the tracking of users.

Data protection and privacy

When information about individuals is combined with details of where they live and work, safeguards are required to monitor, regulate and, where necessary, restrict ways in which the resultant information is used. Disaggregated data are more useful to analysts than aggregated data because of their flexibility for analysis and integration with other data sets, but concern for privacy may arise if individuals can be identified from the synthesis of data sets. These privacy issues become increasingly significant as opportunities for data dissemination become greater. The responsible use of integrated data requires safeguards and data security provisions must be acceptable to all parties.

Data protection

In the UK, the storage of personal data about third parties may be subject to the restrictions of the Data Protection Act, 1998. This Act seeks to ensure the free flow of information to meet business needs, whilst respecting the private lives of individuals. The Act applies to the processing of personal data. The Act does not prohibit personal data from being published but such publications must respect the provisions of the Act. In many property-based GIS initiatives, a distinction is made between 'land and property data' and 'personal' data, with a legislative framework that supports access to former and ensures personal privacy by restricting access to the latter. So whereas the Data Protection Act protects the rights of the individual, initiatives such as the recording of price paid information on the Land Register and NLIS propose open access to land and property data. Consequently there is a need to distinguish between property data and personal data. For example, when buying a house it is necessary to know whether a mortgage is outstanding on the property prior to purchase, but is it necessary to know the amount of the mortgage?

Data Protection involves three interest groups:

- The Data Protection Registrar, who protects the rights of the subject of a data set.
- The User/Data Controller/Data Processor, who uses/stores information about the subject.
- The Data Subject, who usually has no idea that there is a problem!

The 1998 Data Protection Act defines personal data as 'Data which relate to a living individual who can be identified from that information or from that and information and any other information in the possession of or likely to come into the possession of the data user' including expressions of opinion about that individual and any indications of intention of the data user or any other person in respect of that individual. 'Identification from that and any other information...' means that if two data sets cannot identify an individual independently, but can be easily integrated to permit this, then the data in both are personal data.

If data might be personal data, then the user must register with the Data Protection Registrar the sources of the data, the purposes to which they will be put, any disclosures that will be made and to whom these will be made available. The Registrar has six months to reply, during which time the user may carry on using the data unless previously refused permission or currently involved in an appeal with the Data Protection Tribunal. The Registrar will make a decision about what can and cannot be done with the data. However, a user who disputes the ruling can apply to the Data Protection Tribunal for a neutral re-evaluation and ruling. Both the Registrar and Tribunal have wide-ranging powers and may refuse to register the purpose, issue an enforcement notice restricting use, de-register the data so use becomes illegal, enter premises and seize data sets with or without notice and prohibit transfers to third parties. Data sets that do not have to be registered or may be exempt from the Data Protection Act include those listed in Table 9.3.

Freedom of information and access to government information

Freedom of information and data protection legislation vary from country to country, and cultural and social factors are equally variable. In some cases, changes in legislation to release key property data may be required to generate a 'critical mass' of data, thus ensuring the viability of property-based GIS development. There is a corresponding need to ensure that the rights of the individual are safeguarded.

Table 9.3 Data sets that need not be registered or are exempt from the Data Protection Act

---

1 Data collected for purposes of national security
2 Data collected for purposes of crime prevention and taxation
3 Data collected for statutory purposes
4 Data collected for purposes of health, education or social work
5 Data collected for regulation of financial services
6 Disclosures required by law
7 Data that is in the public interest to make available (journalism etc.)
8 Data held for domestic purposes

Tensions arise in government policy because of pressure on public sector data providers to be more commercially minded, coupled with restrictions on their participation in the provision of value-added products and services. There is also a conflict between financial targets for government data providers and increasing pressure on openness of data provision (Coopers and Lybrand, 1996). Onsrud and Reis (1995) suggest a move away from treating publicly held information as a public good but ask whether the public interest will be best served by allowing governments to charge for GIS data. There is little empirical evidence to suggest the best approach, but the question of how far a government department or local authority can go in collecting and selling data before it exceeds its statutory powers must be raised.

In November 2000, the Labour Government passed the Freedom of Information Act in the UK. This Act will be phased in gradually, with a view to full enforcement by November 2005. It gives a general right of access to 'recorded information' held by public authorities and sets out exemptions from that right. The Act appears to be considerably watered down from its original, wide-ranging brief. The legislation has so far achieved little in terms of widening access to land and property data, being more concerned with exceptions to public information access restricting the 'right to publish'. The UK needs to look to the US for an example of an information policy that allows businesses to repackage public domain data.

The Local Government Access to Information Act specifies that public access must be available for papers relating to meeting agendas for local authorities. The Geographical Information Charter – a DTLR/Citizen's Charter Group initiative forms part of the 'Government Direct' move towards more open access and accountability. European Directives are being prepared on a number of themes:

- Rights over databases
- Liability for information services
- Data protection
- Access to public sector data – harmonisation of access to data and of charging regimes across Europe.

Confidentiality – from data secrecy to data dissemination

Confidentiality does not exist as a legal concept in the UK. However, central government departments often have disclosure rules for particular data sets and some have specific restrictions. For example, the 1920 Census Act restricts dissemination of census information to minimum aggregations of fifty households and Census of Employment statistics are gathered and collated so that individual firms cannot be identified.

An efficient property market requires its buyers and sellers to have adequate access to reliable information on which to base decisions, but 'the operation of the property market is constrained by an absence of information, with barriers to data transparency arising from several sources' (Adair *et al.*, 1998). If the quality of information is poor or it is not available in a sufficient quantity then decision-making will suffer. The property market is characterised by features that tend to restrict information flow; transactions occur relatively infrequently when compared with other markets, the holding period for property is relatively long and confidentiality is a factor with most transactions. Comparable evidence in support of property valuations, for example, has always been at a premium and valuers traditionally rely on an informal network of contacts, or the 'jungle telegraph' as it is known, to obtain such information. It is not difficult, therefore, to find references in the property press and academic literature that highlight shortcomings in current data sources, and there have been many calls for the public sector in particular to review data access policies. Widely publicised reports that strongly recommend an examination of data access in the property industry include the Mallinson Report (RICS, 1994a) and the IPD/University of Aberdeen study of property cycles (RICS, 1994b). The Mallinson Report on commercial property valuations highlighted the property profession's protective nature regarding data; explained by the need to maintain a competitive edge, but at the same time, stressed that valuers must work from a wide base of comparable evidence. This produces a conflict between the need for confidentiality of client information and the need for a comprehensive property database for property decision-making. Adair *et al.* (1997) argue that 'the property market, unlike other investment markets, has no formal market place making data collection difficult. The fragmented nature of information sources, inconsistent geographical definitions and difficulties involved in data assembly further complicate analysis'. Paradoxically, given the heterogeneous nature of each property interest (if only in terms of each property's unique geographical location) and the financial significance of each decision, detailed analysis of available information is often required before property advice can be given.

Many market commentators have highlighted the need for a comprehensive property data dissemination strategy. The Mallinson Report (RICS, 1994a) recognised the need for a national property database to improve the valuation process. There have also been calls for property information collected and maintained by the VO to be released into the public domain (DoE, 1987; Marriott, 1989; IPD, 1993). Private sector initiatives have been successful in meeting some of the demand for property data by providing market indices and databases compiled from published sources but access to detailed property data remains elusive.

After publication of Mallinson's findings the RICS General Council endorsed an institution policy that recognises the need to make public and private sector property data available, whilst having regard to the confidentiality of personal information. The policy aims to 'ensure that society benefits from wide access to land and property information, allowing informed decision-making and promoting a simpler and more transparent property market'. It includes an action plan that seeks to remove barriers that prevent data dissemination, to encourage data sharing and the re-use of data, and to promote the adoption of data standards for quality, accuracy, completeness and compatibility of data sets.

Adair *et al.* (1997) found, in a survey of RICS members, that it 'is generally agreed that the wider availability of credible data would be advantageous, ... [with] greater opportunities available in the industry or profession through public access to information'. However, 'data sharing is viewed with some reservation due to commercial interests involved' but 'where data are utilised ... in the analysis of market trends, this is not perceived to diminish commercial advantage'.

Despite these efforts there are very few true primary sources of property information. Most of these belong to private sector organisations that treat them with confidentiality because of their commercial value. Central and local government also collect primary data but they are often aggregated prior to publication, such as official statistics. The market for informal and anecdotal information is, on the other hand, fairly effective and is often relied upon as a source of vital information. The majority of property data sources are, therefore, secondary. These are widely available but can often be incomplete, aggregated, historic, infrequently updated and focused on specific market sectors only.

Public sector confidentiality

Government is a key geographical data collector and manager. Important departments as far as land and property data are concerned are listed in Table 9.4. But government land and property data have been locked away. Very little data are available at the individual property or land parcel level. In public sector organisations, such as the VO and Land Registry, statutory enforcement ensures the accuracy and completeness of data, but government sources are subject to many access restrictions and often lack the currency that is required by many property professionals. Often, the original

Table 9.4 Important government departments as far as land and property data are concerned

Ordnance Survey
Land Registry and Registers of Scotland
Department for Transport, Local Government and the Regions
Department for the Environment, Farming and Rural Affairs
Office for National Statistics
Valuation Office
Environment Agency
British Geological Survey
Local Authorities

purposes for which these organisations were required by statute to collect data are quite different to the marketable uses to which they might now be put. For example, OS mapping was originally used for military purposes and the VO collects property transaction data in support of the rating assessments for residential and commercial properties. The Chorley Report (Department of the Environment, 1987) made strong recommendations for the release of public sector data and significant progress has been made, but problems remain. The price of data is still a substantial barrier, especially for smaller firms. Unlike the OS and HMLR (which have been very proactive), some government departments have made slower progress, particularly, as far as land and property data is concerned, the VO of the Inland Revenue.

In contrast, New Zealand's digital land information is made available at the cost of dissemination and carries no copyright fees. In Europe, Norway, Sweden, Austria, Greece, Portugal and Luxembourg all manage to survive without the government asserting copyright over data collected at the taxpayer's expense. IPRs held by central and local government and government agencies create a major barrier to widespread use of government information (Roper, 2000).

In the UK, government departments were monopoly collectors and providers of certain information, but had little or no understanding of the modern electronic information market. In fact, the highest earning government monopolies were those that collected geographical data, such as the OS, HMLR and the Meteorological Office. In the past it made sense to build information monopolies but this mentality is changing, partly due to changing public sector doctrine that mimics the US, where information is increasingly regarded as a national resource rather than a government asset. Also, technology such as the Internet makes the creation and continuance of information monopolies almost impossible. We now live in an information economy and the stimuli to data release can be classified as legal (such as price paid data on the Land Register), market (such as Government Trading Fund agencies like the OS) and technological (GIS and the Internet, for example). Restrictions to access to public sector data have been organisational rather than technical and these tend to take longer to remove. Government is becoming a facilitator in the e-revolution so freedom of information legislation and a national data dissemination network such as NLIS should encourage innovation and competition in the supply of land and property data.

Private sector confidentiality

Property consultants negotiate and advise on all manner of property issues and this is the source of their data. Agents create property data when deals are struck and transactions undertaken. They are therefore uniquely placed to collect, compile and disseminate this information, yet Feenan (1998) suggests that there is a 'data stalemate' in the property industry, which can be explained by a combination of three factors.

The first is the mistaken belief that data hoarding is the best source of market advantage. Given the barriers to data sharing that exist within firms, it is not surprising that the pooling of data resources between firms has traditionally been regarded with suspicion. There is, therefore, duplication of effort between firms and a lack of reliable national data sources. The IPD is the notable exception to this rule. In the information age competitive advantage may depend upon recognising that property is an information business and the wider dissemination of data is an opportunity for the profession as clients seek advice based on interpretation of the facts. At the centre of a professional service should be an ability to assimilate, interpret and advise based on data rather than an ability to collect data.

The second factor has been the lack of resources dedicated to the complex and costly task of data collection. To date the culture in many firms has been towards progressing the next direct fee-earning task, with the recording of information from the last task being given a low priority. For this reason few firms of property consultants possess large databases of good quality, well-maintained property data. The benefit of good data storage to a firm is that everyone can have access at all times. This reduces duplication of effort in searching for data and also reduces costs.

The third factor is the clients' reluctance to allow private transactional data into the public domain. Confidentiality constraints and commercial restrictions limit public access to property information. Confidentiality clauses inserted into the legal documents relating to many property transactions prevent the release of property information into the public domain. Details of selected market transactions are reported in the property press if the agents involved wish (or are allowed) to publicise the transaction. Unfortunately, many press releases omit vital information, which is required if they are to be used for comparable evidence purposes. Also, the extent of geographical coverage is very poor, especially when categorised by property type. Instead, an informal network of information dissemination exists among property advisors. This issue clearly needs to be addressed if the flow of information is to improve. The success of IPD demonstrates that the confidentiality of property-level data can be adequately protected through the application of effective safeguards on the release of information. Subject to these limitations, a high degree of disaggregation can be achieved where necessary. One solution is for property consultants to publicly advocate an end to confidentiality clauses and to privately counsel clients against their use. Several organisations now include clauses in contracts with clients that permit them to use the information for other purposes.

Recently a number of catalysts have emerged that place improved access to property data high on the agenda. They include:

- The information superhighway and rapid development in information technology present major opportunities for professions that deal with data. As Part II demonstrates, GIS, in particular, is well suited to the management of land and property data.

- The ease with which data can now be collected, assimilated and analysed means that clients now demand more information and reasoning in support of advice. This has been evident with regard to property valuation advice where it is no longer sufficient to provide a description and a valuation figure. Analysis of comparable evidence and an estimation regarding future movements in value are increasingly being requested.
- The availability of some property data is becoming more widespread and more competitively priced. It is now possible to purchase a variety of property and demographic data such as postal address files, large-scale digital maps, census data and rateable values, all of which can be analysed using desktop computers with standard statistical analysis and GIS software.
- The lack of market activity in the early 1990s, also acted as a catalyst for improvements in the dissemination of property data. For valuers, when comparable evidence is hard to obtain, advice is harder to support. This has acted as an incentive for developments such as IPD where a number of firms have shared data to gain a clearer picture of the property investment market.
- Government-led 'N' initiatives have focused a great deal of attention and effort in co-ordinating access to land and property data. The NLIS, NLPG and NLUD are obvious examples.

It is encouraging to note that the quinquennial review of HMLR (Edwards, 2001) suggested that the Land Registry and the VO should collect better information at the time of registration on property types, floor areas, financial considerations, mortgages and lease information. Land Registry application forms and Inland Revenue 'Particulars Delivered' forms for sales and lettings of property would be consolidated to assist publication of useful statistics on various property markets in sales, leases and mortgages. Purchasers and lessees will be required to complete a standard form to cover the requirements of the Land Registry and the VO. It will also assist the VO in bringing transparency to rateable value assessments. The Land Registry is also considering how their data might be combined to assist publication of more informative property price statistics that take into account physical and legal attributes of properties when they are reported on an area-by-area basis.

Data standards

The property industry has witnessed the digitisation of substantial amounts of property data with the aid of increasingly sophisticated technology, including GIS. The public sector, encompassing central government departments and agencies, such as the Department of Transport, Local Government and the Regions, the Office for National Statistics and HMLR, OS and the VO, together with local government, collect and maintain vast amounts of land and property information. It is perhaps unsurprising, therefore, that this sector has led the development of land and property information systems. Development began in a disparate fashion with each organisation or department within an organisation spearheading the digitisation of its own property data. Often such initiatives were championed by individuals and would succeed or fail depending on the person's calibre, motivation and loyalty. The private sector, also, has realised the commercial value of land and property data and companies such as Estates Gazette Interactive, Focus, Experian, Promap and Landmark are all market leaders in the provision of property data to specific sectors of the property industry.

The development of independent information systems inevitably leads to data communication problems. Given that the public sector had a head start, it is this sector that was the first to realise that some standard means of describing property was required if government departments and other public bodies were to share property information effectively.

Standards are required to prevent data duplication, access, retrieval and exchange problems. A lack of standards adds to the cost of GIS projects or prevents key projects from being taken forward. This section describes key issues that arise when land and property data are recorded in a database and, in particular, a geo-referenced database. For example, when describing a property geographically do we refer to its centre (or seed) point or do we delineate its boundary? If we wish to store boundary information, should it be the physical, legal or ownership extent of the building or land parcel? How should interests in multi-occupancy buildings be classified? These issues are fundamental to the ability to integrate property data from disparate sources for property market analysis, research and management decision-making. Some of the issues have been addressed in the British Standard for an NLPG but there is flexibility built in to this standard, which inevitably leads to disparity between property databases created by local authorities. There are also particular concerns when the standard is applied in a commercial rather than public sector environment. These concerns are also discussed in the following sections.

Standards for geographical data

Standards are necessary for the collection and recording of geographical data. Standards increase confidence among collectors and users of data by protecting them against changes in technology. The successful development of GIS requires the creation of digital geographical data and the technological and human resources to handle those data. As GIS technology has matured, the need for methods of promoting data exchange and integration has increased. The drive for standardisation follows on from pressure to promote compatibility between systems and data sets and to allow data users to assess information from external sources.

Standardisation includes:

- Interoperability: the ability to communicate between dissimilar systems in such a way that the system providing the service to the user is transparent.
- Portability: the ability to move a piece of software and its associated data from one manufacturer's machine to another's. This also includes people portability, which is the ability to move people easily between different systems and networks without the need for retraining.
- Scalability: the ability to run the same software or utilise data with acceptable performance on any size of system.

GIS involves more complex considerations than inter-system compatibility. For this reason, standards can be required in the following areas:

- Data definition (e.g. feature coding and classification)
- Data description (metadata)
- Data representation
- Data quality
- Data exchange
- Programming and systems development
- Database query design
- Spatial analytical functions
- Open systems
- Graphics protocols
- Networks and communications
- Systems analysis and design.

De facto standards are those developed by industry and readily accepted as the way to do things. A good example is the DXF format for exchanging digital drawing files or graphics formats such as TIFF and the Microsoft Bitmap. De jure standards are legally defined standards, endorsed and published by the British Standards Institution or a higher level international standardisation body such as the International Standards Organisation (ISO) or Comité Européen de Normalisation (CEN). There is a clearly defined hierarchy of standards organisations that issue de jure standards. Although these standards are legally approved, there is no requirement to use any of them unless specifically called up in legislation. Within Europe, if a new de jure standard is to be created ISO, CEN and national bodies must be looked to first to see if a standard exists already. CEN standards take precedence over BSI standards and ISO standards take precedence over CEN standards.

The rise of the Internet as a forum for discussion, information search and retrieval has had a profound influence on the development of geographical data management strategies. Local, national and international initiatives that were difficult or impossible before the advent of the Internet have now become widespread. National geographic data infrastructures are developing, and wide-ranging international initiatives are following. Such infrastructures have far reaching implications for land and property information management.

A number of countries are exploring the concept of geographical information infrastructures following the inception of the National Spatial Data Infrastructure (NSDI) in the US. NSDI is conceived to be 'an umbrella of policies, standards and procedures under which organisations and technologies interact to foster more efficient use, management and production of geospatial data' (Federal Geographic Data Committee, 1996) and encompasses organisations, technology and information. NSDI depends fundamentally on co-operation among various levels of government, the private sector and academia. Other examples occur in Western Europe and Australasia, including the National Geospatial Data Framework (NGDF) in the UK, the Australian Geospatial Data Infrastructure and the European Geographic Information Infrastructure and related initiatives of the European Commission. In addition, commercial organisations are developing integrative technologies through the Open GIS forum. There is thus an emerging 'Global Spatial Data Infrastructure' (Burrough, 1997). The universal motivating factor behind such policies appears to be a perceived need to optimise the use of spatial information for decision-making in both public and private sectors. This should result in greater profitability for the private sector and improved services for the citizen from the public sector (European Commission, 1995).

The key requirements of such infrastructures are emerging. Critical factors are a sympathetic political and regulatory framework, a genuine desire and willingness to facilitate co-operation between agencies, governments and the private sector, the provision of technologies for communications and data dissemination and the development and implementation of universally accepted standards in primary areas of service provision. The penetration of these factors, and the extent to which they are adopted, will determine the success of the infrastructure.

Burrough (1997) explains why the creation of such an infrastructure is not straightforward. The degree to which issues of interoperability, data exchange and data description impinge upon the user community varies widely across countries, so that a primary concern must be to motivate purely local interests to collect, process and make available their information to the wider community for uses that were never considered when the data were collected. This will require legislation, 'best practice' guidelines, standards and incentives on an unprecedented scale.

It is recognised that the integration of multiple geographical data sets without relevant documentation and appropriate methodologies leads to a reduction in the quality of decision support. However, the need for a supporting infrastructure is coupled with a requirement to be realistic about what can be achieved. Byfuglien (1995) requests clarification of pan-European requirements for the harmonisation of geospatial data, and for fiscal and legal considerations to be studied. Problems are compounded when a 'local' infrastructure is extended to the national or international level. The challenge is not simply the move from NSDI to global ones (GSDI), but from NSDI to an intermediate level (perhaps an appropriate term is 'Multinational Spatial Data Infrastructure', or MSDI, as exemplified by the European ESDI initiative) and then to a global scenario.

Whether LIS and associated land and property data infrastructures are at the local, regional, national, multinational or international scale, large-scale mapping and cadastral data are the foundations of parcel-based LIS, although Dale (1991) suggests that locally derived parcel based gazetteers are also a key to national LIS/GIS projects. Accuracy and geographical completeness are essential, and indicators of precision, accuracy, currency, ownership, authenticity and lineage are equally important. In multi-agency systems, a nodal approach is commonly applied, whereby the GIS database is held and accessed over a distributed network and each data supplier is responsible for accuracy, authenticity and maintenance. An outline of this structure is presented in Figure 9.2.

The integration of geographical data across organisations, countries and continents will require technical standards for data definition and classification, data representation (including data models), data dissemination and

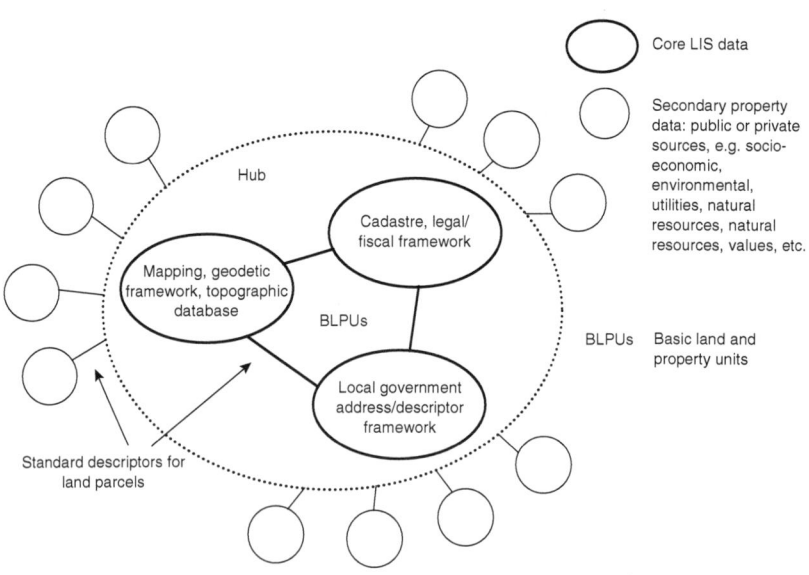

Figure 9.2 Nodal structure for a national LIS/GIS.

Source: Larsson, 1991.

documentation, including common transfer standards and metadata. This requirement has been universally recognised, as the work of groups like the Federal Geographic Data Committee, ANZLIC, the European Commission and ISO illustrates. Development of technical standards is ongoing but, like the infrastructure developments, of which most of them form a component, they are being constructed concurrently at local, national and international levels. In the US, for example, federal government initiatives are moving towards an NSDI (Federal Geographic Data Committee, 1994a) that will facilitate the integration of federal and local government geographical information resources. Similarly, the geographical data standards development work by the Commission of the European Communities (see www.eurogeographics.org for a discussion) demonstrate that the role of standards for geographical information in aiding data integration is very significant.

In the UK, users of property data within a GIS environment are becoming increasingly aware of the need for standards for data specification, description and transfer if the substantial benefits of GIS data integration are to be fully realised. The Interdepartmental Group on Geographical Information (IGGI), central government's co-ordinating body for geographical information, has examined the potential for integrating the plethora of spatial units that are currently utilised within central government. In a paper describing this initiative, Masser and Blakemore (1995) explain that much of the geographical data collected by government might be useful for more than one purpose, but that inconsistency in geographical referencing frameworks, excessive complexity, lack of accuracy and differences in times of measurement mean that integration is often impossible. They suggest that the establishment of geographical data standards will greatly facilitate data sharing initiatives and bring substantial benefits in optimising the use of existing information and reducing data duplication.

The UK approach to standardisation is pragmatic and the main focus is on information standards that will deliver a direct impact on the way in which information is handled throughout the UK. A British Standard can now be developed in six months, from inception to publication, given enough interest in development. Delay often arises in the adoption of the standard nationally. Legislation is the best way to achieve this, but not an easy thing to take forward.

Standards for documenting and describing data

A full understanding of data sets in terms of errors that they might contain, key definitions and limitations is critical to the analysis process. Therefore, another type of standard that is seeing widespread application is concerned with the description of spatial information. 'Data about data', or 'metadata', are now recognised as having a key role to play in the effective transfer of information resources from one place to another. If data are to be exchanged between departments, organisations or even countries, basic information will need to be recorded about them so that others can use them effectively. Metadata describes a data set and will include information on the quality parameters described in the section 'Data access' of this chapter and other information such as:

- Definitions and terminology
- Data models
- Geographical extent
- Temporal coverage
- Quality statements and accuracy measures
- Projections and co-ordinate systems used
- Organisation contact details
- Restrictions on use
- Pricing and availability
- Delivery formats.

The biggest problem with data quality documentation and other metadata are their size and complexity. Typically, a data quality statement can run to 10–15 pages. A metadata document can be as much as fifty pages in length for a single data set. If users are to be expected to apply such information and gain value from it, they need to be able to access it easily and quickly. Web-based approaches and 'intelligent' metadata systems that tailor their responses to the needs of the user are possible solutions. Because metadata are so cumbersome, a lot of work has been carried out to try and make their use easier. The AGI have published Guidelines on Geographic Information Content and Quality (AGI, 1996) and NGDF has produced guidelines on how to apply metadata standards and developed core high level metadata standards and 'discovery' metadata – the bare minimum necessary to decide if information is of value. The Americans have developed a Federal Government standard for describing geographical information (Federal Geographic Data Committee, 1994b), which has been adopted by all Federal government organisations and many other local government and executive agencies under the auspices of their NSDI programme. Development of a similar ISO standard is currently underway.

If data are to be exchanged between departments, organisations or even countries, basic information will need to be recorded about them so that others can use them effectively. Considerable effort has already been spent in setting up a suite of standards to facilitate the documentation of geographical data sets. The major effort has occurred in the US, but there are four key initiatives:

1   US Spatial Data Transfer Standard (SDTS): SDTS resulted in the development of a five-point data quality documentation statement, as well as detailed information about data structures.

2   US NGDI (NSDI) (URL: http://www.fgdc.com).
3   In Europe, there are parallel initiatives under the CEN banner.
4   At the international level, metadata documentation standards are being developed by ISO.

Metadata standards will allow users of geographical information to evaluate the usefulness of data resources from other organisations for their own applications. The issues of data quality and the suitability of data for specific uses are now being raised in the GIS community. If expensive mistakes are to be avoided, this type of information is essential because it will allow potential users to evaluate information resources before they purchase or use them.

Adherence to data description standards will ensure that land and property data can be integrated with other data and used for different applications because users will know something about the data they wish to use. If no standards apply to a particular type of data then it is vital that as much information as possible is recorded about the way in which survey and collection took place. This will allow the user to decide whether the data are suitable for a particular application.

Standards for land and property data

An early barrier to the integration of land and property data was the lack of standard property descriptions. This is confirmed by Adair *et al.* (1997) who comment that 'given the nature of data and the need for data purity, a common industry standard is deemed essential before data can be successfully collected and pooled'. In order to devise a standard by which all land and property can be described, we must define what we mean by a land parcel and a property. It is straightforward to define the extent of a house and garden with a boundary fence and perhaps of a flat too, although the latter may be a little harder to represent on a 2D map. But what about a garage block in a residential area or parking spaces allocated to specific flats? Here several geographical features comprise a composite property. Some applications need to refer to the composite property as a single entity, such as tax assessment, others need to refer to specific features within the composite property, such as building maintenance management. Further complications ensue when the properties concerned are complex entities such as an airport or an industrial estate.

A property-based GIS requires each land parcel to be uniquely referenced in a standard way. This is usually achieved using a UPRN. If land parcels are to be located in space, the UPRN will usually contain a geographical reference. Most countries have therefore adopted UPRNs that relate to their national geodetic frameworks. Standards are required for describing the location of properties in relation to one another, such as an address. One of the major hurdles for a GIS that integrates property data from several sources is address-matching between the data sets. This is particularly so when extensive property data are collected and maintained by different organisations with their own formats for addresses and property descriptions. Non-addressable properties also require standard recording formats for describing location. A UPRN is commonly applied in the identification of such properties but there are problems as different organisations may define the boundaries of non-addressable properties differently and these will need to be reconciled.

Inconsistent geographical definitions are sometimes cited as a problem impeding the analysis of property data. Terms such as the 'M4 Corridor' or 'East London' are used frequently in the property industry to describe development, investment or occupier activity. They have the advantage of being widely recognised but no formal definitions of such areas exist and therefore their delineation is open to misinterpretation. For example, many believe that the property map of Central London is being redrawn and the once distinct boundaries, which separated the West End and the city and East End are being broken down and extended as city institutions move into new developments around Canary Wharf and the Docklands. The river Thames, a natural southern boundary to the city is also no longer perceived as being so as the area between Waterloo and London Bridge becomes the focus of many 'Square Mile' institutions. Such changes have been brought about by space shortages in the city and West End, particularly for larger office requirements. A lack of standard geographical boundaries for property markets hinders data integration. Comprehensive, 'geo-coded' databases would overcome this problem, providing the flexibility to analyse any defined geographical area, but as yet little land and property data are available in this form.

Other issues surrounding definitional standards can also be identified. The preoccupation of many data sources with 'prime' property and lack of evidence for 'secondary' has often been cited as a problem, but there is no consensus as to how these two market segments might best be identified. Again, conclusive analysis will only be possible if the data are collected and stored in such a way as to support the various definitions researchers might choose, be they value based, geographical, functional or a combination of these measures.

Despite the absence of a mandate to develop data standards, representatives from local government have been instrumental in publishing the British Standard for land and property referencing. Many address databases exist within each local authority and standard is of vital importance to the integrated management of address-based and geographical information within local government. Local authorities are therefore spearheading the development of standards for land and property data, driven by their internal requirements for integrated GIS. The Land and Property Gazetteer Working Party, with a strong local authority influence, were responsible for the creation of standard referencing methodologies for streets, land parcels and address information. The Improvement and Development Agency (IDeA) has continued the pioneering work of the LPG Working Party by working with the AGI, the British Standards Institution (BSI) and others on the development of a suite of standards for spatial information that come under the banner of British Standard BS7666 (BSI, 1993). BS7666 is a British Standard for recording property information, which was introduced in 1993. Migration to this standard will present new opportunities for integrated property management. BS7666 promotes the interchange of land and property data on a consistent basis (Rackham, 1995). The standard is in four parts:

Part one: Specification for a street gazetteer. This part specifies the data to be maintained in a gazetteer of all streets in Britain. District authorities have responsibility for the creation, demolition and naming of streets. The standard specifies ways of referencing a street (name, description, route number or unique reference number) and specifies the means of representing its spatial location.

Part two: Specification for an LPG. This part specifies the data to be maintained in a gazetteer of all land and property in Britain. It specifies a BLPU, the itemised contents for each entity and the other entities to be recorded in a LPG. It also specifies the relationship between these entities and externally maintained data such as the boundary of the land and property unit as recorded on a map.

Part three: Specification for addresses. This part specifies a model and structure for an address. It provides a nationally consistent means of structuring address-based information. It will enable the exchange of address-based data and aggregation of other data related to it.

Part four: Specification for rights of way. BS7666 is of major importance to any individual or organisation working with land and property information. In particular it can help to reduce data duplication and maintenance costs, reference multiple applications to the same locations, introduce standard search tools throughout the organisation, provide a single point for maintenance and allow recruitment of pre-trained staff.

The creation of an LPG under part two of the standard draws together parts one and three because the former is a prerequisite and the latter is achievable in parallel. The aim of part two is 'to provide a nationally acceptable method of referencing land and property in order to facilitate the identification, retrieval, integration and exchange of land and property-related data' (BSI, 1993). The purpose of the LPG is to:

(a) provide a standard method of defining land and property parcels such that each parcel definition is justifiable and unique and such that the resulting parcel is meaningful to users;

(b) define a common referencing system in four dimensions (three spatial and one temporal) for land and property;
(c) provide methods of identifying and retrieving land and property information including those already used by the public;
(d) permit the local creation and national integration of land and property information.

(BSI, 1993)

Figure 9.3 illustrates the logical data model for an LPG. The figure shows how the components of the standard are related and how a local and national gazetteer might fit together. Application databases are linked to the LPG via an application reference and the LPI records a brief description of the BLPU. The LPG is, fundamentally, an amalgamation of BLPUs, which are defined as 'contiguous areas of land under uniform property rights' and are identified by UPRNs.

Under the standard, the extent of a BLPU can be defined according to ownership, occupation, physical extent or land use. The standard suggests that 'in the absence of documentary evidence of ownership then the unit shall be defined by its physical features or inferred from occupation or use' (BSI, 1993). The 'uniform rights' used to define the BLPUs are recorded in the provenance code together with provenance annotation. In this way optimum use is made of the information available. The nature of and relationships between BLPUs can be complex. For example, flats may be parts of a block

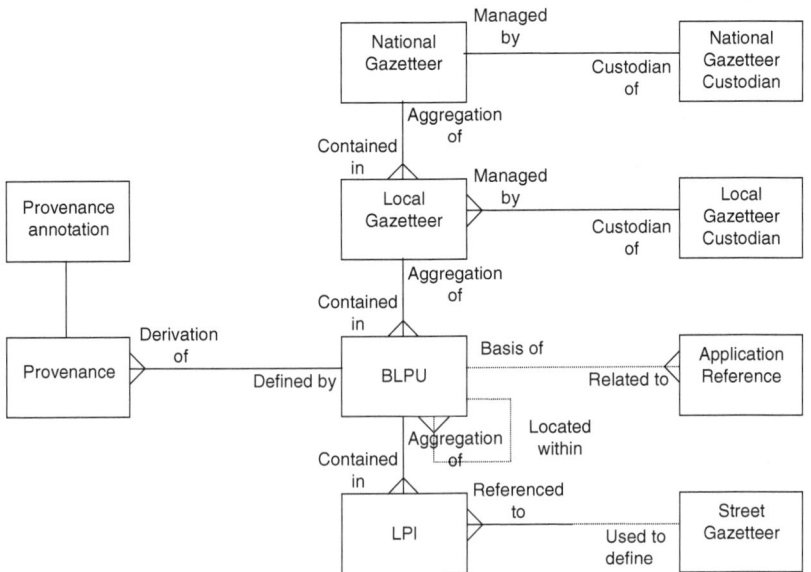

Figure 9.3 Logical data model for an LPG (after BCC, 1997).

and fields parts of a farm. In such cases separate BLPUs for the block and farm can be aggregations of the subdivisions. Horizontal subdivisions such as flats create particular representational problems on maps. The way in which these are handled is at the discretion of the 'local gazetteer custodian' and custodianship and maintenance of the LPG are the responsibility of the local authority.

Overseas case study: BS7666 in practice – the Bristol LPG

The purpose of the LPG is to provide a central core to all GIS applications within Bristol City Council. It will be the primary means of entry to the map base and a common referencing system for all streets, properties and areas of land within the city.

(BCC, 1997)

The Bristol LPG is a land parcel-based database in which BLPU boundaries are recorded. The BLPUs in the Bristol LPG are based on a combination of physical, occupancy, use and legal extents. BCC collected boundary data for every BLPU, which were derived from internal sources including local tax, planning, environmental, electoral register and the land terrier, and external sources including the OS and HMLR.

BCC have provided guidance on the allocation of BLPUs to subdivided property (BCC, 1997):

- Allocate one BLPU for the overall property (building and grounds) – usually in single ownership but at least the subject of some collective property rights.
- Allocate separate BLPUs for each unit (e.g. flat) within a block. These BLPUs may have their own seed points if physically identifiable but more often would overlay the seed point of the main building.
- Car parking spaces, allotments and grave plots are not normally designated separate BLPUs but they could be recorded in BS7666 format in a linked application database.
- It is advisable to include subdivisions in an LPG as they are easier to delete than to insert.

BCC (1997) also provide guidance on valid BLPU types:

- All conventional house, garden, garage combinations where they are not divided by another unit such as a road or path.
- Individual plots, garages, etc. where these are identified (not normally a block of garages in a communal space serving a block of flats).

- Aggregations and disaggregations of BLPUs where an administration, ownership or use brings together or divides the units, for example:

  - individual flats in a block and the communal space;
  - sub-units on an industrial estate;
  - plots on a building site;
  - shops in a retail space;
  - allotments or graves in local authority ownership.

- Public space (roads, paths, open space, parks, pedestrian areas, etc.).
- Geographic features (lakes, rivers, coastal margins, etc.).
- Historical BLPUs.
- Provisional BLPUs (construction sites, planning applications).
- Super BLPUs (collections of units such as administrative areas, business parks or conservation areas).
- Mobile BLPUs (house boats, caravans, etc. – reference mooring or hook-up point).

Each BLPU is linked to an LPI via its UPRN. The LPI includes an address and unique street reference (as defined in BS7666 part one). A one-to-many relationship can exist between BLPUs and LPIs. Thus, the LPIs record a primary address plus any additional address aliases that may exist. So the main components of the LPI are the UPRN from the BLPU, the USRN from the National Street Gazetteer and sufficient elements from the hierarchy of Primary and Secondary Addressable Objects necessary to uniquely identify the BLPU. An Addressable Object is a real world object that has a fixed location and that may be identified and referenced by means of one or more addresses. A Primary Addressable Object Name (PAON) is simply the name given to an addressable object that can be addressed without reference to another addressable object, for example, a building name or street number. A Secondary Addressable Object Name (SAON) is given to any addressable object that is addressed by reference to a PAON, for example 'First Floor'. Both PAONs and SAONs are structured to hold both numeric and character data. The data model for BS7666 used by BCC is shown in Figure 9.4.

BS7666 issues

The following sections examine some of the issues surrounding the creation of a BS7666-compliant LPG and suggest some ways that the current standard might be improved.

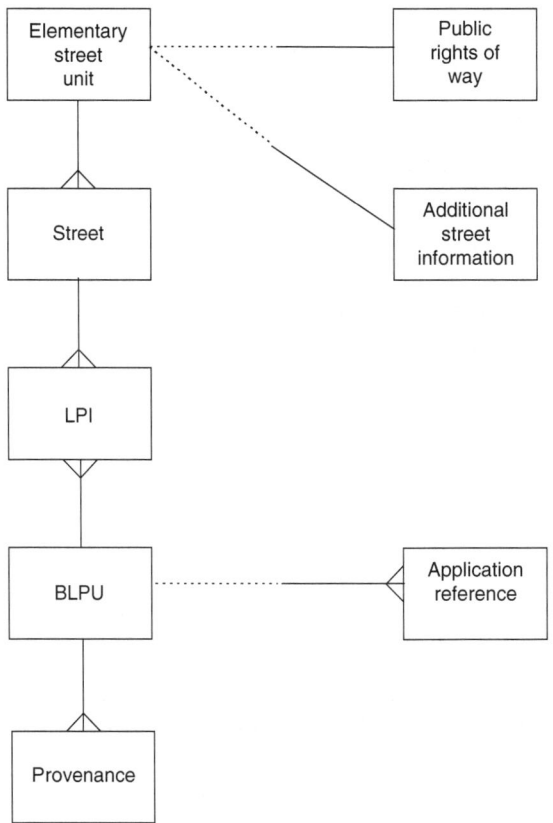

Figure 9.4 Logical data model for the Bristol LPG (after BCC, 1997).

Derivation of BLPUs

The development of an LPG in Bristol has demonstrated that the creation of a gazetteer involves linking data from a variety of sources. The BS7666 standard states that 'in the absence of documentary evidence of ownership ... then the unit shall be defined by its physical features or inferred from occupation or use' (BSI, 1993). However, the data sets that record ownership, occupation and use information in England and Wales suffer a number of disadvantages and allowing each authority to source these data sets without guidance will lead to inconsistencies. Consider each method of defining a BLPU in turn.

LEGAL RIGHTS

Legal rights are recorded by HMLR. The spatial extent of the legal right is illustrated on the Land Registry Title Plan and the information is publicly accessible, for a fee. However, Fairbairn (1993) lists several drawbacks with this information:

- the data set is geographically incomplete (but improving);
- land-related burdens, overriding interests and souvenir plots are not recorded;
- financial burdens are not disclosed; and
- only freeholds and leaseholds longer than twenty-one years are recorded.

HMLR has digitally scanned all of the Title Plans. However, these images are 'island maps' that are unable to include topographical relationships (Tobin, 1995). The Land Registry has therefore embarked upon a programme of full digitisation of both the Title Plans and Index Maps, that will be consistent with OS mapping. Since the end of 2001, new Title Plans are prepared digitally and Index Maps will be digitised by 2004.

OCCUPANCY

Address details of hereditaments (taxable units of occupation) are available from the VO. This level of detail in a property database may be of greater use to property professionals than a cadastre that records only land parcels. The main problem is how to record the spatial extents of these units of occupation and how to maintain such fluid information.

PHYSICAL EXTENT

The UK has no cadastre in which land parcel boundaries are fixed with reference to a co-ordinate system. Instead we have a general boundary system, which means that the line-work on OS maps does not delineate legally enforceable boundaries between land parcels.

BS7666 and OS Mastermap

BS7666 states that 'for the area covered by a gazetteer, the entire surface shall be accounted for by defined BLPUs' (BSI, 1993). This statement suggests that a parcel-based land information system or cadastre should be created. However, due to the way in which the BLPUs may be constructed this would be a cadastre based on general boundaries that have been derived from several sources. BCC admit that 'ideally, a BLPU would have a unique boundary but in practice different boundaries could be associated with a BLPU; therefore an extent provenance is also included in the format, that is, T = from title deed, P = derived from physical features or OS digital data and A = approximated for identification purposes only)' (BCC, 1997).

OS Mastermap is a national set of spatially contiguous polygons derived from topographic features recorded on large-scale digital maps. It would be helpful in terms of boundary reconciliation if the BLPUs defined using BS7666 are related to the land parcels portrayed by Mastermap. In BS7666, individual local authorities are left to decide whether the legal, physical or occupational extent of a land parcel should be used to define its extent. As a result, incompatible local LPGs may result and the process of looking up the derivation of BLPUs will be time-consuming.

Maintenance of BS7666-compliant LPGs

The BLPU and associated provenance information transmits the complex relationships between occupation, use, physical extent and ownership that characterise the property market. Buildings and their occupants create one-to-one, one-to-many, many-to-one and many-to-many relationships with one another.

Take a hypothetical example: the Bristol LPG refers to one BLPU as 1–2 High Street, Clifton. This land parcel is under uniform ownership, occupation and use and comprises two separate buildings. The Rating List records the land parcel as one hereditament. The planning authority, HMLR and all of the utilities regard the parcel as one entity with a single address. The owner decides that one of the buildings, 2 High Street, is surplus to requirements and instructs an agent to dispose of the interest. In drafting marketing particulars the agent refers to the property as 2 Plaza Buildings, High Street, Clifton, perhaps to enhance the image of the property. This address is used by the Western Development Partnership to market the property to a wider catchment and a tenant is eventually found. The agent immediately informs EGi that a deal has been done on 2 Plaza Buildings while the new tenant notifies the Post Office of the new address as Ground Floor, 2 High Street, Clifton. The property has been split and new addresses have been created. The local authority is the custodian of address information in the UK so for a LPG to work it requires the collaboration of all of the above individuals and organisations. This is the only way that external data collectors, users and suppliers will be able to keep up with a constantly changing database.

Geographic information is rarely static and longer term plans for how to deal with obsolete data and how to review and monitor progress in response to user needs must be considered. Only unilateral use of BS7666 will prevent data cleaning and associated costs being incurred each time data are used for another purpose or combined with other data.

Multi-occupancy properties

BS7666 promotes the creation of parcel-based LPG. Many property professionals require access to information on individual property interests

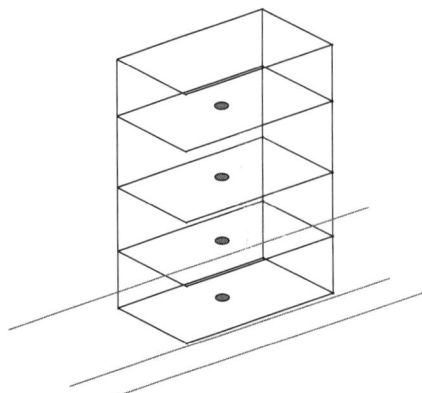

Figure 9.5 Seed points for property interests on each floor of a building.

rather than land parcels. Although a LPG may record a UPRN for each property interest in a building their geographic positions will be concurrent and will thus be displayed as one point only. This is illustrated in Figure 9.5. It not only hinders the display of property interests but also prevents spatial analysis because each interest is referenced to the same geographic position.

One way of representing multi-occupancy properties in an LPG geographically may be an object oriented database. This will allow a location to be simultaneously occupied by more than one object (Tobin, 1995). Another solution might be the use of a database or GIS that is able to display and analyse information in three dimensions. Both of these methods are technologically advanced but the property profession needs to consider the best way to represent multiple occupancy buildings in an LPG geographically.

The UK Standard Geographic Base (UKSGB)

The development of an NLPG provides an excellent opportunity for establishing a standard means of aggregating and disaggregating of property data. The UKSGB is a standard authoritative and central source of information about geographical areas, including their names, codes and boundaries. It comprises a published set of widely used postal and administrative geographical units. It is a single point of contact for descriptions of commonly used geographical data including names, codes and boundaries. Key data sets include the Central Postcode Directory and Postcode Address File from the Royal Mail, and AddressPoint and digital map data from the OS. The information is published on the Internet and the data descriptions are based on metadata standards provided by NGDF (1998).

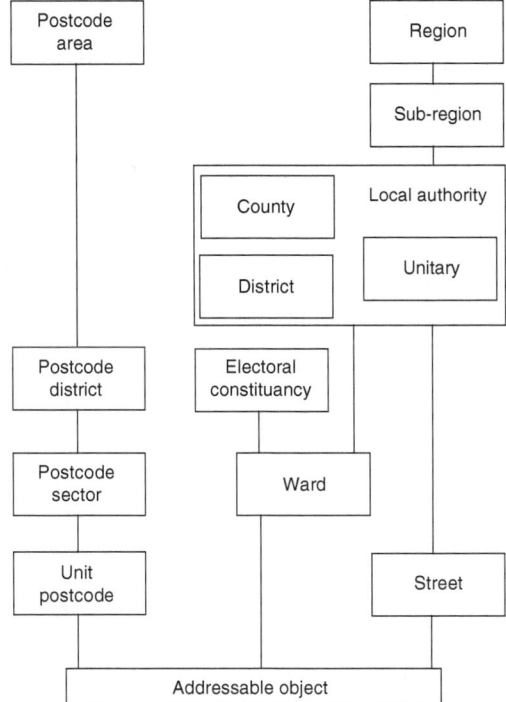

Figure 9.6 Core spatial units in the UKSGB.
Source: UKSGB, 1995, DETR, used with permission.

Figure 9.6 illustrates the way in which the core spatial units are structured in the UKSGB data model. For the standard to be successful, widespread awareness of the UPRN as a means of referencing land parcels and properties is key. But there also needs to be careful consideration of census and other area-based geographies such as land use change statistics. These need to be maintained in a consistent format to enable time series analysis and forecasting. The UKSGB also needs to consider education, labour market and travel-to-work-areas.

With regard to the integration of data from LPGs, a data 'thesaurus' approach could be considered providing linkages between different definitions of the same land parcel. This might be useful where boundaries of land parcels are difficult to define. The overriding consideration is that the data classification must be unambiguous, flexible and adaptable to aggregation for the maximum range of uses. Perhaps the most important issue is the establishment of a single source of geo-referenced address data as a foundation for the UKSGB.

Land use data

The NLPG will create a framework for identifying and mapping land uses on a national basis. It is important then that BS7666 LPGs are constructed in such a way that the data may also be used to populate the NLUD. This will involve the use of a common referencing standard wherever possible. Compatibility between BS7666 BLPUs, OS Mastermap land parcels and NLUD basic spatial units is a crucial issue for local authorities if they are to reap maximum benefit from economies of scale of land and property data collection.

Maintenance of these data is going to be very difficult, particularly in urban areas. This is why similar surveys in the past were not regularly updated. The ability to track history of geography is very important and therefore maintenance must handle transition to any new units created.

Summary

There can be no doubt that the property market is hindered by poor access to comprehensive property data. This is just one of a number of factors that, in economic terms, mean that the property market is imperfect. The property market is unique: every property is geographically distinct, there is no central market place, there is a long lead-in time for the supply of new property and longevity of existing property distorts the supply and demand cycle. The high cost of individual assets, long holding periods, high transfer costs and illiquidity of property assets means that comprehensive, up-to-date, accurate data are vitally important. It is for these reasons that a sea-change in the provision of data in the property profession will ultimately be inevitable.

Historically, comprehensive property data have been difficult to obtain in England and Wales due to legislative restrictions, commercial secrecy and privacy concerns. The expense of map data and paucity of attribute data has hindered the development of GIS applications in the property industry. Despite this the property market is one that operates relatively efficiently, property assets are not substantially overvalued or undervalued for significant periods of time. Much of this is due to the informal network of information dissemination that has developed within the industry, a network that is complex due to the high number of small firms. It is this informal network, with its barriers, restrictions and attitudes that must be tackled if a comprehensive GIS-based property information system is to succeed. The technical issues surrounding the hardware and software requirements are straightforward compared to the awareness, confidentiality, commercial secrecy, training and education issues that are of paramount importance. Probably the most important technical issue to tackle initially is to develop referencing standards for land and property data as a prerequisite to data integration.

The ability to integrate land and property data with other data sets geographically will allow property market analysts and decision-makers to ask questions that would have been difficult and time-consuming to answer in the past. But for this to occur an appropriate framework of standards for defining and representing land and property data that is accessible and universally acceptable must be in place. Once data integration has been achieved, the maintenance of the system and procedures for the updating of data must be carefully considered. But barriers remain – not least the price of data and lack of awareness of potential benefits.

BS7666 was published because the public sector had experienced significant problems referencing and describing land and property data. As more organisations digitise their property data sets the use of the standard will increase. The creation of LPGs is designed to lead towards the development of an NLPG. The NLPG is widely anticipated to be a very useful initiative in preventing data that are purchased from various suppliers being referenced in different ways. There are particular problems with addressing (and some of these were highlighted by the IPD case study in Chapter 8) and postcodes. Postcode geography is not stable and the boundary changes that come as Postcode Address File (PAF) updates are difficult to handle. Unit postcodes are a generalisation of settlement geography designed for speedy delivery of mail. They represent the spatial average of approximately 14–17 properties on a postman's delivery and are geo-referenced by a single point. Their geography is not that of clearly defined areas (Thurstain-Goodwin and Unwin, 2000).

But GIS is not just about systems. There are human, information and commercial issues to be considered. Having defined the task and outlined system and data needs, access to certain information may not be possible due to confidentiality constraints or legislative barriers. Certain information is commercially and personally sensitive and must be handled accordingly; copyright law and the Data Protection Act must be adhered to. The information that is collected or purchased may need to be integrated so that analysis can occur and the data formats may be incompatible. A land and property information manager needs to be aware of the rights that exist in the data that he/she is responsible for. Appropriate authorisation for use needs to be confirmed before data can be applied in particular ways. There are also data protection issues. If a data set might allow for the identification of an individual, it should be registered. The sources of data and disclosure within the system should be carefully considered to prevent identification of individuals.

Organisations such as the Department for Transport, Local Government and the Regions, Bank of England and the Royal Institution of Chartered Surveyors have recognised that there are strong public policy arguments in favour of developing more comprehensive property data, and all are actively engaged in discussions to promote this. Private sector initiatives have been successful in meeting some of the demand for property information by providing market indices and data compiled from other sources, but access to disaggregated property market data remains elusive.

Larger firms may feel that their in-house data resources give them a competitive edge in the property industry. Nevertheless, they are not immune to data pressures: the traditional structure of surveying practices (departments relating to areas of technical expertise such as valuation, agency and rating) is under threat. Clients increasingly demand project-based consultancy (property strategy and information systems for occupiers, for example). This requires improved data flows between departments. Such consultancy requires client confidentiality and is not perceived by clients to sit comfortably with the informal data sharing philosophy within the traditional surveying practice. There are still substantial organisational issues to overcome.

References

Adair, A., Berry, J., Deddis, B., McGreal, S., Keogh, G. and Key, T., 1997, Data sharing in the surveying profession: barriers, trends and implications, RICS Cutting Edge Conference, September, Dublin Institute of Technology.

Adair, A., Berry, J., Deddis, B., McGreal, S., Keogh, G. and Key, T., 1998, Barriers to data sharing in the surveying profession, *Journal of Property Research*, **15**(4), 331–346.

Association for Geographical Information (AGI), 1996, Guidelines for geographic information content and quality, Association of Geographic Information, London.

BCC, 1997, Land and property gazetteer basic design and data model, Bristol City Council and ESRI (UK), October, version 1.0.

BSI, 1993, BS7666: spatial data sets for geographical referencing, British Standards Institute.

Burrough, P., 1997, GSDI and ESDI – views on interoperability and spatial data infrastructures in Europe, URL: http://www.frw.ruu.nl/eurogi/forum/parma.html.

Byfuglien, J., 1995, The European data user's view, Proceedings of the Geodata for All meeting, URL: http://www.frw.ruu.nl/eurogi/forum/geodata.html

Chrisman, N.R., 1991, The error component in spatial data. In Maguire, D.J., Goodchild, M.F. and Rhind, D.W. (eds) *Geographical Information Systems: Principles and Applications*, Longman, London, chapter 12, pp. 165–174.

Coopers and Lybrand, 1996, Economic aspects of the collection, dissemination and integration of government's geospatial information, Consultant's Report for Ordnance Survey.

Dale, P., 1991, Domesday 2000: the professional as a politician, AGI Conference Proceedings, Association of Geographic Information, Birmingham.

Department of the Environment, 1987, Handling Geographic Information, Department of the Environment, HMSO, London.

Edwards, A., 2001, Quinquennial review of HM Land Registry, Land Registry.

European Commission, 1995, GI2000: towards a European Geographic Information Infrastructure (EGII), EC Directorate General XIII, Brussels.

Fairbairn, D.J., 1993, Property-based GIS: data supply and conflict. In Mather, P. (ed.) *Geographic Information Handling – Research and Applications*, Wiley, Chichester, UK.

Federal Geographic Data Committee, 1994a, Content standards for geospatial metadata, FGDC, US.

Federal Geographic Data Committee, 1994b, Strategic plan for the National Spatial Data Infrastructure, FGDC, US.

Federal Geographic Data Committee, 1996, The National Spatial Data Infrastructure, FGDC Internet publication, URL: http://www.fgdc.gov/nsdi2. html

Feenan, R., 1998, Property data and market development. In Wyatt, P. and Fisher, P. (eds) *Property Information Today and Tomorrow*, Royal Institution of Chartered Surveyors, London, pp. 13–18.

Ferrari, E., 1999, From points and polygons to housing strategies, AGI Conference pp. 3.3.1–3.3.5.

Flowerdew, R. and Openshaw, S., 1987, A review of the problems of transferring data from one set of areal units to another incompatible set, NorthEast Regional Research Laboratory Research Report 87/0, University of Newcastle upon Tyne.

Flowerdew, R., Green, M. and Lucas, S., 1991, Analysing house price variations with GIS, AGI'91, Conference of the Association of Geographic Information, AGI, Birmingham, pp. 3.25.1–3.25.8.

IPD, 1993, Department of the Environment consultation paper on commercial property leases: a response by the Investment Property Forum, Investment Property Forum, London.

Larner, A.G., 1996a, The legal and institutional restrictions on the handling of digital land related data in the United Kingdom, unpublished PhD thesis, University of East London.

Larner, A.G., 1996b, Balancing rights in data – elementary? In Parker, D. (ed.) *Innovations in GIS 3*, London, Taylor and Francis, pp. 25–35.

Larsson, G., 1991, *Land Registration and Cadastral System: Tools for Land Information and Management*, Longman, Harlow, UK.

Li, Y., Brimicombe, A.J. and Ralphs, M.P., 2000, Spatial data quality and sensitivity analysis in GIS and environmental modelling: the case of coastal oil spills, Computers, Environment and Urban Systems, Vol. 24, pp. 95–108.

Longley, *et al.*, 2001, Geographic information systems and science, BIG GIS BOOK.

McKeon, A., 2001, What stands in the way of e-commerce? Poor data quality, *GI News*, September, 42–44.

Maffini, G., 1990, The role of public domain databases in the growth and development of GIS, *Mapping Awareness*, 4(1), 49–54.

Marriott, O., 1989, *The Property Boom*, Abingdon Publishing, London.

Martin, D., 1996, Geographic information systems: socioeconomic applications (2nd edition), Routledge, London.

Masser, I. and Blakemore, M., 1995, Final Report of the AGI/Census of Population Roundtable, AGI Publication 1/95, London: Association for Geographic Information.

Mirchin, D., 1997, The European database directive sets the worldwide agenda, *NFAIS Newsletter*, 39(1), 7–12.

Monmonier, M., 1996, *How to Lie with Maps*, University of Chicago Press, Chicago.

NGDF, 1998, http://www.ngdf.org.uk/uksgb, National Geospatial Data Framework.

Onsrud, H.J. and R. Reis, 1995, Law and information policy for spatial databases: a research agenda, *Jurimetrics Journal of Law, Science and Technology*, 35(4), 377–393.

Openshaw, S., 1984, *The Modifiable Area Unit Problem*, GeoAbstracts, Norwich.

Rackham, L.J., 1995, Surveying the address, Proceedings of the AGI'95 Conference, Association of Geographic Information.

Raper, J., Rhind, D. and Shepherd, J., 1992, *Postcodes: The New Geography*, Longman Geoinformation, Harlow, Essex.

RICS, 1994a, The Mallinson Report: Report of the President's Working Party on Commercial Property Valuations, Royal Institution of Chartered Surveyors, London.

RICS, 1994b, Understanding the property cycle: economic cycles and property cycles, Investment Property Databank and the University of Aberdeen on behalf of the Royal Institution of Chartered Surveyors, London.

Robinson, A.H., Morrison, J.L., Muehrcke, P.C., Guptill, S.C. and Kimerling, J. 1995, *Elements of Cartography* (6th edition), Wiley, New York.

Roper, C., 2000, Contrasting data policies: Europe and the USE, AGI Conference.

Rybaczuk, C., 1993, Error and accuracy in spatial databases, Association for Geographical Information, AGI, London.

Simpson, L., Thomasson, E., Kuszyk, K., Prashar, R. and Morin, G., 2000, Small area statistics on-line, *Journal of the Royal Statistical Society*.

Smith, A. and Wyatt, P., 1996, Commercial property analysis and advice: information holds the key, Proceedings of the 'Cutting Edge' Property Research Conference, Royal Institution of Chartered Surveyors, University of the West of England, 20–21 September.

Thurstain-Goodwin, M. and Unwin, D., 2000, Defining and delineating the central areas of towns for statistical monitoring using continuous surface representations, *Transactions in GIS*, 4(4), 305–317.

Tobin, D., 1995, BLPU Extents – an NLIS issue, unpublished MSc dissertation, University of Edinburgh.

Tobler, W., 1979, Cellular geography. In Gale, S. and Olsson, G. (eds) *Philosophy in Geography*, Dordrecht, Reidel, pp. 379–386.

UKSGB, 1995, UK Standard Geographic Base Final Report.

Voss, P.R., Long, D.D. and Hammer, R.B., 1999, When census geography doesn't work: using ancillary information to improve the spatial interpolation of demographic data, CDE Working Paper 99–26, Center for Demography and Ecology, University of Wisconsin-Madison.

Wyatt, P., 1993, A national land information system for Britain – the potential benefits for property valuers, *IRRV Journal*, June.

10 Implementing and managing GIS

Introduction

The successful implementation and management of GIS requires thorough consideration of business requirements, a sound business case and careful planning of the implementation strategy and system management. In this chapter we provide an overview of the most important considerations for land and property sector organisations that are interested in implementing a GIS. The next section considers the implementation of a desktop GIS from the perspective of a small business or single department within a larger organisation. The 'Corporate GIS implementation' section develops the discussion of the implementation process and related issues from the perspective of an organisation-wide or corporate GIS solution. Planning for implementation is discussed, including problem recognition, obtaining management support, the identification of the benefits and costs of GIS and financial analysis of GIS implementation. The issues that arise during GIS implementation are also explored including project definition, system design, installation, maintenance, update and review. The section on 'Implementation issues for national land and property management initiatives' turns briefly to the issues that arise when implementing GIS in support of national land and property information initiatives. In 'Organisation and administrative issues for GIS implementation' we examine some of the organisational and management issues that are raised by the development and use of GIS, and describes some of the lessons learnt by public and private sector organisations when implementing the technology.

Project-led GIS implementation

There are three stages in the process of GIS implementation:

(a) Problem recognition and project definition
(b) System design and installation
(c) Maintenance, update and review.

The scale of the GIS project will influence the amount of time and effort that is spent in considering these issues, but any organisation wishing to consider the introduction of a GIS should spend some time thinking about each one.

Problem recognition and project definition

Aronoff (1989) suggests six types of problem spark interest in GIS:

(a) Existing geographical data are out-of-date or of poor quality. For example, there may be redundant information stored that makes searching for and retrieving data inefficient.

(b) Data are not stored in standard formats and maps vary in quality from one area of the organisation to another. At the same time, different methods of data capture are employed, such as aerial photography, hand drafting, photographic enlargement and photocopies.

(c) Several departments collect and manage similar geographical data and different forms of representation, data redundancy and related inefficiencies are evident in the collection and management of the data.

(d) Data are not shared due to confidentiality and legal concerns.

(e) Analysis and output are inadequate.

(f) New demands are made on the organisation that cannot be met using the data and technology currently available.

Whereas a corporate GIS implementation might involve a great deal of feasibility study and cost–benefit analysis, a small GIS project might be justified on the basis of a short-term payback over the life of a specific project. This makes cost justification more simple but probably harder to achieve in some cases. The main costs are data, human, computer-related and maintenance and update related. Also, a small-scale GIS project can be regarded as a pilot for wider use of GIS within an organisation.

The best way for a small-scale GIS implementation to work is to focus on an application or business process that is geographical and requires the integration of data from different sources. If this is the case then GIS, rather than some other form of information system, might be suitable. Peel (1995) suggests that a successful GIS implementation in the land and property sector requires an application where data integration and spatial analysis are key requirements, such as development control. Furthermore, if software and investment in data can be justified on the basis of a single application then that is a bonus, but it is worth remembering in the cost–benefit analysis that the data and software will also be available for other applications, subject to copyright and confidentiality considerations. The key is to get in-house data in order first because often that is the most expensive part of a GIS implementation for a small business.

System design and installation

GIS implementation can be large or small scale. Small-scale GIS implementation is obviously easier than an organisation-wide or corporate GIS implementation. This is borne out by the many local authorities that have invested in GIS. Most have done so at a departmental level rather than authority-wide. The same is true for the private sector: organisation-wide implementation takes a great deal of planning, resources, investment, skill and commitment, as 'Corporate GIS implementation' section will testify. Consequently, many GIS implementations, especially in an industry characterised by a large number of small firms, have been application-led. In other words, GIS has been introduced to meet a particular business need. For example, IPD introduced GIS to perform a geographical analysis of property investment portfolio performance on behalf of clients, and property consultants JLL and FPDSavills invested in GIS to aid business relocation advice to clients. A primary aim must be for the investment in GIS to generate a sufficient return.

Whereas GIS used to require powerful UNIX workstations, nearly all GIS software is now available on desktop PCs. It is possible to purchase desktop GIS software for under £1,000 but it is important to remember that GIS software on its own is useless – like an empty spreadsheet. It requires data, skilled staff to analyse and interpret the data and produce output in the form of maps and reports. GIS, like any other information system, is data hungry and a problem that the property industry has suffered from for a long time is poor management of data and a strong sense of data secrecy, which inhibits data sharing. Fortunately, the cost of data is tumbling as more data suppliers enter the market and bid prices downwards.

In the long run, application-specific GIS implementations are expensive to undertake. It is generally regarded as an inefficient use of information to collect data once and not update them and use data for one purpose only. It makes better business sense to have a long-term view of data collection and management. Application-specific GIS implementation prevents the realisation of one of the key benefits of GIS use in an organisation – that of adding value to data collected for a specific purpose by using it for another. Consequently a cost–benefit analysis based on just one application may not include all of the potential benefits that a GIS may offer an organisation. Nevertheless, GIS implementation has to start somewhere and it is invariably driven by a specific business need in the first instance.

Maintenance, update and review

The first stage of the geographical analysis of data might include the simple geographical display of proprietary maps (such as large-scale OS mapping, AA road maps or aerial photography) and the overlay of in-house data on top of those maps. This apparently simple process involves the installation of GIS software, the digitisation and geographical referencing of in-house data and the purchase of external map data and possibly other information such as aerial photography from third parties. It is important to think about the scale of the mapping required and the geographical extent of its coverage because both of these factors will have a very significant impact on the cost of external map data. It is also important to think about the licensing and copyright issues surrounding the use, reproduction and publication of any map-based output from the analysis. Most providers of digital mapping and geo-referenced data in the UK require copyright approval prior to reproduction. Perhaps the most significant factor in determining the success or failure of such an undertaking will be the availability of skilled staff. A successful small-scale implementation will require personnel who understand the issues of geography, data management and data availability that enable pragmatic decisions to be taken about the purchase and configuration of data. Such individuals should also be able to make sensible recommendations about how to maximise use of the analytical functionality of a desktop GIS application and will need to be able to provide sensible and pragmatic solutions to problems with limited resources. If the right person can be found, this should maximise the chances of early gain from implementation in a small firm or single department.

The choice between and mixture of external and in-house data in a GIS project may be driven by cost and fitness for purpose with regard to external data and availability, quality and accessibility with regard to in-house data. For example, are in-house data in a GIS-ready format? Have they been geo-referenced? As mentioned in Chapter 9, the Association for Geographical Information (AGI) has published a set of guidelines on geographic information content and quality (AGI, 1996). The guidelines are designed to help those who require geographical information determine that it is fit for its intended purpose and how to decide which aspects of quality are important given different user objectives.

Implementing a desktop GIS in the private sector

The IPD supplies market indices and portfolio performance statistics to the property industry. At IPD, the first problem that the implementation of a GIS sought to address was the quality of the address information about each building held in the Databank. Many addresses of properties held in the IPD databank were of a dubious quality, leading to possible geographical coding errors (both at a regional and local authority district area level), the extent of which was unknown. IPD's local market area statistics were based upon 1974 district boundaries and postal areas within London. The rather historic method of coding properties based on postal towns inevitably led to certain properties being coded within the wrong district area, and possibly even the wrong region. The extent of this problem was also unknown. Other problems were associated with IPD's standard definition of geographical areas. Many data suppliers to IPD find it difficult to associate certain postal areas to specific blocks of streets unless provided with a street map showing the area boundary. This meant that in some cases clients found geographical enquiry of the Databank difficult. Small micro-level analysis of local property markets had previously relied on postcodes. This reliance on postcodes for geographical classification meant that any micro-level analysis within IPD was constrained to Central London (the location of most addresses with postcodes). Similarly clients requesting a geographical analysis of the Databank on areas not standard to IPD found the process difficult and time consuming. Due to the inflexibility of the Databank for geographical analysis the scope for joint projects with other firms was therefore limited.

In the early days, GIS implementation at IPD focused on data integration. This is regarded as the key to successful GIS application. All IPD records are geo-coded using addresses and grid references and a unique reference for each lease is used as the identifier. OS CodePoint™ was used to assign geo-codes based on postcodes and address records are largely BS7666 compliant. The assignment of postcodes had to be carried out at the time of data collection from the supplier. This presented problems because some suppliers assigned incomplete or inaccurate postcodes. If standard addresses were used, they would see benefits in terms of faster transaction times.

A data quality issue arose when assigning property data to relevant local authority districts. Inconsistent referencing of addresses was found to affect performance statistics. In other words, statistics would change depending on which properties were allocated to which district. GIS was used to produce maps showing the location of the local authority boundaries and the location of data points classified by their stated local authority district. It was then possible to identify which properties were incorrectly coded.

The establishment of NLPG will bring great benefits to organisations like IPD who can spend considerable time and effort coding data. Other problems that had to be overcome included the geo-coding of current and historic data from the Databank.

Corporate GIS implementation

Organisation-wide or corporate GIS implementation should not be taken lightly! The process is usually longer term than an application-led GIS implementation and it can be months or years before a corporate GIS is fully operational. Calkins (1997) provides an excellent summary for managers of critical factors that the prospective GIS user should be aware of. Three main points arise from their discussion:

(a) A corporate GIS requires the building of large databases and configuration of hardware and software before it becomes useful. The construction process will involve database design and configuration, may require bespoke applications development to tailor off the shelf software to corporate requirements and is likely to require substantial testing. Such tasks are complex and will require careful planning and proper management if success is to be assured.

(b) GIS is, at present, a technological innovation. The adoption of techno-logical innovations is not always straightforward. Staff may require retraining, new ways of working may need to be developed and new systems will need to be tested and calibrated before they can be used for production work.

(c) One of the key benefits of GIS cited throughout this book has been the opportunity it presents for bringing together and sharing data resources. A necessary condition for maximising the benefits of an information system is for different departments and organisations to co-operate and to pool resources. In this way, the information system becomes an organisation-wide resource and its potential contribution to working practice is correspondingly increased.

Using the same three stages in the process of GIS implementation from the section 'Project-led GIS implementation', this section describes the process of GIS implementation at the corporate level.

Problem recognition and project definition

In order for an organisation to become interested in acquiring a GIS, some-one within the organisation must perceive that the way in which they store, manipulate and analyse information is inadequate for their needs. Equally, there must be an awareness of the capabilities of GIS technology and the potential of GIS to address problems within the organisation.

To obtain a successful outcome, it is critical to gain the full support of the decision-makers who will be required to commit resources to the project. Decision-makers will need to be assured that the GIS project will be devel-oped and managed in a sound manner and that there will be measurable benefits for the organisation resulting from GIS implementation. They will need to know:

(a) What a GIS is and how it is used. Examples of GIS application should demonstrate the use of GIS in a context that is familiar?
(b) What GIS can do for the organisation?
(c) What the indicative costs and benefits of such a system might be?
(d) How long GIS implementation will take and the time that is likely to elapse before the benefits of any new system are felt?

Once a broad base of management support has been obtained and a clear need for GIS has been identified, the next phase of implementation is to for-mally define the GIS project and establish aims and objectives, outcomes and milestones. A carefully managed development plan is essential for success and administration of such projects requires strong leadership. Openshaw *et al.* (1990), in a rare published example of constructive analysis of a failed GIS project, cite clear definition of requirements and the consideration of long-term project management issues as key elements of success.

The project definition and planning stage determines the scale and scope of the implementation process. It will usually involve the following tasks:

(a) Identification of the current role of spatial information.
(b) Production of a project plan or project proposal.
(c) Undertaking one or more feasibility studies.
(d) Production of a detailed business case that builds on the project proposal and includes sufficient information to determine potential costs and benefits.
(e) Identification of potential costs and benefits.
(f) Financial analysis of GIS implementation.

These tasks are considered in more detail in the following sections.

Identifying the role of spatial information

The role of this task is to identify where geographical data are being used at the moment, to explore how these data are applied to the work of the organisation and to suggest potential ways in which GIS might be used to improve the management and analysis of such information.

The project proposal

The project proposal should define objectives, identify requirements and determine user needs and likely GIS products. The project plan should be dynamic, adaptable and refined as better information becomes available. Plans will be very general during the early stages of implementation and will probably just comprise a description of why it is necessary to investigate further and a broad strategy for proceeding. For those charged with developing a project plan, it is important politically to discover who or what is the force behind the interest in GIS. The individuals involved and the significance of the problem are important in determining how to proceed with selling the idea to the organisation.

A sensible approach is to focus initially on proven applications for which there are high levels of demand but to have a long-term plan for multi-purpose systems and for flexibility in future development. Part II of this book illustrated that there is great diversity in applications to which GIS are put. For example, on a yearly basis, over 60 per cent of enquiries to the Swedish Land Data Bank have no direct connection with land transfer or title registration. A willingness to adapt and to think laterally emerges as a key factor for success. When switching from a paper-based system to a computerised GIS, alternatives must be fully considered.

Feasibility studies

Many subjects might be considered for such studies. Examples might be an assessment of the amount and types of map use in the organisation, evaluation of the state of the art in GIS technology, identification of potential GIS benefits and application areas within the organisation. Other possibilities are an indicative study of user requirements, the identification of key potential users or an estimation of the project budget. The goal of such studies is to reduce uncertainty and provide illustrative material to complement the project plan.

Business case

A business case should define the expectations of senior management about the role and performance of the GIS within the organisation and its contribution to business practice. Over optimistic expectations are frequently cited as a reason for failure in GIS implementation. To gain a full understanding of the potential costs and benefits of GIS implementation the impact of a GIS on the organisation and business as a whole should be taken into account. Maffini (1993) suggests that the focus should shift from cost reduction to value enhancement. GIS should not just be regarded as an add-on to existing technology, but as a potential replacement to some existing data management techniques. Maffini (1993) suggests that there are several ways that GIS can contribute to business activities:

- Changing the way in which the collection and processing of data is carried out.
- Using the GIS to facilitate spatial analysis rather than simply as a map output tool.
- Restructuring the business by promoting decentralised units.

Identifying the benefits of GIS implementation

There is only limited evidence to support the measurement of the financial costs and benefits of a GIS. However, in the US, 'those local governments engaged in selling GIS data sets to the private sector have found that land and property data are in more demand than any other layer of information' (Onsrud, 1995). Strassman (1985) found that information systems reduced both direct and indirect costs, but that their contribution to profitability was marginal. Smith and Tomlinson (1992) investigated GIS benefits for the city of Ottawa and found that the time required to search and respond to requests for information was reduced by 50 per cent through GIS use.

We can divide the benefits of GIS implementation into two types: tangible and intangible. Tangible benefits can be clearly defined and are typified by a statement like 'the cost of producing maps will be reduced by 20 per cent'. They include reduced operating costs, staff time savings, cost avoidance and increased revenue. Intangible benefits are less easy to identify and include benefits like improved decision-making, decreased uncertainty or an enhanced corporate image. Such benefits are difficult to value precisely in monetary terms. Because of the difficulty of quantifying intangible benefits, it is often prudent to list them and leave their evaluation to the project decision-making group. Intangible benefits are often tied to expected products. These products might be the same as before but created using GIS techniques or new products that could not be produced without GIS. It is difficult to define some products – map or report is easy to see, a database query is less so.

In order to try and quantify the benefits that a GIS might offer, Parker *et al.* (1988) suggest six classes of value:

(a) Return on investment.
(b) Strategic match – value of GIS strongly aligned to the extent to which GIS supports business strategy.
(c) Competitive advantage – GIS may aid competitive advantage.
(d) Competitive response – competitors may use GIS.
(e) Strategic information system (IS) architecture – value of GIS may be enhanced if fits in strongly with IS requirements.
(f) Management information – provision of effective management information is a strong benefit.

Obermeyer and Pinto (1994) suggest that managers should look for benefits in the following areas:

(a) Piggybacking – some applications can be given a flying start because they can be introduced on the back of another application. For example, a property management company GIS may be justified on the basis of operational management. If the marketing department wishes to introduce GIS in the same company, the basic facilities are already there.
(b) Faster flow of information – improved customer service and faster access to information are substantial sources of efficiency saving.
(c) Easier access to information – quicker access goes beyond time saving in that more readily available information sparks off creativity and enhances the decision-making process.
(d) Opportunity hours – effective management of time can release staff to do productive work.
(e) Better control – managerial control over work flow and work programmes brings benefits of proactive management.

The potential cost reduction benefits of GIS can be estimated using a two-stage process. First, the current cost of using the data that will be placed into the GIS database must be ascertained. Next, the portion of this cost that will be saved by using GIS should be estimated. A fast way to estimate the current cost of map data is to interview the supervisors of departments that will benefit from the GIS with the aim of finding out the amount of time the staff currently spend using mapping data (Korte, 1997). 'Use' is defined here in the broadest sense, and covers data collection, maintenance, analysis, retrieval and distribution. The cost of current use can be derived by comparing the amount of time each employee currently spends on these tasks to the total cost of employing that person (their salary plus indirect costs) and totalling these for the department.

An interview procedure like this is usually carried out formally as part of system planning process. More sophisticated approaches to work and cost assessment can be used and may produce more accurate results but will also involve a much greater level of effort (Korte, 1997). It may be necessary to estimate current costs by examining the cost implications of particular business functions. It is important to not to overlook the cost of tasks that are contracted out and which could be reduced by the GIS. The second stage of cost assessment is to calculate how much the current cost of using data can be reduced by the new system.

Identifying the costs of GIS implementation

The costs of GIS can be broken down into implementation and maintenance costs (Korte, 1997). Implementation costs may include development effort, hardware and software, database creation, data conversion and user training. Maintenance costs are incurred during the daily operation and maintenance of the system. Examples include application software maintenance, incremental data storage expenses, incremental communications costs, software and data licensing, consumables, data maintenance, hardware replacement and upgrade and ongoing training programmes.

The Joint Nordic Project (1987) described sixteen GIS projects in North America and Italy and their costs, benefits and applications. The projects reviewed ranged from automated mapping programmes to fully integrated corporate systems. The results of the study showed that:

1   A digital system used only for computer-aided mapping and updating produced a return on investment (benefit/cost (B/C) ratio of 1:1).
2   If the system is also used for planning and engineering, benefits are doubled (B/C 2:1).
3   Automation of conventional maps can bring a benefit of treble the investment (B/C 3:1).
4   If the system permits the sharing of information among different organisations, this can result in a benefit of up to four times cost (B/C 4:1).
5   Where manual map production processes were inefficient, the benefits of automation gave B/C ratios of upto 7:1, with an average reduction of 50 per cent in map production time.

Overseas case study – financial analysis of the Edwards Airforce Base GIS, California, USA

Edwards Airforce Base began a GIS programme in 1991. Nine departments benefit from the system, which includes forty-nine workstations and over 3 GB of data. GIS database construction was completed in 1995. Data include 1:200 maps of the base area, 1:50 maps of built up areas, digital orthophotographs, cultural and natural resources, utility systems and environmental data.

Distribution of costs was as follows:

Base mapping	34 per cent
Digital orthophotography	3 per cent
Environmental data collection	18 per cent
Civil engineering data collection	9 per cent
Data collection pilot project	5 per cent
System development	6 per cent
Consulting services	4 per cent
Equipment and software	19 per cent
Training	2 per cent

Full cost recovery time was estimated at 1999, eight years after project initiation and four years after the system became operational. Given this evidence, it seems likely that GIS implementation is likely to cause a productivity improvement of approximately 2:1 compared to current activities involving maps and map related data. Some tasks will have greater time savings than others, and so this estimate is probably a conservative one. It is important to remember that cost savings will be achieved in phases. A GIS user typically requires three to six months of training before achieving proficiency. Cost saving might only be 10 per cent in year one, 25 per cent in year two and 50 per cent from year three onwards. Time savings will not necessarily mean reduced employee costs – time and money saved will probably be allocated to other activities. Contract costs may be reduced through GIS usage.

Financial analysis of GIS implementation

Initial project justification might require answers to the following questions:

- Why not continue to invest in the current system?
- Will the system actually save money?
- Will the new system deliver tangible benefits?
- Do users understand what they will get?
- Is the proposal strategically appropriate?
- Does the return exceed the cost?

Alternatives will then need to be evaluated. For example, does the proposed solution fit into the corporate IT architecture, does the proposal fulfil selection criteria in terms of availability, reliability, performance and flexibility and does the technical solution deliver an answer to the organisational problem?

Corporate GIS implementation usually involves a substantial initial investment, which is paid back over time by cost savings and, potentially, increased revenue. Savings may increase steadily and eventually may be larger than the operating cost of the system. The initial investment is thus recovered gradually. The payback period is the length of time it takes to get back the money that was put into the system. The shorter the payback period, the stronger the case for investment.

A Net Present Value (NPV) analysis can be used to examine the financial aspects of GIS. NPV analysis takes account of the devaluation of money. Money received today is worth more than money received in the future (due to opportunity cost and inflation) and the amount of the difference in value increases over time. The present value of future income can be determined by discounting back to the present at an appropriate interest rate. NPV analysis converts all future cash flows to equivalent amounts at a common point in time. The interest rate used to determine these amounts is known as the discount rate. If the NPV computed at the desired discount rate is positive, the investment is justified. If the NPV is negative, the investment will not yield the desired rate of return. The discount rate is usually decided based on the level of risk associated with the investment. The greater the risk, the higher the desired discount rate. The greater the discount rate, the greater the difference between the value of money received now and money received in the future (and therefore the level of risk) is emphasised. The US Office of Management and Budget recommends a real discount rate of 3 per cent for investments like GIS. Korte (1997) provides some examples of the application of NPV analysis in GIS implementation.

Typically, an NPV analysis will be projected for 10–15 years, reflecting the minimum expected life of the GIS. The actual useful life of the system is likely to be much longer. Once maps and related data have been converted from hard copy to digital format, data transfer to new platforms should involve minimal cost. The greatest saving in GIS investment is recovered through savings made in using the data. Integration of operations and user base offers the greatest opportunity for repeated use. A corporate solution, where data are made available to a large number of users, represents the most effective means of maximising return. The sooner users have access to data, the sooner the benefits will begin to accrue.

System design and implementation

If a corporate GIS is to fulfil expectations, and to satisfy anticipated cost/benefit estimates, it must be made operational on time and within budget, provide sufficient functionality and produce outputs of sufficient quality. Creation of a generalised 'model' of what the system might look like, covering its core components and how it might integrate with current business practice, is a useful strategy. Such diagrams are useful when presenting ideas and broad concepts to managers or departmental groups who need to know how the GIS might fit in with their work patterns and what a new system might comprise. The goal of the systems design and implementation process is to deliver these expectations. Design and implementation usually involves the following elements:

- Project definition, development of a business case and preliminary analysis of costs and benefits, which were covered in the previous sections.
- A detailed systems analysis, involving a requirements study and often including pilot projects and feasibility studies.
- A final implementation plan and detailed systems design.
- Rollout of the design, including data capture, procurement of hardware and software and commercial data sets, user training and configuration of databases and subsystems.
- Maintenance and day-to-day operation.

Systems analysis, system design and implementation planning

Organisations are complicated. Even a single department may have many filing systems, databases and staff carrying out a range of different jobs. In large IT implementation projects, it is often useful to adopt a standard method of system design in order to maximise the chance of delivering a successful project outcome. The role of systems analysis and design is similar to that of the architect on a building project – to provide a communication channel between the client (the user) and the builder (systems programmers, hardware, data and software suppliers).

Although it may seem that specifying requirements for an information system is straightforward, users often do not have a clear idea of what they want and need time to rationalise exactly how they wish to proceed. Also, in a large organisation, new requirements may already be in place somewhere, but people might not be aware of their existence. There may be several ways of achieving the requirements, with different cost and performance implications, and a rational decision on how to proceed might need to be reached. Finally, constraints may have been overlooked and interdependencies between departments or subsystems and redundancy and duplication in data sets may not have been considered. Systems analysis and requirements studies clarify these issues and produce a coherent system design and strategy for implementation that contains tangible milestones and a clear specification of what and when the system will have to deliver. The systems analysis process is used to develop a clear picture of existing processes within the organisation including where data are stored, how information moves around the organisation and how key work processes are carried out. This information, usually gained through a series of detailed interview procedures, is used to develop a complete picture of existing work processes that forms the basis of a development plan for the new systems.

Several factors are known to contribute enormously to successful development (Longworth, 1992; Eva, 1994). The most significant is user involvement. By reviewing user requirements at each stage of analysis and design, and involving the users from day one, the risk of producing the wrong system is very much reduced, and problems can be spotted early on. A related issue is quality assurance. By making the user authorise or sign off the system design at various stages of the project, quality is assured. Independent reviewers could also be brought in.

When implementing a new system it is critical that an effective systems analysis procedure is used. The most common systems design tools use structured approaches to systems analysis. They break down the implementation problem into clearly defined, small tasks, and specify the sequence and interaction of these activities. They use data modelling techniques, usually involving diagrams, that express the relationships between data and subsystems in a common format. Structured approaches are useful because they provide a clear specification that everybody can understand. They improve project planning and control because a key element of the structured process is to define milestones and the stages of system construction. Methods such as Structured Systems Analysis and Design Methodology (SSADM), the key stages of which are shown in Table 10.1, provide a clearly defined route through the implementation process.

A well-defined project management system, operating alongside or as part of the systems analysis framework, provides the control mechanism necessary to achieve implementation objectives. Typically, it involves planning (estimation, scheduling, evaluation), monitoring (progress against time and budget, product quality, user involvement) and authorisation (change control, rescheduling, collection of statistics). The benefits of effective project management are that staff know what is happening, what is expected of them and how what they are doing fits in with other staff in the team. The manager can make sure that workloads are appropriately distributed, schedule work and determine staffing levels effectively. Senior management can assess budgetary requirements and look at the value of the system and individual projects to the organisation. In order to be effective, a project management system must be up-to-date. Failure to ensure this will quickly result in the breakdown of the management process.

Pilot projects and their role in system design and implementation

A pilot project permits the informed scoping of costs and benefits, allows for the testing of ideas prior to the commitment of extensive resources, and presents an opportunity for the user community to see how a final product

Table 10.1 GIS implementation issues

---

1 Overemphasis on technology: planning teams are made up of technical staff, emphasise technical issues in planning and ignore managerial issues such as staffing, resource commitment and costs and benefits.

2 No long range planning: planning teams are forced to deal with short-term issues, have no time to address longer term management issues. A GIS implementation usually has a long-term payback, rather than a short-term one.

3 Rigid work patterns: it is difficult for the planning team to foresee necessary changes in work patterns. Some jobs will disappear, other jobs will be redefined, for example, drafting staff reassigned to digitising. Retraining and persuasion will be needed.

4 Organisational inflexibility: the planning team must foresee necessary changes in reporting and management structure, organisation's 'wiring diagram'. Departments, which are expected to interact and exchange data must be willing to do so.

5 The Project Management Team: many GIS projects are initiated by an advisory group drawn from different departments. This structure is adequate for early developments, but in later stages must be replaced by a well-defined management structure for the project to be successful. A project may be derailed if any important or influential individuals are left out of the planning process.

6 Assignment of responsibilities: this is a subtle mixture of technical, political and organisational issues. Typically, assignment will be made on technical grounds, then modified to meet pressing political and organisational issues.

7 Integration of information requirements: management may see integration as a technical data issue, but it is much more than that. A corporate data integration strategy for GIS is likely to require, *inter alia*, departments to work together, the sharing of information resources, development of corporate codes of practice, implementation of standards, centralized/shared data repositories and the unification of existing systems.

---

might look. Usually the last major milestone prior to corporate and technical commitment, it recognises the difference between theory and practice. It is part of the effort to sell the system within the organisation. Pilot systems can be shown to decision-makers as evidence of the immediate value of the project and provide a tangible way of communicating potential to sceptics. Pilots are also useful for verifying estimates of costs and benefits and evaluating systems design alternatives.

Pilot projects may be used to demonstrate proof of concept. A demonstration of limited facilities on a small area, using a system that may not be part of the final product, mainly for development and hands-on experience, provides early visibility of the system to management and users. In some cases, the data used in the pilot may not be part of organisation's operation. In others, the pilot may be a prototype or full-scale model of the future system, designed to identify problems and finalise implementation decisions. A pilot project should be managed and defined as effectively as the major project of which it is part. Objectives must be defined clearly and typically include the evaluation of the proposed system design, the testing of alternatives and technical procedures, some exploration of ways of generating products and the formats of those products.

If support is minimal, the pilot project must be orientated towards building a sound business case for the system. An effective pilot will cost money. To be successful it must justify this cost and the subsequent, larger cost of the full system. If the pilot covers a region within the organisation's service area, this region should be significant or relevant. Pilot project design must consider the current level of experience of the project staff and must allow sufficient training and experience for those involved to permit realistic evaluation of the potential of the system.

The results of a pilot should, as a minimum, provide experience of implementing a GIS project and elicit management approval to proceed with implementation. Ideally, it will reduce risk and improve efficiency in the early stages of a major project. Potentially, it could result in trained staff and users, well-developed technical, managerial and production procedures, an improved implementation plan and enthusiastic support from managers and users.

Rollout of the system design

Once the requirements specification has been completed and a final, physical design has been achieved the system design can be rolled out. This will involve the selection of hardware, software and data suppliers and the provision of detailed specifications to them so that the system design can be implemented.

The rollout process is usually performed through a formal document known as an Invitation To Tender (ITT). The ITT forms the basis of a contract between one or more systems suppliers and the user who has developed the technical specification.

The system design created as a result of the requirements analysis and system planning phase forms the basis for system selection and evaluation, as well as the contractual basis for the ITT. It must specify concrete business requirements, and not abstract ones that are hard to assess or quantify. An ITT is normally sent out to systems suppliers, inviting them to submit a bid for constructing the final system design. It usually comprises the requirements/technical specification, contractual requirements and instructions. Typically, a major project ITT technical specification would include:

- Project background – an explanation of what the project is all about, how it came about and what it is trying to achieve.
- An explanation of how the ITT is structured.
- A description of existing organisations and procedures.
- A statement of IT strategy and the project plan for the subject of the ITT.
- A precise scope for what the tender is asking for and guidance on the type of solution that is required including clear stipulation of any terms and conditions of supply.

- A detailed user requirements analysis, including the specification of future requirements.
- A glossary of technical terms and other jargon.
- Examples to illustrate current practice.

User requirements need to be specified so that they are unambiguous and ideally can be subjected to conformance tests. Two bad examples of requirement definition might be 'the system must be able to produce hardcopy output' or 'users must be able to query the database'. In each case, the requirement is extremely vague and could be met by a wide range of solutions. A better example that defines hardcopy output requirements in much more detail might be:

> The system must be able to product hard copy output of both digital maps and database tables. Predefined templates for map output with fixed scales, border and content must be set up to conform to existing map products. The system must also facilitate the production of maps according to user defined parameters for scale, border, content and presentation. Database tables must be printed in the standard report format specified by the department. User defined templates must also be supported, and suppliers should explain how this will be achieved.

An ITT document, particularly if the project is complex, can be very large. The tenders submitted in response are also likely to include a wide range of information, and it will usually take a significant amount of time to assess them properly. The following are suggested as stages in the review of ITT response documents.

(i) Examine responsiveness of the document:

- Whether or not the document is a substantial response to the ITT.
- Whether or not it meets the requirements and restrictions of the ITT in terms of format.
- Whether it generally complies with the requirement.

  ITT responses that do not meet all three of these criteria would usually be rejected. This may not be the case if there are very few responses, but most organisations recommend rejection at this stage if these three conditions are not met.

(ii) Rank the responses: At this stage, a preliminary read of the documents is carried out and a rough judgement made about which ones look better than others. This is particularly useful if there are lots of bids. For example, if the ITT requested a specification for hardware and software and only software was included, that bid would go to the bottom of the pile.

(iii) Rigorous review: The third stage is to applying the selection criteria and evaluate each proposal for content, compliance with specification and any special circumstances or innovative approaches. This process might include presentations by respondents, visits to supplier sites, benchmarks, visits to user sites where the technology is currently in place, reference checks (telephone interviews with satisfied/dissatisfied customers) or any combination of the above.

(iv) Cost review and re-ranking: Having considered the proposals that will satisfy the needs of the organisation, stage four is re-evaluation of the proposal for cost effectiveness – which does not necessarily mean the cheapest submission! There might be hidden costs and risks, for example warranty disclaimers, expensive support agreements, aggregate pricing (to hide expensive items) and 'free' additions with limited support or guarantees. If in doubt, a lawyer would normally be instructed to make sure that the procurement contract is legally binding and does not include potential loopholes.

(v) Selection and recommendations to management: Having conducted thorough reviews, a recommendation to purchase will be made, usually to the management authority behind the ITT. Because these people are responsible for the procurement, they will be accountable at the end of the decision. The evaluation process must be well documented and presented with a well-organised summary of the process and its results.

(vi) Benchmark testing: A benchmark is a series of tests on hardware, software and other options put forward in a response to tender in order to assess the suitability of a proposed solution. Benchmarks can be used to rate the performance of different solutions against one another, and compare their utility and are discussed in more detail below.

(vii) Selection and negotiation: Once the review is complete, and legal and administrative issues have been dealt with, it is time to notify suppliers. At this stage, it may be necessary to shortlist two or three suppliers and negotiate with them to see who can provide the best deal. If shortlisting is to occur, it should be clearly identified that this will be the case in the original ITT document. The negotiation process can result in greater cost effectiveness, by playing off one supplier against another. However, this should not be at the expense of good relations with suppliers.

(viii) Follow-up: It is usual practice to notify unsuccessful respondents promptly, and to allow them an opportunity to find out why they have not been successful. Decisions must be final, and a proper, well-conducted debriefing will eliminate most attempts at overturning a decision. The need for a well-documented decision-making process is reinforced at this stage, in case challenges to decisions are made.

There are alternatives to the rigid ITT procedure. A Request for Proposal (RFP) allows the supplier more leeway in suggesting a route forward, often used where management would like to see some input from suppliers into how best to move forward.

Procurement policies

Normally, a large organisation will have a standard code of practice for the procurement of new equipment and other resources. In large organisations, such as government departments, these policies are highly formalised, can be governed by legislation, and may be managed by a separate procurement organisation. Procurement policies can cause problems for GIS procurement. Here are some potential problems:

- An IT standard already exists, and GIS requires additional facilities that this does not support.
- The organisation always selects the bid with the lowest price.
- Procurement policy only allows a tender or bid process excluding any changes to specifications.
- The procurement process does not permit negotiation with suppliers.
- The organisation has a 'buy local' policy, but there are no local suppliers with adequate expertise.

In order to make sure that GIS procurement goes forward smoothly, it is important that existing policies and practices are known, and their impact on the procurement of products and services for GIS is understood. Second, procurement administration staff should be involved as early as possible in the purchase process. Policies that obstruct the GIS project were not designed to do so. They came about to support the organisation's existing requirements and an understanding of why they were developed will help to solve contradictions. Special dispensations may occur when the existing procurement framework is entirely inappropriate for GIS acquisition and the organisation may elect to allow a special case.

Benchmarking

Benchmarking is the process of evaluating systems performance and typically forms part of procurement, allowing for the technical comparison of different solutions and the evaluation of system performance in the chosen solution. It is a key element in minimising the risks in system selection. Often, customers do not have precise plans and needs – these are determined to an extent by what the industry has to offer. Customers need reassurance – a real, live, demonstration – that the system can deliver a vendor's claims under real conditions. A benchmark allows the vendor's proposal to be evaluated in a controlled environment. Typically, the customer supplies data sets and a series of tests to be carried out by the vendor and observed by the customer. An evaluation team is assembled and visits each vendor, performing the same series of tasks on each prospective system. Tests examine specific capabilities, as well as general responsiveness and user friendliness and the demonstration is carried out in an environment over which the customer has some control. For example, equipment may be provided by the vendor and data and processes by the customer.

Overseas case study – benchmarking exercise

Donnay and de Roover (1992) describe a suite of tests for GIS performance evaluation based on a set of standard data sets:

- SPOT satellite image
- Land cover layer (raster)
- Contour lines
- Road network co-ordinates
- Land use polygons
- Name and population data file
- Road section attributes
- Paper map.

All data sets were referenced to the Belgian Grid (Lambert projection). Benchmarks were executed on a UNIX workstation with access to a tape drive, a digitiser, a plotter and a printer. The machine required file transfer facilities to a connected PC. Companies were given one week to prepare for the tests so that time-consuming jobs could be dealt with.

The general benchmarking procedure that they followed included three steps:

1   Execute test session
2   Save history files for session and list the functions that were used
3   Generate system statistics: CPU time, I/O parameters, etc.

Because of time constraint, some sessions were prepared in advance. These were assessed in demonstration or batch mode execution. In demonstration mode, tape access and other time-consuming functions were discarded. In batch mode, interactive functions were dropped and system performance statistics for the batch job were retained. History files provided data on how the solution was arrived at and how long it took to achieve the required results.

The following tasks formed the main operational content of the benchmark:

1   Data entry and management – digitising, importing files, automatic generation of features.
2   Generation of network features and linking to attributes.
3   Generation of DEM and derivation of gradient and aspect surfaces.

 4 Generation of a viewshed and printout of results.
 5 Generation of vector layer and subsequent reclassification, selection, merging and overlay.
 6 Direct overlay of raster and vector data and classification of attributes based on overlay.
 7 Analysis of satellite imagery and statistical output generation.
 8 Add attribute data to existing maps.
 9 Shortest path route analysis.
10 Dasymetric mapping – population density.

Data capture

The process of data capture can be long, complicated and expensive. For example, computerising every planning application back to 1947 within a local authority area (some authorities have done this) would lead to the provision of a comprehensive planning application database but could take many years. A balance needs to be struck, and careful consideration of priorities made. Could the GIS include the later planning applications and the older ones be held manually for the rare occasions when it was necessary to refer to them? Could there be benefits in prioritising other data sets that might begin realising benefits sooner, such as mapping areas liable to flooding or covered by restrictive planning policies? The best approach is to begin with simple applications and add data to the GIS that is going to be most useful to users but is not going to be costly to computerise. The GIS can then expand and take on more complex tasks incrementally.

Maintenance, update and review

Full consideration must be given to the life of the GIS project once it progresses beyond the initial development phase. The issues and problems that users of the GIS will face once the system is operational and actively supporting the decision-making process are critical for project success after development. Typically, they might include staff management, workload prioritisation and effective budgeting and project management.

Managing GIS staff

Successful operation of a GIS is dependent on three factors related to the behaviour of staff:

(a) The role of consultants. Outside consultants can be highly effective during the feasibility, cost benefit and vendor analysis stages of a GIS implementation project. However, once implementation is completed, the role of external consultants should be carefully questioned. In-house staff are generally more responsive to user needs than consultants because of their local knowledge base and dependence on success of organisation. King and Kraemer (1985) found that local governments in the US relying on in-house knowledge showed greater success with computing than those relying on external bodies. In-house staff might resent reliance on consultants, particularly if the consultants are apparently more highly valued than them. Consultants might be given more interesting work, for example. A balance between the two groups of workers is therefore essential.

(b) User support vs programmed assignments. Technical professionals usually gain more job satisfaction from day-to-day assistance of users than they do from long-range projects for the benefit of all users. This can result in the neglect of long-term strategic objectives in favour of helping others. Managers must be aware of the potential for conflict between these two roles and assign a reasonable balance.

(c) Staff turnover. Turnover of staff is a critical issue for GIS managers. Most of the early GIS success stories were characterised by long-term continuity of staff. The retention of staff is a result of effective management and heavily dependent on corporate policy and personnel strategies. However, job satisfaction can be enhanced by clearly defined career paths and management opportunities. Staff autonomy, staff development frameworks such as the UK Government's 'Investors in People' initiative, and an understanding of role in the overall mission of the organisation and how individuals fit within this mission can all aid retention.

Managing the workload

The day-to-day operation of an information system involves the application of resources to a range of projects and ongoing maintenance and support services. Equipment will become obsolete, software and hardware will need to be updated, data sets will need to be altered, changes in operation are likely to be required in the longer term. The operational phase of the system will involve project development, creation of work plans to apply resources and schedule project completion and preparation of budgets to ensure that adequate resources for individual projects are available.

Pertinent information about each project should be recorded in a standard format to avoid confusion and to enable all members of staff to understand what is happening. An active GIS support unit in a busy organisation may deal with hundreds of projects in a year, all of which need to be kept track of. Typical information that the project definition should include would be:

- Project title
- Project number (usually for automatic project management purposes)
- Project originator

- Project objectives
- Estimates of time, duration, total cost
- Recommendations (feasibility, priority etc.)
- Authorisation (managerial authority to execute project).

It is standard practice for quality assurance purposes to implement this type of procedure in order to manage workloads effectively. Even if the number of projects dealt with is small, it is still useful for auditing and managerial purposes although the manager should always be aware of the balance of time spent by staff on administration functions at the expense of project activities.

Costs of GIS operation and maintenance

There are eight major cost categories for GIS services and these are listed below.

(a) Staff
(b) Materials and supplies (consumables)
(c) Equipment purchase (includes hardware)
(d) Software
(e) Maintenance of hardware and disaster recovery
(f) Internal services
(g) Data resources
(h) Other expenditure.

We will look briefly at each one.

STAFF

Staff costs include all costs associated with the use of human resources to perform work. Labour costs should include costs to the company for pension schemes and tax contributions as well as gross staff payment costs. A distinction may be made between internal staff and consultants (it may be possible to charge consultant time to particular projects, for example). There is also a distinction between direct and indirect staff costs. Direct costs can be attributed to particular projects and invoiced to them, while indirect costs are associated with time spent on work that is not directly attributable to one project. Examples of indirect costs could include daily backups, user support and staff training.

MATERIALS AND SUPPLIES

This category includes all office supplies, books, diskettes, CD-ROMs, plotter paper rolls, printer cartridges and other day-to-day items required for effective function.

EQUIPMENT

Equipment includes all purchase or lease costs for computers, fax machines, furniture, photocopiers, printers, disk drives, tape drives, telephones, network cabling and other items that form part of the organisation's equipment inventory. Equipment obtained for a specific project might be charged to that project.

Computer equipment and peripherals become obsolete relatively quickly. Replacement costs need to be part of the long-term budget of the GIS office. At the moment, the useful lifetime of a PC is approximately 2–3 years and the replacement of obsolete equipment needs to be managed. It is usually cheaper to replace than to upgrade computer equipment. Where possible, all equipment should conform to corporate IT strategies and standards or other recognised standards if there are no corporate ones.

SOFTWARE

There are two major types of software that can impact on the GIS operation: GIS and related analytical or data presentation software and Operating System software. The GIS manager must ensure that software are used and licensed appropriately. Like hardware, software become obsolete and new releases of strategic software must be managed carefully. Key issues for upgrade include:

- Operating System changes may cause key software packages to cease to function and require additional upgrades.
- Some software may be limited by older versions of the operating system or require the replacement of hardware.
- New releases of software may contain bugs.
- Manufacturers may cease to support old versions of software.
- Hardware compatibility may change.
- Users may need retraining.
- Existing training materials may need revision following upgrades.
- Administrative staff may need new skills.

HARDWARE MAINTENANCE AND DISASTER RECOVERY

Most manufacturers provide the option of a maintenance agreement that provides support in the event of hardware failure. Such agreements can be expensive, but may be essential to permit effective long-term functioning. A related issue is the strategy for dealing with disasters. Computer systems go wrong sometimes and a computer implementation should include a policy for disaster management explaining how serious problems will be dealt with and what contingencies are in place to minimise the impact of a disaster. For example, how would the organisation recover from a major fire that destroyed critical computer systems? As well as physical hardware problems there may be software problems including computer viruses, software crashes, hacking and illegal system access, user errors and accidental removal of key data. A backup strategy is vital. Key considerations include how data are backed up, what is the policy on archiving data and software and how the organisation would recover in the event of a catastrophic failure.

INTERNAL SERVICES

Depending on the nature of the organisation, the cost of services performed by other departments for the GIS facility may be billable (cleaning, power, heating, office space, network access, for example).

DATA

Data may involve a number of separate charging regimes:

- Initial outlay
- Copyright and royalty fees
- Maintenance charges
- Cost of administration
- Outsourcing for updates etc.

Data set usage must not contravene agreements with suppliers and data must be of an appropriate level of completeness and quality.

OTHER EXPENSES

Other expenses include travel costs (conferences, user forums, business trips), car allowances, staff training and payments to professional bodies.

Implementing a corporate GIS in the public sector

A survey of the handling of property data in the East Thames Corridor, which runs from East London to the Essex and Kent coast of southern England, revealed some of the important issues surrounding the implementation of corporate GIS within a local authority environment in the UK (Lopez, 1992). Lopez discovered that many early applications of GIS in local government suffered from common problems:

(a) Lack of finance.
(b) Lack of a corporate information strategy.
(c) Lack of comprehensive GIS feasibility and cost–benefit systems.
(d) Emphasis on technical rather than institutional and data-related issues.
(e) High cost of digital data and manual data conversion.
(f) No mandate to share data.
(g) Problems were highlighted regarding the use of government statistics where certain data were not suitable for local government functions due to inappropriate scale, detail or accuracy.

Information 'islands' developed between which data were rarely shared and independent databases designed for one function were created rather than networked as part of a corporate resource. The survey showed that departments were not sharing data to a large extent and those that were, were doing so on a project-by-project basis. A lack of central guidance exacerbated the problems encountered by individual authorities.

As the level of computing skills within local government increase, users are realising that constraints to information flow are creating obstacles to IT development. It was argued in the survey report (Lopez, 1992) that local authorities should assess their departmental data requirements, work towards corporate data standards and a policy of sharing information before implementing a GIS.

The findings of the RTPI's IT and GIS surveys in 1995 and 2000 revealed only slight (7.5 per cent) growth in the implementation of GIS in local government. GIS appears to be succeeding departmentally, in planning and highways departments usually, but struggling to become a corporate resource. Two problems were cited; a lack of skilled GIS staff and a lack of demonstrable applications to show how GIS might offer benefits to the delivery of core local authority services.

Recent developments are contributing to the development of GIS implementations in local government. The Service Level Agreement, agreed between the IDeA and the OS, allows local authorities to purchase digital map data at a discounted price and central government initiatives for electronic access to government information are gathering momentum. For example, Medway Council has implemented an intranet-based GIS throughout the its PC network. A major driver is the e-government initiative to have all government services online by 2005.

GIS can offer benefits and savings to local authorities in a number of core areas. However, authorities also need to be aware of the costs and some of the necessary adaptations to their own practices that can be attendant on the implementation of a GIS. These can be considered as the following.

Financial costs and benefits

While the general benefits of GIS are widely recognised, it has often proved difficult to predict all the relevant costs and benefits discussed above. Savings from speedier information handling and staff reductions may be relatively easy to measure, but the benefits of well-informed decision-making and higher quality services may be elusive but, nonetheless, significant. Many local authorities advise that start-up costs, including substantial data capture exercises, should not be underestimated. For example, GIS implementation at Milton Keynes Development Corporation involved staff working 10-hour days for 7-day weeks for a period of 25 months to digitise information on 120,000 parcels of land and 40,000 legal transactions. In total, the Corporation estimated that the data capture job alone took 56 person-years in total.

Changes to organisational structure and data handling routines

The introduction of new information system is likely to have strong impacts on the work routines of any organisation. A good IT strategy will aim to minimise unintended effects. Strategies for authority-wide, shared information systems had long been favoured because of the potential for raising efficiency and avoiding duplication and waste. However, such synergy has often failed to materialise because organisational structures and work procedures do not necessarily change according to the potential offered by IT. Indeed, some argue that it is not possible to create a single general information system to meet the needs of all departments across an authority. Many IT strategies therefore favour smaller, departmentally controlled GIS which fit in more easily with existing structures and practices.

It can be easier to implement a GIS on an application-by-application or a departmental basis because it is often difficult to identify information requirements and flows on an authority-wide basis. Also it is easier to persuade a smaller group of managers than the whole local authority (Peel, 1995). Application-led implementation means quick results and acceptance of GIS among users. However, if support is forthcoming a corporate or organisation-wide approach provides better management, co-ordinated data capture, consistent standards and easier data sharing (Peel, 1995).

Staff, employment and skills changes

GIS represents a major investment whose pay-off depends to a significant extent on the skills and competence of the people who use it. The skills required for the successful adoption of GIS include general knowledge of GIS concepts, understanding of user operations, system management and applications tailoring. Skills shortages have been widely reported as a constraint on the adoption of information systems by local authorities. Sometimes, these will be 'bought in' from outside by way of job specifications for new staff appointments. The skills of administrative and cartographic staff, on the other hand, are becoming more dispensable, and redundancies of cartographers and technical and clerical support staff are often given as important cost arguments in support of GIS adoption.

Learning from the local authorities

The main problems experienced with GIS by local authorities concern the cost, availability and (although this is becoming less of a problem) the quality of data; incompatibility of spatial referencing systems; and the management context in which the system is developed. These issues are closely linked to financial constraints that limit the range of solutions. There have been cases where GIS had been implemented for a whole authority and then abandoned or reduced to a single department, because of such problems. In other cases, the maintenance of the geographical referencing database has been divided between departments, leading to the loss of consistency.

While there is a strong argument for taking advantage of the economies of scale achieved by a corporate GIS, short-term cost considerations are powerful arguments for providing single departments with stand-alone facilities. Some of the more successful GIS implementations have begun in a modest way, with the commissioning of one or two workstations in a single department, producing identifiable benefits (even if these are only good quality paper copies of maps!) from day one. In this way, users are convinced of the value of systems and are more likely to argue for their development and extension.

Cost–benefit analyses for GIS, or even comprehensive cost estimates, have proved difficult and rather unreliable. It is widely assumed that data capture is one of the biggest cost items in setting up a GIS, but there may well be challenges to that assumption as local authorities make use of cheap digitised maps and existing computerised data sources. The experience of Plymouth City Council suggests that data capture costs were only in the region of 20 per cent of total costs.

Some local authorities report data-related problems in GIS implementations; lack of compatibility between data sets and the cost of data capture have been mentioned, differing levels of accuracy and precision of many independently gathered data sets are a major obstacle to their integration. Certain data sets are simply not available, incomplete, or not reliable. The problems are compounded when a local authority decides to hold historic data on its GIS. Whilst freshly generated data (such as planning applications) can be recorded on a GIS on an ongoing basis, possibly as part of the registration procedure, the computerisation of historic manual data sets, or the integration of obsolete computer databases, presents quite a different task. One large urban local authority decided to digitise all of the planning registers covering the period 1947–1990. The records (which obviously had deteriorated with age) were held in bulky ledgers, which had to be photocopied and then sent off-site to a data capture agency. It is probably legitimate to wonder exactly what use the authority would have made of the older records.

Implementation issues for national land and property management initiatives

Many countries have now implemented (or are in the process of implementing) national frameworks for land information management. Countries that have provided a framework of sympathetic political support and government assistance have generally been more successful than others. Given the magnitude of national LIS initiatives, financial viability may only be possible if public/private sector partnerships are employed for the development, maintenance and use of data sets and systems. The political climate may influence how such partnerships are perceived, but increasing use is being made of them. Additionally, co-operation is essential and feedback is crucial, particularly given the long-term development time-scale for most LIS projects.

For developing countries, advice and support is important. A contractual framework and commitment (human and financial) from data providers is also necessary. A widely adopted approach has been the formation of a new government body to manage relationships between data suppliers and users and to control access and data supply using IT (the FGDC in the US, or the NLIS Agency in the UK, for example). This is often accompanied by a second body to co-ordinate development, publish metadata and administer standards. The management structures of the Western Australian LIS and the UK National Geographic Framework[1] (NGDF) initiative reflect such an arrangement. A related policy has been to merge land registration and mapping agencies so that cost recovery for the latter function can be absorbed by the former, as is the case in New Zealand. Where possible, this should be achieved so as to minimise disruption to existing government structures when creating a new agency/department.

For any multi-agency project like a national LIS there is a need to ensure that participating organisations are in a position to co-operate and commit resources and that there is the political will to achieve this. A related issue is the requirement for restructuring existing governmental and legislative frameworks in order to make progress. This can encompass the creation, combination or removal of departments, staff training, or the redrafting of organisational mandates. While the aim has tended to be to minimise disruption to existing frameworks, the creation of a new department or agency may offer vitality and focus. Initiatives in Kenya and Sweden suggest that this idea has validity, and the UK NLIS Feasibility Study (Local Government Management Board, 1997) calls for the formation of a new agency to take forward development. The role of the private sector in the administration and provision of integrated land resource data and services may also have important organisational implications. In the UK, primary spatial data are being developed through a mixture of private and public sector initiatives. Government agencies like the OS often subcontract or have partnership arrangements with 'value added' commercial resellers. Given the expense of geographical data infrastructures as evidenced by NSDI, it may be that public/private sector joint ventures are the only way by which sustainable development of an NLIS can be realised.

If possible, LIS development should be self-financing, for example, using registration fees or property taxation. However, many of the lessons remain the same: government and private sector support are required, data quality is important, standards are needed, and training and education are required. The development of an LIS that comprises a fiscal cadastre offers a revenue-generating capability. Also, buyers and sellers of property may be prepared to pay a fee for public access to information that has been collected and maintained by the state, such as physical characteristics and legal details of interests in property. It is worth mentioning that in some countries this information is provided free of charge or at a very low price that represents the cost of distributing the information.

Organisation and administration issues for GIS implementation

GIS is more likely to require organisational changes than other innovations, because of its role as an integrating technology. The need for changes, co-operation, breaking down of barriers, for example, may have been used as key arguments for investing in GIS. Organisational change is often difficult to achieve and can lead to failure of the GIS project. Organisational, political and institutional issues are much more likely to be the cause of failure in a GIS project than any number of technical issues. Resistance to change has always been a problem in technological innovation – the early years of the industrial revolution bare testimony to that.

Obermeyer and Pinto (1994) categorise the major risks for GIS implementation projects as follows:

(a) Organisational risk – any project requires a mix of skills and organisational conditions present for it to be successful. In a GIS implementation, important skills may be unavailable. Skill requirements need to be clearly identified and a plan of action to remedy any deficiencies will need to be drafted and executed. A multi-user GIS requires appropriate staffing to ensure effective management of resources and operational efficiency. The system planning team may not recognise the necessity of these individuals. Management may be tempted to fill these positions from existing staff without adequate attention to qualifications. In addition, personnel departments may be unfamiliar with nature of positions, qualifications required and salaries. Managers must establish clearly defined job descriptions, performance evaluation criteria and career development plans and a consistent methodology for assigning and managing work.

(b) IS infrastructure risk and technical uncertainty – a GIS project introduces new technology into business. Is the required level of technical support available? Is the GIS project breaking new ground in terms of untried technology? The fact that technology works elsewhere does not mean that it will work in a particular organisation. The more untried the technology, the greater the risk.

(c) Definitional uncertainty – information requirements are notoriously difficult to define. How certain is the business that requirements have been identified and that the GIS specification is appropriate for identified objectives?

Location of GIS within an organisation

Even though GIS may be an organisation-wide tool and seen as a decentralised resource, co-ordination of the GIS operation is still necessary to ensure efficiency and cost effectiveness. This will help avoid redundancy in the collection of data and ensure that expensive hardware and software is being used efficiently. The location of the GIS and its staff within an organisation will reflect how GIS was introduced to the organisation, the management structure and organisational policies and mandates. The location of the GIS manager and support staff will be seen as the location of the GIS unit within the organisation and will affect the way the GIS staff interact with the rest of the organisation. Somers (1990) suggests that there are three basic options for the location of the GIS:

(a) Operational department location. For example Planning and Development, Pipeline Management, Property Services. GIS often develops from a small system obtained to deal with specific needs that grows to support activities outside the mandate of the original department. The advantage is that it is very responsive to original users' needs. The disadvantages are that a departmental focus makes it difficult for other users to have their needs and priorities recognised, it may not have higher-level management support and may lead to 'territorial' squabbling.

(b) Support department location, such as IT, Computing Services. In these locations, GIS is seen as a service operation like payroll, personnel and data processing and will be supported by the organisation as such. The advantage is that there is objectivity of system design and management. Disadvantages are that it is remote from the users of the GIS, it may not be responsive to the needs of users and priorities of department may be different than users' priorities. Centralised computing services with large staff complements are disappearing because new, distributed workstation hardware requires less support.

(c) Executive or management level location. Here, the GIS team is located at management level. The advantages are high-level visibility, support and attention, objectivity. Disadvantages are distance from the real operations of the organisation and users may feel GIS support staff are out of touch with their needs.

There is no one ideal placement, but successful implementations usually reflect the way in which the organisation operates. If the top elected official is a 'hands on' manager with interests in the day-to-day operation of many facets of the organisation, responsibility for the system is likely to be directly under that person's control. If the management structure of the organisation is more decentralised, it is likely that GIS management will reflect this too, and the system will be under the control of a different department or group of departments (Huxhold and Levinsohn, 1995).

Change requires leadership. Initiation of GIS strategy requires a 'project champion' within an existing department. Such leaders take great personal risk – there is ample evidence of past failure of GIS projects and the initial 'missionary' is an obvious scapegoat. Corporate uptake requires commitment of top management and of individuals within departments. Despite reasoned economic, operational, political advantages of GIS, the technology is new and outside many senior managers' experience.

At a more pragmatic level, methods of data capture, storage and manipulation may change dramatically in a GIS environment, including the need for recording geographical location, database construction/structures, analysis/manipulation methods and map-based rather than tabular output.

Defining the GIS professional

In 1994, the RICS Education and Membership Committee stated 'the growing importance of IT in areas of practice such as GIS is liable to have a significant effect on the needs of many employers. Overall the technological revolution which will affect the profession in the period up to the turn of the century will have many far reaching implications both in the way in which business is conducted and in the type of skills which will be needed to survive' (RICS, 1994).

In Australia, the Australian and New Zealand Land Information Council (ANZLIC) commissioned a survey to identify the education and training needs of the land and geographic information community. The results correlate with similar work undertaken by the AGI (Capper and Unwin, 1995). Some of the major themes to emerge from the ANZLIC survey and the work of the AGI include the following:

* Training must include fundamental spatial concepts (scale, errors, data management, basic cartography).
* There could be adaptation of existing programs, for example, undergraduate and postgraduate modules to short courses, pre-qualification structured learning or continuing professional development.
* As IT develops and more data become available in digital form, core disciplines associated with surveying (cartography, valuation) are being replaced by more analytical, data-related disciplines. Rapidly changing technology means that continuing professional development will be crucial.
* There is a need to develop a national competency standard for land information management, similar to the core curriculum developed by the American National Centre for Geographical Information and Analysis (NCGIA).

A lack of training and education is often seen as a barrier to the use of GIS by property professionals. Many organisations are concerned that new technologies require substantial investment in education and training. Many property people are not familiar with analysing property information geographically on a computer, and the employment of a full-time GIS expert is not regarded as feasible in many UK property firms at the moment. Small property firms will not wish to invest the substantial amounts required to obtain the capabilities of large GIS packages. Such practices do not have the resources for investment in hardware, software, training or the employment of dedicated IT personnel.

The integration of GIS skills into property education could occur at a number of levels; undergraduate, postgraduate, during professional training and as part of continuing professional development. At the undergraduate and postgraduate levels, universities now provide GIS courses. It is within this framework that GIS education must fit. Table 10.2 suggests curriculum content for GIS teaching on property courses in the UK based on a hierarchy of skills taken from Dixon (1995) and the AGI (Capper and Unwin, 1995).

At the professional and vocational levels the AGI recently published a draft report that proposes a professional development framework for those employed within the geographic information (GI) industry. The framework is based on that employed by the British Computer Society and consists of a

---

Table 10.2 GIS course content (after Dixon, 1995 and Capper and Unwin, 1995)

Level 1:
- general ICT skills; hardware, software, networks, data sharing, data transfer
- introduction to GIS and its relevance to surveying
- definitions and terminology; GIS, LIS, cadastres, digital mapping, remote sensing, image processing
- key players in the UK; OS, Land Registry, VO, the Office for National Statistics and private sector organisations
- compare and contrast with GIS abroad.

Level 2:
- elements and principles of a GIS, GIS software, peripheral devices (scanners, digitisers, plotters, printers)
- data models, raster, vector, topology, spatial concepts
- databases; attribute, spatial, relational, integrity
- computer-based tutorial introducing GIS capability
- geographical analysis; introduction.

Level 3:
- geographical data sources; data acquisition; availability, cost, format
- data input and integration; standards, quality, currency, accuracy
- data management; database update, extension, data conversion, import and export
- GIS applications, for example;

    - Retail   location analysis, economic appraisal, etc.
    - Utilities   maintenance and management of plant
    - Environment   impact assessments, land quality statements, contaminated land, etc.
    - Local Government   planning, property management, etc.
    - Health care   epidemiology, pollution monitoring, location of services, etc.
    - Transportation   least-cost path, efficient network analysis, etc.
    - Financial Services risk analysis

- begin computer-based project

(continued)

Table 10.2 (Continued)

Level 4:
- GIS project work flow
- project definition – data acquisition – hardware – software – data input – assimilation – manipulation – analysis – reporting – maintenance – update – feedback – improvement
- data analysis techniques
- possible reporting procedures
- computer-based project continues.

Level 5:
- data management; metadata, quality assurance, integrity, quality, currency
- data availability, accuracy, completeness, timeliness, etc.
- legal issues; data protection, confidentiality, privacy, copyright issues
- computer-based project continues.

Level 6:
- advanced analysis; retail site location…
- other problems; error propagation, data accuracy
- examples of geographic and land information management and systems
- NLIS demo.
- computer-based project completed.

Table 10.3 Skills matrix

Level	DB	DA	DM	GA	DV	HI
1	DB-1	DA-1	DM-1	GA-1	DV-1	HI-1
2	DB-2	DA-2	DM-2	GA-2	DV-2	HI-2
3	DB-3	DA-3	DM-3	GA-3	DV-3	HI-3
4	DB-4	DA-4	DM-4	GA-4	DV-4	HI-4
5	DB-5	DA-5	DM-5	GA-5	DV-5	HI-5
6	DB-6	DA-6	DM-6	GA-6	DV-6	HI-6

Notes
DB = Design and Build; DA = Data Acquisition; DM = Data Management; GA = Geographical Analysis; DV = Data Visualisation; HI = Human Issues and: 1 = Skilled Entry; 2 = Initially Trained Practitioner; 3 = Fully Skilled Specialist; 4 = Team Leader/Senior Specialist; 5 = Senior Manager/Consultant; 6 = Principle Manager/Director.

matrix in which the columns are roles within the GI industry and the rows are skill levels. Table 10.3 illustrates the matrix suggested by the AGI (Capper and Unwin, 1995). Individuals progress through this matrix by moving along the rows or up to another level. The AGI Draft Report contains a detailed description of the requisite skill requirements for each cell of the matrix.

With an understanding of the organisational and technical issues that surround GIS, the property professions will be able to offer advice to clients that is based on an objective analysis of GI and its relevance to property.

Summary

In this chapter we introduced the process of implementing GIS and issues of ongoing maintenance after implementation. We saw that GIS implementation can be undertaken at the project or application level and would typically involve the purchase of a desktop GIS together with in-house and external data. At the other end of the spectrum implementation of a corporate GIS can be a complex, long-term, multi-stage process. Effective project management, coupled with a rigorous, structured design methodology will help to ensure success. Importantly, numerous studies have found that many of the serious problems that stem from GIS implementation are caused by political, organisational or cultural issues rather than technical constraints.

We explored the role of a systems implementation team and suggested that their role, and the role of a GIS project manager in particular, is multifaceted. Political, technical and managerial skills are all required to bring an implementation project through to a successful conclusion and to ensure the successful long-term operation of the GIS facility. At the same time, close attention must be paid to user requirements and ongoing consultation undertaken to ensure that user concerns are considered fully. Ideally, everybody who is to be involved with the project should support it to ensure success. This will require a sustained campaign of liaison between managers, implementation team and the user community. Commitment and support at management level is critical and planning should be for the long term. Many GIS facilities do not recover their costs for some years after implementation, and this is particularly true in the case of large, corporate solutions.

Finally, we saw that GIS may cause far-reaching changes to an organisation's working practices. The location of a GIS facility within the organisation may affect how it operates and how effective it is. Specialist staff may be required to perform the work that the new technology enables and existing staff may require training to maximise their ability to operate alongside a new information systems environment.

Note

1 A NGDF, being co-ordinated by OS, will link each of these national developments. The aim is to ensure that all geospatial data resources and their presentation will be accessible to the user community through the use of standard metadata protocols.

References

Aronoff, Stan (1989), *Geographic Information Systems: A Management Perspective.* WDL Publications, Ottawa.

Association for Geographical Information (AGI), 1996, Guidelines for geographic information content and quality, Association of Geographic Information, London.

Burn, J. and Caldwell, 1990, The management of information systems technology, Alfred Waller.

Capper, B. and Unwin, D., 1995, Professional development for the geographic information industry, Association of Geographic Information, London. Chapter 5 discusses the role of computers and information services within the organisation. Chapter 13 covers the issue of resistance to change in depth, and looks at strategies for overcoming the problem.

Calkins, H., 1997, Geographic Information System Development Guides, New York State Archives, http://www.sara.nysed.gov/pubs/gis/gisindex.htm

Dale, P.F., 1994, Ethics and professionalism, Association of Geographic Information, London.

Dixon, T., 1995, IT skills and education in the surveying profession, College of Estate Management.

Donnay, J. and de Roover, B., 1992, Benchmark issues for the evaluation of hybrid GIS, Proceedings of the 1992 European GIS (EGIS) Conference, Vol. 1, p. 619.

Foley, M.E., 1988 Beyond the bits, bytes and black boxes: institutional issues in successful LIS/GIS management, Proceedings, GIS/LIS 88, ASPRS/ACSM, Falls Church, VA, pp. 608–617.

Hussain, D.S. and Hussain, K.M., 1992, *Information Management*, Prentice Hall.

Huxhold, W.E. and Levinsohn, A.G., 1995, *Managing Geographic Information System Projects*, Oxford University Press.

Joint Nordic Project, 1987, Digital Map Data Bases, Economics and User Experiences in North America, Publications Division of the National Board of Survey, Helsinki, Finland.

King, J.L. and Kraemer, K.L., 1985, *The Dynamics of Computing*, Columbia University Press, New York.

Korte, G., 1997, *The GIS Book: Understanding the Value and Implementation of Geographic Information Systems* (4th edition), OnWord Press.

Local Government Management Board, 1997, Feasibility Study for a National Land Information System, Local Government Management Board, London.

Longworth, G., 1992, *Introducing SSADM_v4*, NCC, Blackwell.

Lopez, X., 1992., Data sharing and spatial information flows within the East Thames Corridor. South East Regional Research Laboratory.

Maffini, G., 1993, The role of public domain databases in the growth and development of GIS, *Mapping Awareness*, **64**(1), 49–54.

Obermeyer, N. and Pinto, J.K., 1994, *Managing Geographic Information Systems*, Guilford Press.

Onsrud, H. (ed.), 1995, *Sharing Geographic Information*, Centre for Public Policy Research, New Jersey, USA.

Openshaw, S., Cross, A., Charlton, M. and Brunsdon, C., 1990, Lessons learnt from a Post Mortem of a failed GIS, Proceedings of the Second National Conference and Exhibition of the Association for Geographical Information, Brighton, AGI.

Parker, M., Benson, R. and Trainor, H., 1988, *Information Economics: Linking Business Performance to IT*, Prentice Hall.

Peel, R., 1995, Fact-track to GIS, *Mapping Awareness*, 30–33.

RICS Education Policy, 1994, Trends in recruitment, education, training and development: a strategy for action, A discussion document prepared by the education and membership committee, March.

Smith, D. and Tomlinson, R., 1992, Assessing costs and benefits of GIS: Methodological and implementation issues, *Int. Journal of Geographical Information Systems*, **6**(3).

Smith, D.R., 1982, Selecting a turnkey geographic information system using decision analysis, Computers, *Environment and Urban Systems*, 7, 335–345. This book provides a complete reference on project management procedures and processes. You might like to read Chapter One, and work through the book as a supporting text for all of the materials covered in this course.

Somers, R., 1990, 'Where do you place the GIS?', *GIS World*, 3(2), 38–39.

11 Future prospects

Introduction

By now you should now have an understanding of the principles of GIS, an appreciation of its functionality and an awareness of the range of applications in land and property management which have benefited from its use. You should also be aware of the major UK initiative to create an NLIS and appreciate some of the information issues that arise when using a GIS to aid land and property decision-making and management. In this last chapter we will draw together the topics that we have covered, summarise our findings and, finally, look towards the future of GIS in land and property management.

The use of GIS in land and property management

The ability of computers to handle geographic data has improved significantly in recent years and GIS are well suited to the integration and management of local, national and international land and property data. Consequently the various sectors of the property 'industry' represent significant potential markets for GIS. In Part I of the book, we introduced the principles that underlie GIS functionality. We looked at some of the most commonly used GIS techniques for visualising and analysing GI, focusing on how these methods might be applied to the land and property sector. We considered the methods by which objects can be positioned on the surface of the earth, and how the information we record about them can be digitally encoded. We introduced vector and raster data structures for storing GI, describing how geographical relationships can be encoded using them. In Chapter 3, we went on to look at methods for visualising and analysing digital geographical data. We considered methods ranging from simple inventory and database query through to more complex overlay procedures to combine data sets, network analysis to investigate accessibility and visualisation procedures including 3D representation of the built environment.

Many people who work in property do so because they want to deal with the built environment. They are interested in the planning, development, construction and management of physical entities, together with the geography of the built environment and the financial, networking and other effects of property decisions. GIS provides a platform on which to model some of these decisions in a way that is intuitive using maps, plans and networks. Given better and more competitively priced data from important data suppliers (a wish that is slowly being granted) and the adoption of the NLPG (reliant on industry-wide support), GIS could become a formidable tool for property information management and analysis.

A GIS is useful simply as a display tool. However, as Fung *et al.* (1996) point out, a GIS is more than just a tool for producing or designing maps. The technology offers potential to analyse large data sets, identify potential development sites, narrow property investment choices based upon a large combination of criteria and view the choices geographically in relation to adjacent properties. A database may be queried and a property that meets a set of required conditions can be identified and mapped together with its attribute information. The technology offers a range of methods to analyse geographical data by manipulating map layers, individually or in combination, to solve geographic problems.

A GIS helps the decision-maker to understand the concept of location. Given the importance attached to location as an influence on all land and property decisions, surveyors, planners, developers, valuers, environmental managers and property managers should regard GIS as a major opportunity for improving the decision-making process. Drummond (1994) notes that GIS can help property professionals create maps that illustrate factors including value, development potential or investment performance. Data may be property-specific or relate to the area such as census and economic data. For example, maps can identify properties for sale, help determine a realistic asking price by highlighting prices within the same geographic area or category as a client's property, or help to choose a retail site by showing vehicle and pedestrian traffic and the potential customers within a 5-kilometre radius. The advantage of GIS over other types of data analysis software is its ability to focus on geographically defined market areas and to integrate and visualise data geographically.

To illustrate how spatially referencing property information is fundamental to property-related decisions, Table 11.1 divides the property cycle into five stages and describes how geo-referenced property information can be used at each stage.

Many people and organisations have invested in GIS because they handle GI. Just about every company has recourse to GI at some point. Many organisations that have implemented a GIS have found that one of its main benefits is improved management of their own organisation and resources. Because a GIS has the ability to link data sets together by geography, they facilitate interdepartmental information sharing and communication. By creating a shared database one department can benefit from the work of another – data can be collected once and used many times.

Table 11.1 How geo-referenced property information is used in the property cycle

Activity	Role of geo-referenced property information
Development opportunity	
Site identification	Select suitable sites (size, layout, aspect, planning, land use, rates, access, neighbouring uses, etc.)
Planning policy	Socio-economic analysis and geo-demographic analysis of population and infrastructure
Development feasibility	Market analysis and economic appraisal of the locality
Development feasibility	
Financial feasibility	Analysis of selected site with regard to market catchment, proximity of suppliers and competitors
Development site feasibility	Planning, legal and competitor searches, government grant availability, regional economic policy, etc.
Environmental survey	Environmental policy support, impact assessments and land quality statements
Design	
Land and hydrographic survey	Digital cartography, boundary disputes, marine development and mineral exploration
Infrastructure design	Identify routes for transportation and utilities' services
Development plans	Local authority planning support; integrate socio-economic, housing, environmental, transport and property data spatially to determine optimum land use and where land use regulation is required (e.g. conservation areas)
Production	
Planning consents	Local authority planning support; notification of applications, planning constraints and histories, local land charges, etc.
Construction programme Infrastructure provision	Contaminated land, geological survey network analysis, pedestrian and traffic flows
Occupation and ownership	
Property investment	Decision support, geographical diversification of a
Valuation and appraisal	property portfolio Impact of location on value; accessibility, proximity to the market, competitors and suppliers, selection of comparable evidence based on locational factors
Property management acquisition and disposal	Management of property assets by large organisations. Conveyancing and agency support (match client's locational requirements with property available)
Regeneration and redevelopment	Analysis of opportunities – back to start of the cycle

Geographical Information Systems applications in property in the UK began in the public sector, in particular, within local government and the utilities. The functions for which these organisations are responsible involve a great deal of data maintenance rather than analysis and GIS was seen initially as a digital replacement for paper maps. Most other organisations are not responsible for large repositories of map data and therefore the decision to invest in GIS must be based on other criteria. The use of GIS in the private sector is at an early stage. Users of GIS in property include retailers, property agents, investment managers and analysts, valuers, tax assessors, corporate property managers, mortgage brokers, lenders and insurers. Uses include map display and presentation, location analysis, rating and valuation, and property market analysis.

Part II looked at some of the ways that GIS has been used in a range of land and property management functions. It provides examples of best practice and experience of 'trailblazers' in various market sectors. When considering the implementation and use of new technology in business, there is no substitute for learning from other peoples' experiences! The five chapters in Part II began with basic mapping and property information management applications. They then gradually introduced applications that involve more analysis and advanced use of GIS. The problems faced when implementing large land and property IS such as the NLIS, the NLPG and the NLUD, as well as local authority property management systems, tend to be organisational and political. As discussion progresses through the chapters of Part II to more analytical applications the problems tend to become more technical in nature.

Take-up of GIS has been slow in the field of land and property management, perhaps due to two reasons in particular. First, cost justification in the public and private sectors for the creation of LIS, property management systems and GIS applications is more difficult when compared to GIS implementation for network management in the utility sector. Second, the selection, installation and management of a GIS for land and property management requires specialist skilled staff or consultants, who are either in short supply or expensive. The complexity of system requirements and the cost of digitisation of paper-based plans and records make the decision to implement a GIS-based property management system a difficult one. However, there are drivers that are encouraging the use of GIS for land and property management: the legislative requirement for government to move towards the electronic delivery of services, the increasing activity surrounding the 'N' initiatives – NLIS, NLUD and NLPG.

Table 11.1 illustrates the variety of professions that can relate to each stage of the property cycle. These professions overlap and need to for the cycle to work, but there is also the need to maintain competitive advantage. This restricts the amount of information sharing and consequent business opportunities that this might offer. There are several professional bodies that represent the built environment professions so, politically and legally, the property 'industry' may and often does not speak with one voice. The

RICS has strongly supported the development and uptake of GIS and the sharing and dissemination of land and property data amongst the property professions, as have the RTPI. Others, such as the Law Society, which lobbied against the reinstatement of price paid on the Land Register, have different agendas.

Individuals and businesses all have a requirement for land and property to some extent and therefore property advisors act on behalf of a great variety of clients. Consequently, the need for mapping and geographical analysis is client as well as task-dependent. Retail occupiers and other location-specific businesses may be more inclined to pay for geographical analysis in support of their property advice. Others, such as the mineral extraction industry or storage facilities may find sophisticated geographical of market opportunities less appropriate. But property is increasingly regarded by occupiers, be they tenants or owners, as a key asset/liability. Changes to accounting rules mean that the long-term asset value/liability charge of property must be reflected in company accounts. For many business sectors property is often the second highest expenditure/outgoing after staff. Given that urban rent theory proves that occupiers pay more for land and property in certain locations (such as city and town centres of other dense urban areas), the use of location analysis along the lines described in Chapters 8 and 9 is valid. The property professions are therefore uniquely qualified to be users and innovators of a technology that is designed to manage and analyse geographically referenced land and property information.

New 'information' markets in land and property management

It is difficult to quantify financially the implications that developments in digital mapping, GIS and the Internet will have on the property profession. It is generally agreed that income or fee generation from brokerage may reduce if the use of Internet-based property listing services increases significantly. What is more difficult to predict is the number and financial value of new professional services that may germinate from the use of these emerging technologies. This uncertainty that surrounds possible threats and opportunities to the future of land and property management functions manifests itself as a reluctance to share information. Unfortunately this sharing of information and data is exactly what technologies such as GIS depend upon. As Adair *et al.* (1997) comment 'the net outcome of this shifting balance of surveying business is considered to be unclear at this stage and information sharing therefore remains a concern for the surveying profession'.

Information about land and property is fundamental to an industry that is concerned with the natural and built environments. In March 1994, the RICS Education and Membership Committee stated that 'The growing importance of IT in areas of practice such as GIS is liable to have a significant effect on the needs of many employers' (RICS, 1994). So, to try and address the reluctance to share data, the RICS has been involved with a number of initiatives that encourage wider access to land and property data: the Domesday 2000 Initiative and NLIS are two examples. The RICS has produced a policy paper on the availability of land and property information that stresses the importance to surveyors of initiatives that attempt to improve access to land and property data. Experience in Australia supports this view. Sharma *et al.* (1996) summarise the position neatly with the following statement. 'In 1995 there is not one of Australia's 17.5 million people whose life is not influenced in some way by GIS technology, nor one of its 7.7 million square kilometres of land that is not administered or managed through the use of GIS'.

Enthusiastic demand for the information generated by national land information management initiatives has been seen repeatedly. The Swedish Land Data Bank, the US National Spatial Data Infrastructure and UK government data dissemination services like the National Statistics 'Neighbourhood Statistics' project or the Land Registry's online house price information service have all seen widespread uptake. This suggests that, if the reluctance of some parties can be overcome, there are wideranging opportunities for the brokerage and application of land and property information via the distribution channels of the new information sector.

The NLPG, and BS7666 on which it is based, were introduced because local authorities experienced huge problems referencing and describing the land and property data that they are required to collect and maintain in support of their statutory functions. Each component of BS7666 provides a valuable resource for information integration and the undertaking by local authorities to adhere to the standard is a significant step forward. The national adoption of BS7666 will ensure that each unit of land and property can be unambiguously identified, which enables data to be shared with confidence and allows data held by different organisations about the same property to be integrated. It is this process of combining data that will unlock the latent value in those data. Construction of a BS7666-compliant NLPG is progressing. Once complete, it will be a substantial resource, not only as a prerequisite for the NLIS, but also in providing a referencing framework that enables the sharing of land and property data and the development of new applications and business opportunities in the private sector. The OS's MasterMap initiative is a valuable parallel resource that will make 'intelligent' topographic mapping data including basic information about the extent of land parcels and buildings available to the land and property sector. Exactly how the two will interface remains to be resolved.

As progress is made towards an NLPG and an NLIS, an increasing volume of property data will become available. Property professionals will have access to these data and it may be a case of deciding which markets will be the focus for professional activity in the future. The two traditional markets of real estate agency (brokerage) and property consultancy may be joined by a third, an information market. Of course this market will mean new competition, from publishers and specialist data providers, but the property professions must decide where they would like to operate in the future. Already, property professionals are finding employment and careers in areas where new technologies such as GIS are used to manage land and property information. But the property industry has not yet fully appreciated the benefits of collecting, storing, analysing and presenting land and property information using a GIS. Even where GIS implementation has been widespread, in the local government sector for example, the use of GIS has largely been to automate operational procedures that have an inherent geographical dimension.

Part III examined how data sets should be managed in order to maximise their value and to inform users as to whether they are fit for the purpose intended. In this way data sets can be traded with confidence in a land and property information market. The assembly of data and the linking of data sets has legal implications. As in all aspects of professional work using data resources, care must be taken over the terms and conditions of copyright, licensing agreements for software and data and the potential caveats caused by data protection legislation. Perhaps the time has come for a new breed of property professional, the 'land information manager' envisaged by Dale (2001), who can understand and appreciate issues of land and property management in the context of computerised geographical information resources and can apply those resources in working towards more effective decision-making.

The UK land and property management professions have worked very well in spite of the lack of readily available disaggregated data on land use, value and ownership. With historical roots that often lie on the eighteenth and nineteenth centuries, professional bodies have created informal networks of information exchange that have benefited both client and advisor. But, in the information economy of the new millennium, things are changing rapidly and significantly. For example, the Land Register of England and Wales, established in 1925 and which includes details of legal ownership, is now publicly accessible by anyone – not just the landowner – and the price paid at the last transaction also appears on the Register since the late 1990s. The Registry publishes average price paid by property type down to the level of postcode sectors on its website, making the value mapping of residential property at quite local levels and time-series tracking an achievable aim in the near future.

So far, big changes to traditional land and property activities such as conveyancing, development and planning have not occurred. These areas may see great changes over the next few years. This book has described many of the initiatives that are going on out of the clients' eye – changes to back-office functions and the way in which analyses are carried out in support of client advice. In the public sector the Land Registry and the OS have invested substantial funds in the development of electronic land registration and searches and large-scale digital mapping. In the private sector, companies like Environmental Systems and Research Institute (ESRI) UK with their

Searchflow electronic conveyancing service and Landmark's environmental investigation service are GIS-based. But the GIS is behind the scenes. It is the quality of the geographical analysis that is evident. Expansion and integration of initiatives such as these is largely dependent on a single catalyst – the completion and widespread adoption of the NLPG.

National initiatives for land and property information

In almost every society, there is a need to manage property information for administrative and planning purposes and the creation of an NLIS is an aid to that management function. There are clear advantages to be gained from a national service that disseminates comprehensive and reliable land and property data quickly, cheaply to those who require them. At the very least such a service should include data on ownership and legal interests, use, development control and planning and property values and prices. These data could be combined with social, environmental, economic and other data.

The process cannot be hurried – Switzerland has proposed to upgrade its cadastral system, which some already regard as the best in the world, and it is estimated that this process will take thirty years and around 2 billion ECUs. The Swedish Land Data Bank took over ten years to complete. In all cases, the only way that a multi-purpose NLIS has been able to succeed is with unrestricted access to data. Such systems are doomed if the data that they contain are not comprehensive, accurate, up-to-date, regularly maintained and reasonably priced.

The technical ability to create an NLIS in Britain is in place. Chapter 5 described the NLIS initiative currently under way and Chapter 10 described some of the issues that need to be resolved before an NLIS can become a reality, many of which are now being addressed. They include:

- the adoption of standards in order to integrate the data sets that are maintained by different organisations;
- the introduction of legislation to release key property information, such as transaction data, into the public domain;
- the protection of personal privacy and commercial sensitivity;
- the availability of sufficient finance;
- public and private sector support;
- the satisfactory allocation of responsibility for data ownership, integrity, maintenance of data and liability for incorrect data; and
- the establishment of a pricing policy for access to and extraction of land and property data via the NLIS.

The Conveyancing Application of NLIS, described in Chapter 5, is helping to address some of the above issues and the Bristol-based pilot has provided valuable empirical results of NLIS operation, which has potentially far-reaching implications for the property market.

It is possible to develop an NLIS that does not contain a detailed survey of land parcel boundaries (as in the UK). In such cases, a parcel identifier will suffice for land registration and property taxation purposes. However, research suggests that the mapping of boundaries is necessary in the long term (although this has not occurred in the UK yet). The choice of property identifier is crucial if an NLIS is to be extended to a true multi-purpose cadastre. In the UK there are three land and property identifiers that can legitimately be described as national:

- the UPRN used in the NLPG and maintained by local government;
- the TOID used in MasterMap digital map data and maintained by the OS;
- the Title Numbers used in the Land Register and maintained by the Land Registry.

The relationship between these three identifiers is very important if we are to achieve a 'joined up geography' relating to land and property information management. In the UK the merger of the Land Registry, OS and the VO would help resolve the discrepancies between these identifiers and encourage a single government department to take responsibility for the provision of public sector land and property information in NLIS, but such a merger would have far-reaching policy implications. Currently, there are overlaps of functions between the three owners of the above identifiers. For example, when creating the topographic units in MasterMap the OS inserted many thousands of inferred links to separate land parcels that are under separate ownership. This resolves the problem of having a single polygon on a map representing the open plan gardens belonging to, for example, a dozen houses. But ownership is not a topographic feature and, strictly speaking, should not be delineated by topographic units. In fact, ownership boundaries are not legally delineated in the UK at all (except in a few rare cases) but the recording of ownership title and description of legal extent of ownership is the responsibility of the Land Registry. It would seem logical, therefore, for the OS and the Land Registry to collaborate on the creation of these inferred links.

Another example is the charging policy for the NLPG. The NLPG is an amalgamation of LLPGs created and managed by local authorities. Via the Local Government Information House, local authorities are able to recoup expenditure on maintaining LLPGs by charging users. But the OS owns the National Street Gazetteer, which is an integral part of the NLPG. What happens if the Land Registry and the VO, whose address databases helped compile the NLPG, want revenue too? Without co-ordination and integration of central and local government departments and agencies, organisational barriers may prevent what is technically feasible. Dale (2001) suggests that now is the right time for NLIS, NLPG and NLUD initiatives to be integrated into a single national multi-purpose cadastre. This cadastre should be based on general boundaries with the basic spatial units as recognised in the NLPG, rather than OS Mastermap land parcel identifiers, because of the former's ability to record property interests that exist in three dimensions.

The NLUD initiative has been on the drawing board for a long time. Perhaps it has languished as a purely public sector inititative, driven (rather slowly) by a central government department. A solution might be for property professionals in the private sector to pool land use data with colleagues in local authority planning departments and the VO using the NLPG as the foundation. The NLPG is superior to OS MasterMap for the purposes of property management because it can handle land and property interests in 3Ds, for example, a shop with a flat above. Such data sharing, as a commercial venture, would be a good example of a Public Private Partnership.

So there is an overall trend towards multi-purpose NLIS (Larsson, 1991). However, the question of whether international systems are yet achievable arises. Dale suggests that the challenge for the UK NLIS is to incorporate geographical analysis as well as simple data inventory. Currently, few multipurpose LIS with advanced GIS capability are operating successfully on a national scale. Despite this, there are already moves towards a global data infrastructure. For a global geographical data infrastructure to arise, technical standardisation will also have to be international, requiring organisations with a global focus. A recent example is the OpenGIS Consortium (OGC), whose aim is 'to facilitate transparent access to heterogeneous geodata and geoprocessing resources in a networked environment'. OpenGIS has already been cited as an enabling technology for the US NSDI by the Federal Geographic Data Committee (FGDC, 1994), and OGC liaises closely with ISO. Although 'new technology ... has greatly increased the potential for developing (multipurpose) systems' (Larsson, 1991), management problems, in particular those associated with standards, IPRs, political and legislative frameworks must be considered fully if a global land information infrastructure is to be realised. Whether a truly global approach is possible or practical will become clear during the first years of the new millennium.

New methods of data visualisation and exploration

Over the next few years we may see the disappearance of GIS as a separate piece of ICT as it becomes integrated with more mainstream database and spreadsheet software. A GIS does no more than geographically 'enable' a database of land and property information. Therefore, geographical analysis of this information should be an extension of typical database and spreadsheet analysis. Products like MapPoint or InfoMap are already steering towards this goal, although these packages typically lack much of the specialist functionality of full-blown desktop GIS software. Often the results of geographical analysis are analysed further and reported in spreadsheet, statistical and database environments, respectively, and a closer coupling of these technologies is probably inevitable. When deployed in tandem, such technologies could greatly enhance the integration of property and business data, allow for high quality property performance measurement and more proactive property management.

Geographical Information Systems have made it easier and cheaper for users of geographical data to explore and analyse complex geographical phenomena. As the technology has moved from laboratory to desktop, this capacity has certainly increased and will continue to do so as computer systems become more powerful and our capability to manipulate and visualise large amounts of data simultaneously improve. Three-dimensional visualisation, complex modelling and interlinking of database systems are now standard components of the GIS software arsenal. We must hope that alongside these new tools the GIS industry and the university sector that develops them and trains new GIS practitioners will also provide suitable instructions and caveats on use. The tools of the future will only empower the GIS user if their limitations and practical applications are fully understood. In general, the movements of the GIS industry in this regard are encouraging. The documentation of GIS data sets is now a standard component of most desktop systems and there is clear recognition that an awareness of the limitations of data is paramount if effective decisions are to be made. At the same time, care must be taken to preserve the integrity that was built into old maps. As the potential for data manipulation becomes more flexible, effort is needed to improve the quality of data and the methods whereby they are subjected to changes in scale.

GIS and the Internet

The Internet is a phenomenally successful communication tool and a major contributing factor to the freeing up of land and property information. It is revolutionising the delivery of data and services. Indeed, GIS functionality will increasingly be accessible via the Internet. Adair *et al.* (1997) state that 'there is consensus that the advances made in information technology with almost limitless data transfer and handling capacity, geographic information systems and the communicative power of the Internet would lead the way forward'. Examples of the deployment of Internet-based geographical analysis and display tools have been included throughout Part II. Perhaps the most sophisticated of these can be found in Chapter 7 where 3D models (developed using GIS tools) can be viewed interactively and in real time using a virtual reality browser over the Internet.

The practicalities of Internet GIS have been limited by problems of bandwidth and processing power to deliver GIS data structures rapidly. However, these problems have largely been overcome and many of the major GIS software houses have produced solutions designed to serve map data across the net. The future of mapping for land and property information may well include virtual environments including animated floor plans and 3D graphical models, linked to complex distributed databases and all served via the Internet in a fully interactive, real time framework. The components of such a scenario are already in place.

The majority of property professionals work in small practices and it is here that latent demand for web-based property information services may exist. For example, the Land Registry records approximately 250,000 – 300,000 transactions each quarter. In 1994, for example, there were 1.27 million conveyances recorded in England and Wales. Before the conveyancing process and subsequent transaction each property must be marketed and valued. It is in these processes that web-based property applications offer significant potential. However, the implementation of web technology that incorporates digital mapping and GIS in an industry that is characterised by a large number of small firms is not without its problems. For example:

- there are very few large firms to make substantial investment in IT;
- awareness of potential benefits of IT are difficult to disseminate – this is substantiated by feedback from NLIS presentations;
- few personnel in small- to medium-sized firms are employed as dedicated IT specialists, therefore IT should be easy to learn and use for all practitioners;
- it may take a long time for a particular IT implementation to reach critical mass;
- pay-back on IT investment by small firms will have to be quick or implementation costs should be phased or by subscription (cf. Promap and Focus).

Some of these points are taken up by Adair *et al.* (1997) who state that 'the structure of the market includes large numbers of small local firms, the perception being that this encourages a degree of general secrecy and only a limited form of information sharing based on informal networks'.

A mobile future

As well as the potential for increased data sharing due to web technologies, an important development is the enhanced capacity for mobile information management presented by improvements in mobile communications technologies and portable computing platforms. The use of such methods in the property sector has been hindered by a static data capture process, requiring agents to return to the office to retrieve data. With the development of cost-effective real time tracking technologies like the GPS and the availability of palm top computing platforms linked to mobile communications systems it is now possible to record the geographical location of property resources alongside other key information in near real time and to transmit these data directly back to a corporate headquarters. The full impact of real time data capture and retrieval is still to be realised, but these technologies represent important opportunities to review and perhaps revolutionise working practice.

Conclusions

Land and property professionals, along with everyone else, face a world that is changing socially, economically, politically and environmentally at an ever-increasing pace. In particular, the following global challenges are pertinent to the subject matter covered in this book:

- The development of ICT in terms of its speed, capacity, portability, format and availability, the mushrooming availability of digital data (mapping and property data), reductions in the cost of ICT and data, and client demand for more rigorous analysis mean the world is becoming more 'data-centric' (McKeon, 2001).
- Economic reform of companies and their relationship with national governments and continental and international parliaments and trade organisations, as well as their effect on society and the environment.
- Globalisation of trade, culture and professions. With population increase and industrialisation continuing to exert pressure on the land it is more important than ever that information about geographical patterns, relationships and trends can be collected, stored and processed efficiently.
- Development of the built and natural environment that is sustainable. Pressure on land and property resources has never been greater. Countries at all stages of economic development are faced with difficult decisions concerning the allocation of resources for economic output, environmental needs, planning, housing and infrastructure provision. These decisions are growing in complexity because of the need to match diverse requirements to an increasingly limited land resource.

Looking back over the past fifteen years or so at the use of GIS in land and property management, a series of events have combined to produce significant opportunities for the future:

- The realisation of cost effective hardware and software technologies to handle mapping and geographical analysis.
- Technology-led application of GIS to land and property management functions.
- Rapid development of technological capability of GIS software to meet burgeoning demands of users.
- Realisation of data limitations in terms of availability, cost to digitise in-house and/or purchase from a third party.
- Substantial land and property data digitisation activity, especially in national mapping and land registration agencies.

- Realisation that the lack of executive level awareness of the potential of GIS in land and property management, together with a shortage of appropriate skills, were restricting the development of GIS applications in land and property.
- Motivation to overcome awareness and skills issues via national initiatives that encourage data to be shared and organisations (in the public and private sectors) to work together in new land and property information markets.

The list shows that, technologically, few barriers stand in the way of the widespread application of geographical analysis techniques to land and property data. There are unprecedented opportunities for deploying these techniques to improve decision-making. Data issues are widely recognised and great efforts have been made to address them. In many countries, these efforts are still underway. At the same time, awareness amongst owners and users of data is improving but there is still some way to go. The organisational and political factors that can affect the deployment of GIS technology remain significant barriers to effective implementation, and there is still substantial scope for improving the way in which information resources are documented and their quality reported.

Dale (2001) argues that the NLIS, NLPG and NLUD in the UK are being launched because of deeper concerns for the environment and the commercial needs and opportunities for development. A similar purpose lies at the heart of the US NSDI initiative and parallel developments in Australia, New Zealand and Europe. There are many ways that data about land and property can be used to enhance our understanding of the land market and to support more sustainable development.

Our exploration of the use of GIS in the property sector demonstrates that laudable efforts have been made to develop technology capable of managing, analysing and presenting land and property information in a geographical context. Some of the ways in which this technology is now being used in the property sector have been described in this book and there appear to be considerable opportunities for the future. It remains to be seen whether the wider land and property management sector will embrace these opportunities and the changes in work practice that they will require and move towards an environment of information-rich property analysis and better quality decision-making as a result.

References

Adair, A., Berry, J., Deddis, B., McGreal, S., Keogh, G. and Key, T., 1997, Data sharing in the surveying profession: barriers, trends and implications, RICS Cutting Edge Conference, September, Dublin Institute of Technology.

Dale, P.F., 2001, Geomatics: yesterday, today and tomorrow, Proceedings of a technical meeting held at University College London, 21 September.

Drummond, W., 1994, Extending the Revolution: Teaching Land Use Planning in a GIS Environment, *Journal of Planning Education and Research*. **14**(4), 280–291.

Fung, T., P.C. Lai, H. Lin and A.G.O. Yeh (eds), 1996, *GIS in Asia*, Selected Papers of the Asia GIS/LIS AM/FM and Spatial Analysis Conference held in Hong Kong on 28–30 March 1994 (GIS Asia Pacific, Singapore).

Larsson, G., 1991, *Land Registration and Cadastral System: Tools for Land Information and Management*, Longman, Harlow, UK.

McKeon, A., 2001, What stands in the way of e-commerce? Poor data quality, *GI News*, September, 42–44.

RICS, 1994, Trends in recruitment, education, training and development: a strategy for action, Discussion document, Education and Membership Committee, Royal Institution of Chartered Surveyors.

Sharma, P., Puller, D. and McDonald, G., 1996, Identifying national training priorities in GIS: an Australian case study, GIS Conference, Columbia MA.

Glossary

AM/FM Automated Mapping/Facilities Management. This is a specific application of GIS to the management of utility networks such as pipes, cables, etc.

Area Homogeneous extent of the Earth bounded by one or more line features to form a polygon.

Attribute An item of text, a numeric value or an image that is associated with a particular spatial feature.

Buffer A zone of user-specified distance around a spatial feature and used to establish the proximity of features, for example, to find areas of development land less than 2 kilometre from a motorway.

Cadastre A data set containing information related to land ownership, use and value. This usually takes the form of maps and descriptions of uniquely identifiable land parcels. For each parcel, legal information such as ownership, easements and mortgages are recorded along with ownership and values for taxation purposes.

Cell The basic element in a raster data set.

Centroid The centre point of a polygon, often used to attach attribute information to an area.

Connectivity The topological identification of connected arcs by recording the from- and to-node for each arc. Arcs that share a common node are connected.

Contiguity The topological identification of adjacent polygons by recording the left and right polygons of each arc.

Co-ordinates Numbers representing the position of a point relative to an origin, for example, National Grid co-ordinates.

Database A logical collection of related information, managed and stored as a unit.

Data dictionary A catalogue of all data held in a database.

Data model A formal representation of the real world.

Data set A collection of logically related data items.

DEM Digital Elevation Model (or Terrain Model). A data model used to represent a 3D surface, often based on a grid with a height value for each cell, or on a set of irregular triangles (see TIN).

Digitising Conversion or encoding of existing maps from paper form into digital form as x, y co-ordinates, for example, via a digitising tablet or scanner.

Entity A collection of objects (people, places, etc.) described by the same attributes.

Entity relationship diagram A graphical representation of the entities and the relationships between them.

Geo-code The attribute in a database used to identify the location of a particular record, for example, a postcode.

Geographical data The locations and descriptions of geographic features – the composite of spatial and attribute data.

Geographical database A collection of spatial and attribute data organised for storage and retrieval in a computer system.

Geo-reference Relationship between page co-ordinates on a planar map and real-world co-ordinates.

Gravity model Used to model the behaviour of populations, the underlying assumption is that the influence of populations on one another is inversely proportional to the distance between them (cf. gravitational attraction from Newtonian physics).

Land line Large-scale digital map data available from the OS (now replaced by Mastermap).

LIS Land Information System for the management, analysis and presentation of information relating to land, including ownership and legal rights. Often an automated development of the cadastre.

Line A set of co-ordinates that represents a linear geographical feature such as a road, stream or railway.

Map A representation of the physical features of a portion of the Earth's surface graphically displayed on a planar surface.

Map projection A mathematical model used to convert 3D reality into two dimensions for representation on a map, or within a 2D GIS database.

Map scale The measure of reduction between the representation and the reality, be it a map or a spatial database. Scale is usually represented as a representative fraction of distance, for example, 1:50,000, 1 unit of distance on the map representing 50,000 units in reality.

National Topographic Base The formal title of the OS's national map archive.

Network A model representing the interconnected elements through which some form of resource can be transmitted or will flow. In GIS this is represented as a series of nodes connected by arcs, each or which has attributes representing flow characteristics, for example, a road or pipeline network.

Node A basic spatial entity within the vector data model, which represents the beginning or end of an arc. Also, a node may be formed when a number of segments join. For example a node might be represented in a road network as a highway intersection.

Point   A spatial entity that represents the simplest geographical element. Represented in the vector data model as a single x, y co-ordinate, and in the raster as a single cell. The point may have associated attributes, which describe the element it is representing.

Polygon   A representation of an enclosed area defined by an arc or a series of arcs that make up its boundary. Polygons may have attributes describing the area they represent.

Raster   A data structure composed of a grid of cells. Groups of cells represent geographical features; the value in the cell represents the attribute of the feature.

Relational database   A method of structuring data as collections of tables that are logically associated with each other by shared attributes.

Scanning   A data capture technique, which digitises information from hard copy into digital raster data.

Spatial analysis   Spatial analysis is the process of applying analytical techniques to geographically referenced data sets to extract or generate new geographical information.

Thematic map   A map, which communicates a single theme or subject. For example, a population density map and political boundary map. This contrasts with a topographical map which is a general purpose map containing landscape features such as rivers, roads, landmarks and elevation.

Topology   The spatial relationships between connected or adjacent geographical objects. Topology is used to apply intelligence to data held in the vector data model. For example, topological information stored for an arc might include the polygon to its left and right, and the nodes to which it is connected.

Vector data   A data model based on the representation of geographical object by Cartesian co-ordinates, commonly used to represent linear features. Each feature is represented by a series of co-ordinates which define its shape, and which can have linked information. More sophisticated vector data models include topology.

Index

3D modelling 190, 193, 196

abscissa 23
accessibility 202, 208, 212, 276
accuracy 32, 286–7
aerial photography 139, 193
angle of grid convergence 24
Aon Corporation 233
Appraisal Institute 272
Architecture Foundation 195
ArcView GIS 193
area profiling location 206
areas *see* geographical features
areas of Ancient Woodland 151
attribute combination, methods of 75
attribute data 39
Audit Commission 123, 124, 126
Aylesbury Vale District Council
    127, 128

Barbican 267
barometric pressure map 35
Basic Land and Property Unit (BLPU)
    92, 318–21
basic spatial unit 101
benchmarking 349–51
Benfield Grieg Ltd 232, 233
Birmingham City Council 131, 174
breakpoints between map classes 60–1
Bristol City Council 105, 118,
    131, 180
Bristol offices 257
British Geological Survey 187,
    189, 231
British National Grid 18, 21, 24–7
brownfield sites 159, 165
BS7666 92, 105, 106, 127, 128, 129,
    136, 139, 176, 181, 246, 319–24
buffer zones 68, 72–4, 76

building society 207
business case for GIS 338–42

CACI 211, 214, 218
cadastre 96, 102, 103
Canary Wharf 185
car dealerships 207
Cardiff City Council 133
cartographic symbols 28
Catalist 208
catchment area 202, 206–8, 210–12,
    214, 220
Central London Office Forecasting
    Service 245, 246
central statistical area 162, 163
Centre for Advanced Spatial Analysis
    (CASA) 161, 191, 193, 196
Character Area 151
Charles Planning Associates 183
Chartered Institute of Public Finance
    and Accountancy 123, 126
choice of colours on maps 60
choropleth map 34, 57–9
City of Vallejo, California 134
class interval definitions on maps 59–62
combinatory geographical overlays 76
combining data geographically *see*
    overlay analysis
Common Agricultural Policy 149
competition analysis 207, 208
completeness 288
Computer Aided Design (CAD) 8,
    140, 191
Computer Aided Facilities
    Management 140
computer hardware 11–12
confidentiality 303–8
conformal projection 22
consistency 288

consumer spending 210
contaminated land 187
continuous characteristics 30
contour lines 30, 36, 38
contour maps 34–6
contributory rule 75
conveyancing 5, 109, 111, 117, 133
co-ordinate systems *see* spatial
   referencing
copyright *see* Intellectual Property
   Rights
corporate GIS 335
corporate property database 180
Corporation of London 193
costs and benefits of GIS 338–42
Cotswold District Council 168
council house sales 129
council tax 221
Countryside Agency 151
crime modelling 234
currency 288
customer profiling 207, 219

data: access to 298, 302–8; definition
   of 5; for Geographical Information
   Systems 13
Data Base Management Systems 7
database information 14
data capture 43, 178
data integration 284; factors
   affecting 284–5
data misuse 295–6
data modelling 7–13
data ownership 298
data protection 301–2
data protection legislation
data quality 284, 285–8; assessment
   of 297; description of 313–15; and
   the display of land and property
   data 294; of land and property
   data 289–94; measurement
   of 286–8; and the National
   Land Information Service 297
data standards 284, 309–18
data structures 39, 76; definition
   of 39; organisation of 39; and
   overlays 76; raster *see* raster data
   structures; vector *see* vector data
   structures
data visualisation
datum 30
Debenhams 206
Defence Estate 139, 140
demography 201

Department of the Environment
   (DoE) 170
Department of the Environment,
   Transport and the Regions
   (DETR) 165
development 5
development agencies 219
development control 181
Digital Elevation Model (DEM) 186
digital imagery 14
digital mapping 179
digital maps 13, 32; scaling of 32
direction of flow 43
discrete characteristics 30–1
distribution network 208
Domesday 2000 105
dominance rule 75
Donaldsons 214
drive-times 202, 211, 212, 214, 218,
   219

Eastern Thames Corridor 126
Easting 24
East Lindsey District Council 182
EGi London Office Database 221, 245
EGPropertyLink 113
English Heritage 91, 155, 156
English Nature 151, 187
English Partnerships 109, 165, 185
Environment Act, 1995 164
Environment Agency 153, 154, 179, 187
environmental risk 187, 188
equator 18–19, 24
equidistant projection 22
Ernst & Young 219
error propagation 294
errors 290
Essex Sites and Monuments
   Records 156
Estates Gazette 113
Euclidean distance 202

facilities management 140, 141
feasibility studies for GIS
   implementation 337–8
financial service providers 205, 234–5
flooding 154, 186, 188, 232, 233, 234
focus 117, 245
food retailers 204, 205
Forestry Commission 148
forward planning 181
FPDSavills 220
freedom of information 302–3
functionality 8, 13

General Boundaries Rule 100
generalisation of cartographic data 31–2
GeoBusiness Solutions 200
geodemographic analysis 201, 204
geographical analysis 64–80
geographical features 29; area
features 29; line features 29, 42;
point features 29
geographical information: definition of
5; and visualisation 80
Geographical Information Systems 4,
8–10, 14–16, 18; core functionality of
11–12; and databases 15; definition
of 6; hardware and software for
11–12; implementation of 331–58;
and information presentation 54–64;
maintenance of 333
geographical relationships 64
geoid 19–20
geology 189
geo-referencing *see* spatial referencing
GIS professionals 362–4
Glenigan 221
Global Positioning Systems
(GPS) 17–19, 21, 148, 149
Global Video 209
Gloucestershire County Council 133
Goad plans *see* retail Goad plans
graticule 21, 24–5
gravity modelling 204, 210, 212
greenfield sites 159
Greenwich Meridian 18–19, 21–4
Grid Co-ordinate Systems 18, 23–6;
origin of 23
grounds maintenance 131, 180, 181
ground water protection 154

Hackney Building Exploratory 195, 196
hazardous materials 188
Henderson Global Investors 234
Highways Agency 136
Historic Landscape Assessment 154
Historic Towns Survey Project 156
Home Energy Conservation Act 131
Horsham District Council 179
housing allocations 169, 181
housing association 134, 135
housing database 179
housing developers 185

implementation of national land and
property initiatives 358–60
Improvement and Development
Agency 108, 165

Index Map 100
information 5
information and communication
technology (ICT) 6–7
information management 283
information systems 6–7
insurance 231, 233
intangible features 31
Integrated Administration and Control
System (IACS) 149
Intellectual Property Rights 298–301
Intelligent Addressing 94
Intelligent Space Partnership (ISP) 258
interaction rule 75
Inter-Departmental Business
Register 162
Internet 377
Internet GIS 190
investment perfomance
measurement 221
Investment Property Databank 214,
240, 249
Invitation To Tender (ITT) 346–9
Ireland 113
Isle of Wight 156
isochrone 35
isopleth 34

Japan 113
Dr John Snow 88
Jones Lang Lasalle (JLL) 95, 211, 214,
218

KPMG 106, 107

labour market areas 220
Lambert Smith Hampton 256
land 4
land charges administration system
130, 180
Land Cover Map 155
landfill sites 188, 189
land information system (LIS) 9–10,
101, 102
Landline 89, 155
Landline® data sets 45
land management 5, 150
Landmark Information Group 187
land parcel 96, 127
land and property identifiers 92
land register 96
Land Registry 91, 92, 98, 105,
108–10, 118, 127, 129, 137, 149,
180, 185, 221, 246

land taxation 103
land tenure 96
land terrier 124–31, 139, 140, 179–81
land use 187
latitude and longitude 18, 21, 22, 25;
　definition of 18
leisure operators 219
liability 295–6
licensing agreements *see* Intellectual
　Property Rights
lineage 287
lines *see* geographical features
link and node structure 42–3
listed buildings 180
Lloyds Chemists 206
local authority property management
　121, 124
local government 177
local land charges 105, 109, 118,
　180, 181
local land and property gazetteers 92,
　127, 128, 129, 131, 176, 177, 181
local planning authorities 176, 177
location value response surface 277
London Borough of Barking and
　Dagenham 128
London Borough of Hackney 87
London Borough of Kingston 87, 180
London Transport Property 139
longitude *see* latitude and longitude
Lothian Regional Council 169

maintaining geographical information
　systems 352–5
MapInfo 211
MapPoint 211
map production 176, 177
map projections 6, 21–6
map scale 27
market research 110, 201, 202
Marks and Spencer 206
Mastermap 90
Mercator 24
Meridian 20–1, 24
metadata 313–15
Metropolitan Life Insurance
　Company 236
Ministry of Defence 139
mobile data capture 378
modelling 7, 13; and GIS
　databases 37–8
motorway corridors 255
Multimap 113, 214
multiple regression analysis 204, 273

National Assembly for Wales 154
National Association of Realtors 110
National Buildings Data Set 91
national grid *see* British National Grid
national initiatives for land and
　property information 374
National Land Information System
　(NLIS) 104
National Land and Property Gazetteer
　(NLPG) 92, 103, 109, 315–24
National Land Use Database
　(NLUD) 109, 160, 164, 165
National Nature Reserve 151
National Parks 122
National Survey of Local Shopping
　Patterns 210, 211
National Topographic Database 89, 185
Natural Area 152
Network Rail 139
new information markets 371
NLIS channels 109
node 42, 43
non-food retailers 205, 206
Norfolk County Council 182
northing 24
Nottinghamshire County Council 152
Nuclear Industry Radioactive Waste
　Executive 189

Occupier Property Database (OPD) 258
Office of the Deputy Prime Minister
　(ODPM) 160, 161, 165
office location planning 217, 221
Office for National Statistics 162
ordinate 23
Ordnance Survey 89, 92
organisational issues for GIS
　implementation 360–2
outlet strategy 205
overhead transmission wires 188
overlay analysis 69, 72–4; combinatory
　overlay 76; and data structures 76;
　line on line overlay 69; line on
　polygon overlay 69; point in
　polygon overlay 69; point on line
　overlay 69; point on point
　overlay 69; polygon on polygon
　overlay 69; practical examples
　of 71–4, 76–80; raster overlay
　76–80; and set theory 69–70

panoramic imaging technology 196
paper maps, limitations of 36–7
parallels 20

pedestrian modelling 258, 266
petrol stations 208
pilot projects 344
planning application systems 177, 179
planning policy 160
planning procedure 169
points *see* geographical features
polygons 43
positioning *see* spatial referencing
postcode 6, 139, 201, 204, 208, 211
Postcode Address File (PAF) 327
Powys County Council 179
Pre-Build Information Service 185
Precision 32, 287
precision farming 149
procurement policies for IT equipment 349
project definition 332
Promap 188, 211
property 4–5
Property Advisors to the Civil Estate (PACE) 91, 135
property consultants 208
property data: coverage of 290; strengths and weaknesses of in the UK 289
property development 182
property information 6; and the property cycle 369
property investment 235
property management 4
property management systems 7, 124, 127, 131
Property Market Analysis (PMA) 240
Property Market Information Service (PROMIS) 240
property portfolio management system 130
property valuation 270
property value maps 267

radioactive waste 230
radio masts 258
radon 188, 189
raster approaches 39
raster cells 47–9
raster database 76
raster data sets 76
raster data structures 45–6, 48–9; benefits and limitations of 49; comparison with vector data structures 50
raster layers 46, 77–8

raster orientation 47–8
raster resolution 47
real estate 4
Realtors Information Network 110
regional planning 169
Registers of Scotland 108, 137
registration of deeds 96
registration of title 96
relief map 30
representation of points, lines and areas 40–4
representative fraction 27
Residential Property Price Report 246
retail development 258
retail Goad plans 204, 208, 211, 240
retail location analysis 202, 204, 206, 209, 210, 211
retail rent forecasting 210
right-to-buy 129
risk management 233
Riverside Housing Association 134
River Thames 186
Royal Institution of Chartered Surveyors (RICS) 105, 107, 123, 124, 126
Royal Mail/Post Office 139, 179
Royal Town Planning Institute (RTPI) 122, 176, 178
Rugby Borough Council 181
rural land management 148

Safeway 205
Sanderson, Townend and Gilbert, Newcastle 208
Saskatchewan, Canada 110
satellite imagery 185
scale 27
ScotLIS 106
Scottish Environmental Protection Agency 187
Scottish Homes 169
shopping centre 206, 211, 212, 214, 216
shopping centre developers 186
shortest path analysis 263
sieve analysis 182, 185
Singapore LIS 106
site appraisal 183
Sites of Special Scientific Interest (SSSI) 151, 187
socio-economic mapping 34
software 12
South Staffordshire Housing Association 134
space planning 140, 141, 145

spatial autocorrelation 292–3
spatial information: aggregation of 292; autocorrelation of 292–3; errors in 291
spatial referencing 17–27, 38, 51; numeric spatial references 5; symbolic spatial references 6
Special Area of Conservation 151, 180
sphere 19–20, 24
spheroid 19–20, 24
SSR Realty Advisors 236
standards for land and property data 315–28
storms 232, 234
Strathclyde Regional Council 131
subsidence 188, 231, 234
Sunkist 150
surface modelling 30
Sutton London Borough Council 133
Swansea City Council 181
Switzerland 273
symbolisation 32
systems analysis 343

tangible features 31
Terence O'Rourke 183
thematic maps 33–5, 54–5; area class 34–5; choropleth *see* choropleth maps; isopleth 34–5; thematic visualisation 80
title plan 98
topographic maps 32–3, 46, 51, 54–5
town centre boundaries 160, 161
town centre index 162
town planning 170, 172
Toyota 207
trade potential report 205
Transport for London 260
Transverse Mercator projection 24–5
tree preservation orders 180

Unique Property Reference Number (UPRN) 92, 103, 124
University of Bristol Healthcare Trust 141
University College, London 161
University of Minnesota 145
urban design 190, 191, 193
Urban and Economic Development Group (URBED) 161

Vale of White Horse District Council 131
valuation 5
Valuation Office 92, 105, 118
vector approaches 39
vector data structures 39–45, 76; benefits and limitations of 45; comparison with raster data structures 46
virtual city 190, 195
Virtual Environments for Urban Environments (VENUE) 191
Virtual Reality Modelling Language (VRML) 191, 193
Virtual Urban Information System 195
visibility analysis 263
visualisation 80; of the built environment 81–3; of geographical data 80, 376–7; in three dimensions 80–2

Wakefield Council 183
West Oxfordshire District Council 179
Willis Risk Management Consultants 186
Wolverhampton 193
Woodberry Down 196